WITHDRAWN

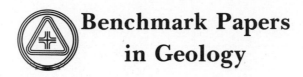 **Benchmark Papers
in Geology**

Series Editor: Rhodes W. Fairbridge
Columbia University

Published Volumes

ENVIRONMENTAL GEOMORPHOLOGY AND LANDSCAPE CONSER-
VATION, VOLUME I: Prior to 1900 / Donald R. Coates
RIVER MORPHOLOGY / Stanley A. Schumm
SPITS AND BARS / Maurice L. Schwartz
TEKTITES / Virgil E. Barnes and Mildred A. Barnes
GEOCHRONOLOGY: Radiometric Dating of Rocks and Minerals / C. T. Harper
SLOPE MORPHOLOGY / Stanley A. Schumm and M. Paul Mosley
MARINE EVAPORITES: Origin, Diagenesis, and Geochemistry / Douglas W.
Kirkland and Robert Evans
ENVIRONMENTAL GEOMORPHOLOGY AND LANDSCAPE CONSER-
VATION, VOLUME III: Non-Urban Regions / Donald R. Coates
BARRIER ISLANDS / Maurice L. Schwartz
GLACIAL ISOSTASY / John T. Andrews
GEOCHEMISTRY OF GERMANIUM / Jon N. Weber
ENVIRONMENTAL GEOMORPHOLOGY AND LANDSCAPE CONSER-
VATION, VOLUME II: Urban Areas / Donald R. Coates
PHILOSOPHY OF GEOHISTORY: 1785–1970 / Claude C. Albritton, Jr.

Additional volumes in preparation

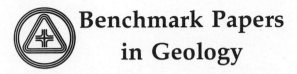

Benchmark Papers in Geology

——— A *BENCHMARK* ® Books Series ———

PHILOSOPHY OF
GEOHISTORY:
1785–1970

Edited by
CLAUDE C. ALBRITTON, JR.
Southern Methodist University

Dowden, Hutchinson
& Ross, Inc.

Stroudsburg, Pennsylvania

82276

Copyright © 1975 by **Dowden, Hutchinson & Ross, Inc.**
Benchmark Papers in Geology, Volume 13
Library of Congress Catalog Card Number: 74–10559
ISBN: V–0–471–02052–4

All rights reserved. No part of this book covered by the
copyrights hereon may be reproduced or transmitted in any
form or by any means—graphic, electronic, or mechanical,
including photocopying, recording, taping, or information
storage and retrieval systems—without written permission
of the publisher.

Manufactured in the United States of America.

Exclusive distributor outside the United States and
Canada: John Wiley & Sons, Inc.

74 75 76   5 4 3 2 1

Library of Congress Cataloging in Publication Data

Albritton, Claude C   1913-   comp.
    Philosophy of geohistory, 1785-1970.

    (Benchmark papers in geology, v. 13)
    1. Earth sciences--History--Addresses, essays,
lectures. I. Title.
QE11.A42   550'.9   74-10559
ISBN V-0-471-02052-4

Acknowledgments and Permissions

ACKNOWLEDGMENTS

AMERICAN ASSOCIATION FOR THE ADVANCEMENT OF SCIENCE—*Science*
    The Origin of Hypotheses, Illustrated by the Discussion of a Topographic Problem
    The Value of Outrageous Geological Hypotheses

GEOLOGICAL SOCIETY OF AMERICA—*Bulletin of the Geological Society of America*
    Role of Analysis in Scientific Investigation

PERMISSIONS

The following papers have been reprinted with the permission of the authors and the copyright holders.

AMERICAN ASSOCIATION FOR THE ADVANCEMENT OF SCIENCE—*Science*
    The Method of Multiple Working Hypotheses

GEOLOGICAL SOCIETY OF AMERICA
    The Fabric of Geology
        Rational and Empirical Methods of Investigation in Geology
        Simplicity in Structural Geology
    Uniformity and Simplicity
        Critique of the Principle of Uniformity

MANCHESTER LITERARY AND PHILOSOPHICAL SOCIETY—*Manchester Literary and Philosophical Society Memoirs and Proceedings*
    The Discovery of Time

PRENTICE-HALL, INC.—*Essays in Evolution and Genetics in Honor of Theodosius Dobzhansky*
    Uniformitarianism. An Inquiry into Principle, Theory, and Method in Geohistory and Biohistory

ROYAL NETHERLANDS ACADEMY OF SCIENCES—*Koninklijke Nederlandse Akademie van Wetenschappen, afd. Letterkunde, Med. (n.r.)*
    Catastrophism in Geology, Its Scientific Character in Relation to Actualism and Uniformitarianism

UNIVERSITY OF CHICAGO PRESS—*Journal of Geology*
    Geologic Time

Series Editor's Preface

The philosophy behind the "Benchmark Papers in Geology" series is one of collection, sifting, and rediffusion. Scientific literature today is so vast, so dispersed, and, in the case of old papers, so inaccessible for readers not in the immediate neighborhood of major libraries, that much valuable information has become ignored, by default. It has become just so difficult, or time consuming, to search out the key papers in any basic area of research that one can hardly blame a busy man for skimping on some of his "homework."

This series of volumes has been devised, therefore, to make a practical contribution to this critical problem. The geologist, perhaps even more than any other type of scientist, often suffers from twin difficulties — isolation from central library resources and an immensely diffused source of material. New colleges and industrial libraries simply cannot afford to purchase complete runs of all the world's earth science literature. Specialists simply cannot locate reprints or copies of all their principal reference materials. So it is that we are now making a concentrated effort to gather into single volumes the critical material needed to reconstruct the background to any and every major topic of our discipline.

We are interpreting "Geology" in its broadest sense: the fundamental science of the Planet Earth, its materials, its history, and its dynamics. Because of training and experience in "earthy" materials, we also take in astrogeology, the corresponding aspect of the planetary sciences. Besides the classical core disciplines such as mineralogy, petrology, structure, geomorphology, paleontology, or stratigraphy, we embrace the newer fields of geophysics and geochemistry, applied also to oceanography, geochronology, and paleoecology. We recognize the work of the mining geologists, the petroleum geologists, the hydrologists, the engineering and environmental geologists. Each specialist needs his working library. We are endeavoring to make his task a little easier.

Each volume in the series contains an Introduction prepared by a specialist, the volume editor — and a "state-of-the-art" opening or a summary of the objects and content of the volume. The articles selected, usually some 30 to 50 reproduced either in their entirety or in significant extracts, attempt to scan the field from the key papers of the last century until fairly recent years. Where the original references may be in foreign languages, we have endeavored to locate or commission translations. Geologists, because of their global subject, are often acutely aware of the oneness of our world. Its literature, therefore, cannot be restricted to any one country and, whenever possible, an attempt has been made to scan the world literature.

To each article, or group of kindred items, some sort of "Highlight Commentary" is usually supplied by the volume editor. This should serve to bring that article into historical perspective and to emphasize its particular role in the growth of the field. References or citations, wherever possible, will be reproduced in their entirety; for by this means the observant reader can assess the background material available to that particular author, or, if he wishes, he too can doublecheck the earlier sources.

A "benchmark," in surveyor's terminology, is an established point on the ground, recorded on our maps. It is usually anything that is a vantage point, from a modest hill to a mountain peak. From the historical point of view these benchmarks are the bricks of our scientific edifice.

Rhodes W. Fairbridge

Preface

Courses on the history and philosophy of the geological sciences are rarely offered in American universities. Whatever most students may learn about these subjects is usually the result of "outside reading" or independent study.

So far as the history of geology is concerned, independent study is well served by several general works and by a growing number of histories restricted to special fields. A massive general history is being planned by the International Union of Geological Societies.

By contrast, there is no standard work on the philosophy of geology, nor even any consensus as to what this nebulous field encompasses. However, the journal literature on such topics as the scope and interrelationships of the geological sciences, the explication of geological terms and ideas, the classification of geological phenomena, and the logical methodology of geohistory is vast and diffuse.

Thus it has seemed to Dr. Fairbridge and to me that a useful purpose would be served by binding together reproductions of geophilosophical writings that should be of general interest to students. Readers are advised that the present selection is biased on three counts. All these writings were originally published in English, the only language most American students read with ease. Books of similar scope and size could be drawn from writings solely in German, Russian, or French, and probably other languages as well. Second, the selections relate mainly to historical geology rather than to physical or economic geology. The only excuse is that historical geology seems the most germane to our understanding of the world and our place in it. Finally, and unabashedly, these selections represent some of my favorite reading, a good part of the body of reflective works that I hope all my students will come to know.

To all the living authors and to the copyright holders who have graciously approved these reprintings, my hearty thanks.

Claude C. Albritton, Jr.

Contents

Contents by Author

Introduction

What happened on earth before man was here? This question has vexed reflective minds of all generations. In the ancient theological traditions the answer usually has been given in the form of creation epics. The Babylonian and Hebraic accounts of creation are notable examples; and the latter is of particular interest to historians of science because it influenced geological and biological thought late into the nineteenth century.

Until the latter eighteenth century there was no path to guide the scientifically minded antiquarian in his search for the past. This is not to deny that much earlier in time a few scholars had used the evidence provided by rocks and fossils to show that remarkable changes in the position of land and sea, or in climate, had taken place during the unremembered past. The names of Xenophanes of Colophon, Herodotus, Strabo, Chen Kua, the Islamic Brothers of Purity, Chu Hsi, Leonardo, Steno, Hooke, Buffon, and Werner come to mind as prophets of historical geology. But the guiding principles of geohistory were not bound together in a coherent theory, free of the strictures of natural theology, until after 1785.

The writings here assembled follow a chronological order except for the first and last papers. In the first essay Toulmin relates the revolution in historical geology sparked by Hutton, Playfair, and Lyell to a general awakening of interest in history during the eighteenth century. The ideas here sketched are amplified and supplemented in a book (with June Goodfield) also entitled *The Discovery of Time*.[1]

Although the "Abstract" that constitutes the second chapter of this book is undated and anonymous, Victor A. Eyles has established beyond reasonable doubt that James Hutton (1726–1797) is the author and that the year of publication was 1785.[2] This abstract contains all the essential conclusions found in Hutton's *Theory of the Earth* published three years later, and these conclusions are more clearly stated here than in the longer work. Hutton has been widely acclaimed the founder of modern geology. His concept that geologic time is virtually limitless in duration prepared the way for Lyell's uniformitarian approach to earth history and for Darwin's evolutionary theory. In demonstrating that granite is not part of the earth's primitive crust but an intrusive rock, he helped to overthrow the Neptunist school of thought led by Werner. His concept that the earth functions essentially as a heat machine remains an essential component of modern geophysical theory. All Hutton's geological writings, now extremely scarce in the original printings, have been reprinted and are in circulation today. The most illuminating biography of Hutton is the one written by his friend and defender, John Playfair.[3]

Hutton's revolutionary views, especially those concerning the antiquity of the earth and the role of terrestrial heat in the consolidation of sediments, were not well received by most of his contemporaries. The idea of a world without beginning or end was unattractive to the Neptunists, whose theory offered comforting parallels, however tenuous, with the Mosaic account of creation. Neptunist accounts of the past, moreover, had the virtue of being historical in format and thus in accord with the contemporary intellectual movement described by Toulmin. By contrast, Hutton's endless cycles of destruction and rejuvenation of continents seemed ahistoric. The murky prose in expanded versions of Hutton's Abstract was itself an impediment to the propagation of his ideas.

Evidently Hutton could make clear in conversation concepts that became clouded when he reduced them to writing. And the man who apparently listened most intently to what Hutton had to say about geology was John Playfair.[4] Following Hutton's death, Playfair set about the task of preparing a treatise "drawn up with a view of explaining Dr. Hutton's Theory of the Earth in a manner more popular and perspicuous than is done in his own writings." After five years of work Playfair completed his treatise, published in 1802 under the title *Illustrations of the Huttonian Theory of the Earth.*[5]

The *Illustrations* are not merely an explication of the Huttonian theory. Sometimes in the text but more often in the 389 pages of notes that follow, Playfair adds concepts of his own design, related to the significance of unconformities, origin of granite, glacial transport of erratics, and concordance of stream junctions. Nor did Playfair hesitate to be critical of Hutton's views. In discussing the origin of calcareous rocks, for example, he allows that Hutton "has sometimes expressed himself as if he thought that the present calcareous rocks are all composed of animal remains. This conclusion, he continues, "is more general than the facts warrant; and from some incorrectness or ambiguity of language, is certainly more general than he intended."

Our third chapter consists of excerpts from the text of the *Illustrations*. Playfair's "law of concordant stream junction" is found on page 102, and his account of the geological cycle begins on page 127. The section on consolidation of strata by heat is omitted.

In the early part of the nineteenth century most geologists attributed the major discontinuities in the paleontological record, as well as the contortion of strata so commonly displayed in ranges of mountains, to catastrophic dislocations of the earth's crust. The energy released during these supposedly brief but violent episodes in the ancient history of the earth was imagined to exceed by many orders of magnitude the most destructive floods, volcanic eruptions, and quakes of historic times. Charles Lyell challenged this view in 1830, with the publication of the first volume of his *Principles of*

Geology.[6] According to Lyell, inferences regarding whatever changes on the earth may have taken place in the past should be drawn solely on the basis of changes now in progress. In Lyell's view it is not only unnecessary but absurd to believe that a mountain range has been thrown up in a single massive upheaval: given enough time a series of small upward dislocations, accompanied by the kinds of quakes we experience today, can accomplish the same result. Given enough time and the proper topographic situation, a stream, gradually and by small increments, may so deepen its original shallow valley as to form a great canyon. The ensuing debate between the Uniformitarians, led by Lyell, and the Catastrophists, in whose camp nearly every other reputable geologist of the 1830s was marshalled, continued throughout the nineteenth century; and the issues arising out of this contest are still discussed today.

Papers 4 and 5 pit the Uniformitarian Lyell against the Catastrophist William Whewell, in writings published during the same year.[7] Lyell cleverly defends his position by telling why the Catastrophists have been so long in seeing the light. They have, he proposes, not properly comprehended the vast duration of geologic time. Compressing an almost endless array of geological events into a short span of time necessarily makes the unfolding of these events seem catastrophic. Other prejudices relate to our general ignorance of what goes on beneath the surface of the earth or below the ocean.

Whewell's defense of Catastrophism is an appeal partly to authority and partly to evidence. The evidence cited includes the tilted and contorted coal measures of England, alternations of fresh and salt water species of animals in the Paris Basin, and the elevations of the continents as expressed in high ranges of mountains. The marked differences in kinds of fossils on opposite sides of certain formational boundaries suggest episodes of destruction of life followed by creations of new life.

The longer part of Whewell's paper is given to an attack on Lyellian uniformitarianism. Here he concentrates on the part of Lyell's method that was most vulnerable: the proposition that the causes of changes during the geologic past have always operated at the same levels of energy as today. To Whewell, as later in Lord Kelvin's view, the assumption that the terrestrial heat machine is moving at the same rate now as for all geologic time is gratuitous and unlikely. Dynamically speaking, Whewell could find no reason to assume that the present is the key to the past.

In the closing lines of his chapter on the two antagonist doctrines of geology, Whewell pled that in 1872 it was yet too early for making final assessments of the uniformitarian and catastrophist views. He urged geologists to gather more and more facts about the earth before drawing conclusions about causes. Whewell went so far as to suggest the cultivation of a subscience entitled "geological knowledge of facts," which he placed at the same level of importance as "geological dynamics."

Looking back on the factual reformation in geology promoted by Whewell, from a perspective of 17 years, T. C. Chamberlin observed:

> The advocates of reform insisted that theorizing should be restrained, and efforts directed to the simple determination of facts. The effort was to

make scientific study factitious instead of causal. Because theorizing in narrow lines had led to manifest evils, theorizing was to be condemned. The reformation urged was not the proper control of theoretical effort, but its suppression. We do not need to go backward more than twenty years to find ourselves in the midst of this attempted reformation. Its weakness lay in its narrowness and restrictiveness. There is no nobler aspiration of the human intellect than desire to compass the cause of things. The disposition to find explanations and to develop theories is laudable in itself. It is only ill use that is reprehensible. The vitality of study quickly disappears when the object sought is a mere collection of dead unmeaning facts.

This quotation is from Chamberlin's famous essay "The Method of Multiple Working Hypotheses," read before the Society of Western Naturalists in 1899 and first published in *Science* the following year.[8] A favorite reading among scientists in general and geologists in particular, it has been reprinted in various journals and is still available in reprint format from the American Association for the Advancement of Science.

Chamberlin is here interested in the logical methodology of natural science. He distinguishes between the method of the ruling theory, the working hypothesis, and what he calls the method of multiple working hypotheses, which he advocates as the best of the three. In applying this favored method, Chamberlin explains: "the effort is to bring up into view every rational explanation of new phenomena, and to develop every tenable hypothesis regarding their cause and history." He extends the application of this methodology from scientific investigation into educational procedures, and even into the realm of coping with the practical problems of life.

In the next paper G. K. Gilbert applies the method of multiple hypotheses to the problem of the origin of Arizona's Meteor Crater.[9] The hypothesis he is finally led to favor is that the crater was formed by an explosion of subterranean steam generated by volcanic heat. This view is now discredited by overwhelming evidence which indicates an origin by impact and explosion of a large meteorite. But Gilbert could not have known in 1896 that projectiles traveling in excess of certain critical velocities explode on impact and do not remain embedded in their targets. All this is not much to the point for the purposes of this collection, however; Gilbert's aim in the essay is to explicate, by reference to a case history, his "hypothesis for hypotheses": the proposition that hypotheses are born of analogies. This he does with his customary clarity of exposition.

In "The Value of Outrageous Geological Hypotheses" William Morris Davis emphasizes the speculative character of earth science, expresses concern regarding what he sensed as a growing complacency among geologists with the current theory of their science, and stresses the fact that new hypotheses are usually regarded as outrageous at the time of their announcement.[10] Davis's view is that the formulation and testing of hypotheses that go against the intellectual grain of the time are necessary to the advancement of science. When he predicted that "violence must be done to many of our accepted principles," he was being more prophetic than he could have known. This essay was written in 1925, the year following the publication of Walther Penck's *Morphologische Analyse* [for an English translation, see *Morphological Analysis* of *Land Forms* (London: Macmillan & Co. Ltd., 1953), 429 p.]. Penck's work was a telling challenge to Davis's cherished "geographical cycle," and led eventually to the modification or abandonment of many elements in Davisian geomorphic theory.

Davis's concept of the geographical cycle has often been criticized as deductivist in character, and to that extent unreal. In 1933, Douglas Johnson, one of Davis's distinguished students, undertook to show that deduction, no less than induction, plays an important part in the method of multiple working hypotheses.[11] Johnson recognizes seven stages in investigations of the kinds conducted by geologists. Although both inductive and deductive reasoning are involved to varying degrees at each stage, induction is especially prominent at the critical step of inventing multiple hypotheses, and deduction is equally so when the time comes for hypotheses to be verified or eliminated. Johnson concludes that the method of multiple working hypotheses is the best guarantee of success in scientific research, but concedes that no methodologic device, "however perfect, can wholly deprive the human intellect of its capacity for making mistakes."

Three decades passed between the publication of Johnson's paper and the appearance of Mackin's essay on rational and empirical methods of investigation.[12] During this interval computers had made a remarkable impact on the scientific scene, first in meteorology and then on other branches of science. With the capability at hand of manipulating vast accumulations of numerical data, geology became more quantitative. Experimentation with mathematical models simulating the operations of natural processes became feasible. In geomorphology interest shifted away from the history of landscapes to the dynamics of the erosive processes—a development Whewell would have applauded. With these changes some geologists, notably Mackin, felt that there was also a swing toward empiricism in earth science. As Mackin explains, he is not so much concerned here with the issue of qualitative versus quantitative, as with the issue of rationality versus blind empiricism. The case history he used to make his points derives from his own investigations in depth of fluvial erosion and deposition.

In the eleventh paper, C. A. Anderson analyzes the workings of the methodological principle of simplicity in historical geology, especially in the area of structural history.[13] In essence he concludes that geologists should not multiply (inferred) historical events without necessity, even though the number of minimal events which the evidence at hand requires should be fewer than the investigator senses they must in fact be. Adherence to this precept gives some guarantee that further elaboration of earth history, as new facts turn up, will proceed in an orderly way.

M. King Hubbert's "Critique of the Principle of Uniformity" is the first of three essays devoted to issues that grew out of the debate between the Uniformitarians and the Catastrophists of the nineteenth century.[14] He traces the history of uniformitarian ideas, as developed in the works of Hutton and Lyell, and later extended into the biological realm by Darwin. In his analysis of the principle, he concludes that its viable element—the postulate that the laws of nature are invariant with time—is no longer peculiar to geology but has now become the common denominator of all science. And he goes on to formulate the following interesting definition of history: "History, human or geological, represents our hypothesis, couched in terms of past events devised to explain our present-day observations."

Simpson's essay is probably the most penetrating account of the issues and ambiguities of uniformitarian thought yet published.[15] Following a review of the development of uniformitarianism beginning with the writings of Hutton, he proceeds to identify and analyze "principles and issues that do still have some interest within the general topic of uniformitarianism." The relationships between that particular ism and naturalism, actualism, historicism, evolutionism, and gradualism are explored in the order listed. Turning then to the problem of drawing historical inferences, he identifies the senses in which the present is the key to the past, and vigorously defends the proposition that the historical sciences have principles of their own that set them apart from the nonhistorical sciences. Although Simpson is a confessed uniformitarian (given his own definition of that term), he is willing to give the catastrophists their due: in his conclusion he praises Conybeare and other opponents of Lyell for their insistence on a historical or cumulative model of the earth.

Few modern scholars profess to see much that was very good in catastrophist views of the eighteenth and nineteenth centuries. Reijer Hooykaas is an exception.[16] He has long insisted that strict adherence to uniformitarian tenets tends to force past phenomena into a preconceived frame built upon events occurring in our time. In the present essay he proceeds to show, by reference to original writings, that supernaturalism is not a common denominator of catastrophist theories and that many catastrophists of the past were also actualistic in their approach to earth history. Methodologically he finds actualistic catastrophism superior to Lyellian uniformitarianism, on two counts: it led to historical rather than ahistorical reconstructions of the past, and it proceeded from observation to theory without imposing limits on the rate of past changes a priori.

The collection ends as it began—with Time. David Kitts, in a prize-winning essay, critically examines the bases for our conception of geological time.[17] After analyzing the methods by which geologic events are arrayed in chronologic order, beginning with the principles of superposition and cross-cutting relationships, he turns to the complex problems of measuring intervals of geologic time and synchronizing geologic events over broad geographic frames of reference. Among other matters, he urges the abandonment of the term "absolute time" as pretentious and misleading, and argues on both theoretical and practical grounds that in the correlation of geologic events only approximate simultaniety is possible. He questions that "time planes" can be established on a worldwide basis. Correlation, he concludes, "is a method which permits the ordering of spatially separated historical–geological events in the relation 'earlier than'—'later than.' "

Notes

[1]Stephen Edelston Toulmin (*b*. 1922) is a graduate of King's College, The University of Cambridge. At various times he has been a faculty member at Cambridge, Oxford, Leeds, Brandeis, and Michigan State University. At present he is Provost at Crown College of the University of California at Santa Cruz. It was while Dr. Toulmin

was Director of the Unit for the History of Ideas at the Nuffield Foundation in London (1960–1965) that the book *The Discovery of Time* was published.

[2]James Hutton (1726–1797) spent most of his years in his native Scotland. At Edinburgh and Paris he prepared himself for the practice of medicine, but he turned instead to scientific agriculture. At the age of 42 he retired from farming, moved back to Edinburgh, and spent the rest of his life in scientific pursuits. His famous essay "Theory of the Earth" was published in 1788, in the first volume of the *Transactions of the Royal Society of Edinburgh*. In 1795 this essay was published in expanded form as a work of two volumes. An incomplete manuscript of a third volume was discovered after Hutton's death and published in 1899.

[3]Playfair, John (1805), Biographical Account of the Late James Hutton, F.R.S. Edinburgh: *Trans. Royal Soc. Edinburgh,* Vol. V, Pt. III, pp. 39–99. (Reprinted in James Hutton's *System of the Earth,* 1785; *Theory of the Earth,* 1788; *Observations on Granite,* 1794; together with Playfair's *Biography of Hutton,* New York: Hafner Press, 1970).

[4]John Playfair (1748–1819) trained for the clergy at the University of St. Andrews, and between 1773 and 1782 he was minister at the Scottish communities of Benvie and Liff. His skill as a mathematician had already attracted the attention of academicians when he was only 18 years old. In 1782 he moved to Edinburgh, where he moved quickly into the circle of intellectuals of which James Hutton was a member. In 1785 he joined the faculty of the University of Edinburgh as professor of mathematics. Throughout his long tenure at the University he maintained an active interest in geology, especially in the investigations of Hutton. In 1805 he moved to a chair in natural philosophy. Widely traveled in the British Isles and Europe, Playfair was a keen observer of geological phenomena in the field. His writings have been widely acclaimed for their clarity and grace. Both his biography of Hutton and his *Illustrations* are in print as facsimiles.

[5]Playfair, John (1802), *Illustrations of the Huttonian Theory of the Earth* (Facsimile Reprint, with an Introduction by George W. White, Urbana, Ill.: University of Illinois Press, 1956).

[6]Charles Lyell (1797–1875) was born on the family estate of Kinnordy, Forfarshire, Scotland. At the age of 19 he entered Exeter College, Oxford, where he became attracted to geology through the lectures of William Buckland. Nevertheless he trained for the law in accordance with a paternal wish and was admitted to the bar in 1825. He abandoned law for geology two years later. Lyell's fame rests largely on three works: *The Principles of Geology* (3 vols., 1830–1833), *Elements of Geology* (1838), and *The Antiquity of Man* (1863). The first two of these went through multiple editions, and the *Principles* is currently available in facsimile. A recipient of many honors from scientific societies, Lyell was knighted in 1848 and created a baronet in 1864. He is buried in Westminster Abbey.

[7]Philosopher, scientist, and historian of science, William Whewell (1794–1866) was associated with the University of Cambridge throughout most of his varied and distinguished career. While a fellow and tutor at Trinity College he was elected to fellowship in the Royal Society at the age of 26. He taught courses in mineralogy from 1828 to 1832. His early writings concentrated on mechanics and dynamics. Later he became in-

terested in the history and philosophy of science. His fame rests primarily on his *History of the Inductive Sciences* (3 vols., 1837) and his *Philosophy of the Inductive Sciences* (1840). In his writings he was a formidable antagonist of uniformitarian geology, but after 1850 his investigations turned mainly to moral theology.

[8]Thomas Chrowder Chamberlin (1843–1928) is best known for his coauthorship of the planetesimal hypothesis, which proposed that the earth was originally an aggregate of small solid particles congealed by freezing of matter thrown out from the sun. This hypothesis offered an alternative to the idea that the earth has cooled from an original molten state, the premise on which Lord Kelvin had based his calculations for a scale of geological time far shorter than geologists of the late nineteenth century felt was needed to accommodate their lengthy chronicle of historical events. He served as Chief Geologist for the Wisconsin Geological Survey (1876–1882), and for five years beginning in 1887 was President of the University of Wisconsin. In 1892 he became head of the Department of Geology at the University of Chicago and served in that capacity until his retirement in 1919. There he founded the *Journal of Geology* and served as its editor until the time of his death.

[9]Grove Karl Gilbert (1843–1918) was one of the six senior scientists appointed to the U.S. Geological Survey when that agency was established in 1875. Prior to that time he had done extensive geological work in the American West as a member of the Wheeler and Powell surveys. In the course of his studies in Utah he discovered ancient Lake Bonneville, and he considered his monograph of 1890 on the history of this formerly great lake to be his best work. He is also remembered for his *Geology of the Henry Mountains*, which contains the first description of laccoliths, and for his *History of the Niagara River*. An extensive account of Gilbert's life and work is found in William Morris Davis's biography published in 1927 in the *Biographical Memoirs of the National Academy of Sciences* (Vol. 21, 5th Memoir, v + 303 pp.).

[10]Best known for his application of the evolutionary idea to the study of landforms, Davis (1850–1934) is generally acclaimed as the person who molded geomorphology into a coherent discipline. Following a long tenure at Harvard, he resigned in 1912 and thereafter concentrated his studies on the American West. There he demonstrated the predominance of fluvial processes in shaping the landscapes of arid regions. His bibliography of more than 500 items includes, in addition to contributions to geomorphology, many works in meteorology and substantial biographies of John Wesley Powell and G. K. Gilbert. A full account of Davis's life and work has recently been published by Chorley, Dunn, and Beckinsale [*The History of the Study of Landforms or the Development of Geomorphology*, Vol. 2, *The Life and Work of William Morris Davis*. Methuen, London, 1973 (U.S. distributor, Harper & Row, New York) xxii, 874 pp., illus.].

[11]Douglas Wilson Johnson (1878–1944) first became associated with Davis when he moved to Boston after receiving his doctorate from Columbia University in 1903. First as Instructor at the Massachusetts Institute of Technology and then as Assistant Professor at Harvard (1907–1912), he studied and worked with Davis. In 1909 he edited Davis's influential *Geographical Essays*. Johnson is well remembered for his studies of shorelines and shore processes, the source of the example he chose as a hypothetical case history for the essay reproduced in this book. Other subjects to which

he made significant contributions include the origins and development of marine terraces, submarine canyons, stream systems, and rock fans and pediments in arid regions. He was also a pioneer in the development of military geology. In 1912 he returned to Columbia, where he remained a faculty member for the rest of his life.

[12]Joseph Hoover Mackin (1905–1968) came under the influence of Davisian geomorphology during his graduate studies with Douglas Johnson at Columbia in the 1930s. Even after much of the theoretical basis for Davis's system had been either challenged or discredited, Mackin proclaimed himself a member of the "Davis Protective Society." What Mackin cherished in the work of Davis and his disciples was their reasoned approach to geological problems in general and their concern with the history of landscape development in particular. Although his main work was in geomorphology, he also was active in engineering geology, having served as consultant for many hydroelectric dam projects. He is a former Chairman of the Division of Earth Sciences of National Academy of Sciences–National Research Council, and member of the Atomic Energy Commission (1955–1957).

[13]Charles Alfred Anderson (*b.* 1902) is presently Research Associate of the Earth Sciences Board, University of California at Santa Cruz. His professional career began in 1928, when he was appointed to the faculty of geology at the University of California at Berkeley. In 1942 he joined the U.S. Geological Survey, from which he retired in 1972. His principal research interests have been with the genesis and distribution of metalliferous deposits in the American West. For five years, beginning in 1953, he served as Chief Geologist of the Survey. He was elected to the American Academy of Arts and Sciences in 1956 and to the National Academy of Sciences the following year.

[14]Marion King Hubbert (*b.* 1903) is Research Geophysicist with the U.S. Geological Survey. Previously he has worked in a variety of capacities, including service as a research geophysicist with Shell Oil and Development Companies, as a professor at Columbia and Stanford Universities, and as researcher in mineral resources for the Board of Economic Warfare during World War II. He is a member of the National Academy of Sciences and of the American Academy of Arts and Sciences. Dr. Hubbert's scientific work has included investigations in exploration geophysics, petroleum geology and engineering, structural geology and the physics of earth deformation, and in the physics of underground fluids. It has also included studies of the world's energy and mineral resources and of the philosophy and history of geology. A past President of The Geological Society of America, Hubbert has received that society's two highest awards: the Arthur L. Day Medal in 1954 and the Penrose Medal in 1973.

[15]George Gaylord Simpson (*b.* 1902) began his career in vertebrate paleontology as Field Assistant to the American Museum of Natural History in 1924. There he rose to the rank of Chairman of the Department of Paleontology and Geology in 1944. In 1959 he joined the faculty at Harvard, and in 1971 retired there as Agassiz Professor of Vertebrate Paleontology. He is presently Professor of Geoscience at the University of Arizona. For his contributions to the study of recent and fossil mammals, and for his contributions to evolutionary theory, he has received numerous awards, including the Darwin Wallace Medal of the Linnean Society of London, the National Medal of

Science, the Penrose Medal, and the Darwin Medal of the Royal Society of London.

[16]Dr. Hooykaas (*b.* 1906) is a Professor of the History of Science at the University of Utrecht, The Netherlands. He is also Vice President (for Europe) of the International Committee on the History of Geological Sciences. He has written extensively on the history of chemistry, mineralogy and crystallography, evolutionary theory, and geology. The essay here reproduced is an extension of a longer work, *The Principle of Uniformity in Geology, Biology, and Theology* (Leiden: Brill, 1959, 1963). Professor Hooykaas is also concerned with the interaction between science and religion. In a recent book, *Religion and the Rise of Modern Science* (Grand Rapids, Mich.: Eerdmans, 1972), he defends the thesis that modern science is in good part a product of the Judaeo-Christian influence on Western thought. As he puts it, "whereas the bodily ingredients of science may have been Greek, its vitamins and hormones were biblical."

[17]David Burlingame Kitts (*b.* 1923) is Davis Ross Boyd Professor of Geology and the History of Science at the University of Oklahoma. At Columbia University, where he earned his doctorate in 1953, he was a student of George Gaylord Simpson. His research interests include investigations of Cenozoic mammals and studies in the philosophy of geology and evolutionary theory. In 1964–1965 he was a visiting fellow in philosophy at Princeton University, where he studied with Professor Carl G. Hempel. It was there that he wrote the present essay, for which the faculty of philosophy awarded him the J. Walker Tomb Prize.

1

Copyright © 1961–1962 by the Manchester Literary and Philosophical Society

Reprinted from the *Manchester Lit. Phil. Soc. Mem. Proc.*, **105**, 100–112 (1962–1963)

The Discovery of Time

By STEPHEN TOULMIN

ABSTRACT

By now most people are aware how far human attitudes were transformed by the astronomical discoveries beginning around A.D. 1550, as a result of which our conception of the size and scale of the cosmos has been increased almost without limit. There has, however, been another similar transformation almost as striking in its effects, involving a corresponding expansion in our ideas about the antiquity of the universe. This " discovery of time " has taken place almost entirely since A.D. 1800, and the mid-nineteenth century debates about evolution were only one small but particularly noisy episode in a much larger intellectual revolution. This change in our time-scale has by now so permeated our whole view of nature and life that we find it difficult to get properly inside the minds of our forefathers—even as recently as the eighteenth century.

This lecture will try to make clear just how our common-sense attitudes have been influenced as a result of this new awareness of the historical dimension of the world.

As a way of introducing the problem with which this lecture is concerned, let me quote Sir Thomas Browne, the seventeenth-century essayist. My first passage comes from the section of his *Christian Morals*, in which he justifies a philosophical attitude towards death by arguing that the normal span of human life gives one a " fair share " of the whole history of the cosmos:

> " He is like to be the best judge of Time who hath lived to see about the sixtieth part thereof .... In seventy or eighty years a Man may have a deep Gust of the World ... a curt Epitome of the whole course thereof."

A little further on in the same essay he returns in passing to this same question—of the Age of the World:

> " The World it self seems in the wane, and we have no such comfortable prognosticks of latter times, since a greater part of Time is spun than is to come."

Looking back to his earlier *Religio Medici* we find several allusions to the same topic, for instance:

> " The World grows near its end .... The last and general fever may as naturally destroy it before six thousand, as me before forty."

These passages may, at first glance, strike a twentieth-century reader as mysterious. But they at once become self-explanatory if we recall that for Sir Thomas Browne, as for other educated men in the seventeenth century, one great fact about the world's history was a matter of common knowledge: namely, that the whole story of the cosmos from the Creation to the Last Trump could be comfortably contained within a time-span of 10,000 years. The most popular estimates looked back to Julius Africanus' analysis of the Jewish Chronicles in the third century A.D.: this set the span of temporal history at 7,000 years, the last 1,000 being reserved for the promised Millenium which would follow the Second Coming of Christ.

The whole course of history, on the general seventeenth-century view, comprised six cosmic "days" of 1,000 years each: four such "days" had been completed at the time of Jesus' birth, so that a few hundred years at most remained before the allotted 6,000 years should be completed. These doctrines were not just the numerological fancies of Archbishop Ussher: on the contrary, a large body of Christian opinion took them as un-questioned axioms. No less a man than Johannes Kepler—or so Mr. Arthur Koestler tells me—demonstrated from the dating of the Crucifixion eclipse that four years had at some time been lost from the chronology of the Christian Era, so putting back the birth of Christ to 4 B.C. and the supposed date of the Creation from 4,000 to 4,004 B.C. And in Sir Thomas Browne we have a witness whose views can safely be taken as representative: though he was a diligent observer, a great reader and a fine writer, there was very little really original in the general line of his thought. He gives us, rather, a reliable and elegant expression of presuppositions widespread among the literate public of his time.

Once we have seen the assumption he here takes for granted, we can recognise in it the starting-point for one of the greatest transformations in the whole history of human thought. This "short time-scale" (as I shall call it) was shared by many scientific thinkers of the time, even including the most distinguished; yet between their day and ours it has been utterly displaced and abandoned. By now, one cannot seriously question that the cave-paintings of central France and northern Spain date back well before 10,000 B.C., and the whole story of Man's existence on Earth is known to be only one recent (and comparatively brief) episode in a cosmic history lasting not a few thousand, but thousands of millions of years. In its own way, this drastic expansion of our time-scale constitutes as dramatic and significant an intellectual revolution as the earlier expansion of Man's ideas about the size and layout of the universe. Yet up to now it has been much less clearly recognised and mapped.

G

Some worthwhile parallels can be drawn between the two intellectual transformations. By this time, the course of the "Copernican revolution" in astronomy and dynamics—a revolution which was completed only in the late seventeenth century, through the work of Isaac Newton—has been fully analysed and reconstructed: we all of us appreciate how, between 1500 and 1700, astronomers and physicists pushed back the spatial boundaries of the cosmos to many times its earlier dimensions, and in so doing displaced the Earth from its former humble (but at any rate significant) position at the centre of things. The immediate effect of that expansion had been to throw doubt on the uniqueness of our terrestrial globe, and we find Bruno and Fontenelle at once speculating about a possible "plurality of worlds". Yet this was only the first half of a double revolution, and the second instalment (I shall argue) is still in progress to-day. Although in 1700 the uniqueness of Here—i.e. our Earth—had been questioned, the uniqueness of Now—i.e. the epoch of cosmic history in which we find ourselves living—had not yet seriously been called in doubt; and, whereas before A.D. 1650 Galileo's telescope had already broken the imagined boundaries of space, the corresponding limits of time have been pushed back *only in the last 150 years.*

How has this change come about? In order to understand the course it has taken, we should try to put ourselves back into the intellectual position of our forefathers, and ask on what evidence they based their conception of the cosmic time-scale, and the course of the world's "life-history". In doing so, we must always remember that, in order to form ideas about things one cannot inspect for oneself, it is necessary to rely *either* on other people's testimony *or* on some secure mode of arguing to the required conclusions from data actually at one's disposal. How were the men of 1700 placed, in these respects, when they asked questions about the remote past? The answer is: they were in little better a position than their forerunners in 400 B.C. When the Greek philosophers began to speculate about the origin and development of the universe, they possessed no historical testimony or records for events before 2,000 B.C., and Christian scholars had pushed back this date to 4,000 B.C. only by accepting in a very literal-minded way the sacred chronologies worked out by Julius Africanus and Eusebius. No other established mode of argument was available in A.D. 1700 for drawing conclusions about the remoter past from observations made at the present time.

In speculating about the origin of the world, accordingly, the first thinkers were driven back on to considerations of an entirely general and philosophical kind. Some were inclined, like Plato, to accept a " Big Bang " story, in which cosmic history was supposed to have begun at some moment in the past when the Creator established the present Order of Nature, imposing on the formless raw material of things the Laws of Nature which still apply to-day. Others followed Aristotle in rejecting this idea, arguing as a matter of principle that nothing meaningful could be said about " the Beginning of Time " and supposing, on the contrary, that the world had existed through all Eternity in the same Steady State—heavens, earth, sea, living things and even the races of men, all displaying the same general aspect at every stage. Others again were drawn to a Cyclic Cosmos, in which the wheeling course of Fate carried the whole world of material things through a recurrent sequence of creation and destruction, which repeated itself every few thousand years. A few dissident thinkers alone (chiefly Ionians and Epicureans) turned their backs on abstract, philosophical arguments, and tried to reconstruct in a more empirical way a picture of the successive stages by which the world gradually developed into its present form.

Once we have acknowledged how very little evidence our forefathers had to go on when reflecting about the course of cosmic history, and how short a period of time this evidence covered, we may find it easier to admit that their belief in the *fixity* of the Natural Order was an understandable, a rational— even a *scientific*—interpretation of all they knew. So far as it went, the evidence they possessed tended to support the hypotheses that the heavens, the earth and organic species had been fundamentally unchanging throughout the 5,000 or so " accessible " years: certainly it lent scarcely any support to the opposite views. Indeed, for a full century after the publication of Newton's *Principia* in 1687, it appeared that his discoveries had reinforced the accepted, static view of nature, and the very phrase " laws of nature " came to play an important part of its own in the argument. Whereas physical scientists in the twentieth century pay no conscious attention to the legal metaphor implicit in this phrase, the men who established the concept around the year 1700 took it extremely seriously. In their view, the Laws of Nature were the principles of design by which the plan of the Divine Creation was governed. In making discoveries about the

Order of Nature, scientists were revealing the rational specifications by which the Divine Architect had laid out his Creation—the immutable edicts to which He had required inanimate things to conform. It seemed only respectful to assume that the Deity had made the natural order to a fixed, stable and pre-ordained plan, and Leibniz' strongest objection to the Newtonian theory of gravitation was its apparent failure to guarantee the stability of the planetary system.

Scientists in the Protestant parts of Europe were, if anything, drawn towards a fixed and fundamentalist view of nature even more strongly than their Catholic colleagues. Throughout this period " natural theology " had a great attraction for them, and their works are full of naïve appeals to the Argument from Design, since they devoutly assumed that the Creator's blueprint could be reconstructed directly from a detailed study of the workings of Nature. Though this attitude provided a powerful motive for experiment and observation, it also gave an inevitable slant to the theoretical interpretations which followed. By the 1730's, Linnaeus could undertake the all-embracing task of categorising the entire *System of Nature* into a single fixed set of taxonomic pigeon-holes. Thus the immediate effect of the seventeenth-century " scientific revolution " was to freeze the Order of Nature, and so make the task of the nineteenth-century revolutionaries (Lyell and Darwin) not less, but *more* difficult.

Meanwhile, the first stirrings towards a change were beginning, at the opposite end of Europe and in a very different field of inquiry. At Naples, in the heart of Catholic Italy, a professor of law called Giambattista Vico was working out his own novel conception of the development of human society. It was Vico, indeed, who launched the modern idea of *historical development* into European thought. By an act of sublime intellectual folly he attempted to have the best of two irreconcilable worlds—orthodox Christianity and Epicurean philosophy. Recalling the disastrous consequences of the Tower of Babel, he put forward the following original argument: that, after the resulting dispersion into peoples of many different tongues, the Jewish people alone lived in an unbroken social order, of which the Old Testament provides a true history. All the other peoples were scattered over the face of the earth, and thrown back into absolute savagery—the " nasty, brutish and short " existence which Hobbes called the " state of nature " Gradually, down the centuries, the

Gentile races were compelled by necessity to *re-acquire* the arts of social life, language and technology, going through the successive stages already hinted at by Lucretius. Farmers succeeded hunters, city-dwellers succeeded farmers, and all aspects of social development progressed through a corresponding succession of phases. Thus, the twin ideas of " *periods* " and " *development* " found their way into modern historical and sociological analysis. (Historians in the previous century, e.g. Clarendon, had followed Thucydides rather than Lucretius, limiting their surveys to their own epochs or—at most—reconstructing an idealised picture of antiquity to serve as a moral example to their contemporaries.)

It is hard to realise just how original this conception of Vico's was. Nowadays, we are so accustomed to thinking of the present state of things—whether in nature, in society, or in the individual human life—as the temporary product of a continuing process of development, that we forget how recent this habit of thought is. Yet it appears that no serious thinker in western Europe before Vico accepted anything like the modern historical idea of " development through period ". Vico, of course, had forerunners in the ancient world (notably Lucretius), but their views had never been more than enticing heresies: Vico was the first man in the history of thought to insinuate this conception of development into the general body of ideas. Through Montesquieu, Shaftesbury and their successors, the idea of historical periods and phases—each with its characteristic modes of life, language and thought—made its way into the European consciousness, and before long Voltaire scandalised his contemporaries by asserting—as Vico himself never did— that the history of the Jews must be treated on a par with that of the Gentiles, since, after all, the Jews were one more pastoral people living in a Mediterranean environment. [*Note added after the meeting:* It is in fact uncertain how far (if at all) Vico's argument was read and understood by any of his contemporaries. Montesquieu may have had his book, but he does not cite it explicitly: his own *Esprit des Lois* treated cultural complexes as " organic ", and to that extent made use of the notion of " anachronisms " or " incongruities of period and location ". But for him the determining factors were *environmental*, rather than *historical*. If, as seems quite likely, the implications of Vico's analysis were appreciated only after 1800, then he becomes (so to speak) the " Mendel of history ". His most significant

disciple, by whom the central ideas of the historical approach he
pioneered were securely launched into the tide of European
thought, was the French historian Michelet, who first read Vico
in 1824.]

By the second half of the seventeenth century, a new sense of
" period " was making itself felt all over the human scene,
sometimes in quite unexpected ways. Two examples will
illustrate the general tendency. The first involves styles in
painting. If one looks at medieval paintings of supposedly
antique scenes (e.g. those by the Dutch Primitives) one finds the
characters represented, as a matter of course, in medieval
costume. A fourteenth-century painter did not trouble to adapt
the clothing of his figures to the supposed dates of the events
portrayed: he saw no reason to do so. After all, clothes were
clothes. Solomon, Holofernes, Augustine of Hippo, or a
living merchant—why should one paint them in anything but
clothing familiar to fourteenth-century men as such? During the
seventeenth century the passion for antiquity had its effect on
painting also, contemporary figures being painted in imitation
togas or Roman armour. But, from a certain point in the
eighteenth century, a new scruple affected painters: they began
to match clothes to subjects, and became anxious as never before
to avoid anachronisms. From this time on, indeed, the very
word " anachronisms " acquired a new sense. Earlier it had
referred to " mis-datings ", but now it was extended to embrace
also " out-of-period incongruities "—those which Coleridge later
distinguished as " practical anachronisms ".

At about the same time, the sense of period made itself felt
on the stage also. Traditionally, productions of Shakespeare had
been dressed out of one and the same wardrobe, regardless of
period: nobody had thought it incongruous to present Julius
Caesar in seventeenth-century clothing. Now, however, a new
and exciting alternative presented itself. For the first time, it
occurred to theatrical producers to present Shakespeare's
ancient history plays—as a matter of deliberate policy—*in the
costume of the original period*. At the time, this move had all the
force and shock of an *avant-garde* gesture. Men experienced the
same excitement on being shown Julius Caesar dressed in the
supposed costume of *his own time* that theatregoers in the 1930's
experienced when (reversing this move) they were shown Julius
Caesar in the clothing of *our own time*—namely, in Fascist
uniform.

During the second half of the eighteenth century, this new sense of " development " and " period " gradually reshaped the whole common sense of Europe. Whereas, earlier in the century, history and social science had owed much to the stimulus of Newtonian ideas, the human sciences were now to repay their debt to natural science. From 1770 on, scientists began to look at the Order of Nature with new eyes—beginning with the Earth. Guettard, for instance, at last recognised the mountains of the Auvergne as *extinct volcanoes*. To-day their appearance seems unmistakable, since our preconceptions do not blind us to the possibility of drastic geological change. Up to 1770, however, their resemblance to active volcanoes in other parts of the globe had gone unremarked: hitherto it had been difficult to produce unanswerable examples showing that the crust of the earth had a very different aspect at earlier stages in its history. From this time on, however, the pieces were gradually collected from which, between 1800 and 1830, a brand-new geological jigsaw picture was to be constructed.

At first, the new historical tasks which scientists were beginning to set themselves encountered widespread scepticism. In his poem *The Task* (1785), William Cowper mocked the enterprise as a waste of time:

> " Some write a narrative of wars, and feats
> Of heroes little known; and call the rant
> An history . . . . Some drill and bore
> The solid earth, and from the strata there
> Extract a register, by which we learn
> That he who made it, and reveal'd its date
> To Moses, was mistaken in its age . . . .
> Great contest follows, and much learned dust
> Involves the combatants; each claiming truth
> And truth disclaiming both. And thus they spend
> The little wick of life's poor shallow lamp,
> In playing tricks with nature, giving laws
> To distant worlds, and trifling in their own."

The more candid geologists themselves confessed that they had a long way to go before they could make confident deductions about the remoter stages in the Earth's history. The Edinburgh geologist, James Hutton, was among the most distinguished and original geological thinkers of the period, and contributed strikingly to our ideas about the agencies by which the earth's crust is progressively built up and worn down. Yet he could see in the geological evidence (as he put it) " No Vestige of a Beginning: no Prospect of an End ", and found himself driven back on to an Aristotelian picture of the Earth as surviving from age to age in a Steady State—the effects of constructive and destructive forces eventually cancelling out.

Only between 1810 and 1830 did the modern geological picture of the Earth's history at last crystallise out rapidly. This happened for three reasons. First, geologists turned away from theorising to a piecemeal study of the Earth's present structure. Secondly, they were now in a position to apply reliable chemical methods in the study of surface rocks and their properties, drawing analogies between the internal chemistry of the Earth and the processes studied in the laboratory by men such as Lavoisier and Dalton. Thirdly, William Smith was in course of establishing the principles of " stratigraphy ", by which the sequence of distinct rocky layers (or " strata ") is used as a clue to the *temporal* succession of geological periods—the period during which these superimposed strata were assumed to have reached their present form. From 1830 on, men at last had at their disposal reliable ways of arguing far backwards in time, making deductions from the present state of the Earth to plausible conclusions about its former state. Finally, if one accepted Sir Charles Lyell's " uniformitarian " principle, and supposed that all the existing strata acquired their present forms through the action of forces comparable in kind and magnitude to those we see at work to-day, a further—and more drastic— conclusion was inescapable: the successive phases of past geological time had lasted far longer than any previous generation had conceived.

The years from 1810 to 1850 were a period of intellectual turmoil. This was particularly so in Britain, where a literal-minded natural theology had taken firmest hold. The story of this turbulent age has been well told by Charles Gillispie in his book *Genesis and Geology*, and there is no virtue in going over the same ground here. All that is worth adding is this: that Darwin's work on the *Origin of Species* simply completed the phase in this new historical revolution which Lyell had begun. Despite the fact that, at a stroke, he had extended the life-span of Nature from 6,000 to 6,000,000 years, and despite his deep insight into the geological development of the Earth, Lyell himself had never been able to see the existing organic species—like the existing rock-strata—as the temporary end-product of a continuing process of development. Until the appearance of Darwin's book, indeed, Lyell continued to support the doctrine of fixed species, and it says much for him that he was prepared to change his mind in public, by adopting the new view of species in the later editions of his *Principles of Geology*.

Despite Lyell's acceptance of fixed species, Darwin's teacher Professor Henslow had given him a copy of the *Principles of Geology* to take with him on H.M.S. *Beagle*, with strict instructions that he was on no account to accept its basic conclusions. Darwin foresaw clearly the reception awaiting his ideas, and this helps to explain why he delayed so long before publishing his theory of " evolution by natural selection ". The violent theological debate which raged around the theory of evolution was, accordingly, only a continuation of the previous violent debate provoked by the new geological views about the formation of the earth.

" Before the Hills in Order stood
Or Earth receiv'd her Frame . . . . "

In a mere fifty years, the stable earth and the immutable species of living things—those things which, ever since the beginnings of science, had provided men with the intellectual framework for their system of nature—had been shown to be the mutable end-products of historical development. As a result, our own human epoch has turned to out be, not " a curt Epitome of the whole course " of cosmic history (as Sir Thomas Browne saw a man's life), but a single brief phase which appears almost lost in the unbounded expansion of our time-scale.

When I called this lecture *The Discovery of Time*, I had in mind that vast expansion in our conception of cosmic history. Since the middle of the nineteenth century, men have no longer been able to regard even the most constant aspects of the world as parts of the changeless framework of Divine Creation, or as immutable from the beginning of things right up to their ultimate dissolution: it has become necessary, instead, to see them all as transitional products of continuing processes. No man, it is now clear, can hope ever again to see "about the sixtieth part " of the cosmic story, or comfort himself with the thought that a single lifetime gives him a representative sample of all that will ever happen. This disillusionment has had its sad side, as Tennyson saw. But it has had its merits also. If the past history of the world has been so very, very long, presumably the future also will not terminate either abruptly or soon. So *The Pursuit of the Millennium*, which (as Dr. Norman Cohn reminds us) was a recurrent obsession throughout earlier centuries, has lost much of its plausibility.

Where the preoccupation of natural scientists in the early eighteenth century was with the Fixed Order of Nature, the watchword of natural philosophy in the years after 1850 became " evolution ". Cultural history, anthropology, comparative ethnology, even comparative religion soon joined zoology and geology in the ranks of the evolutionary sciences. Yet, for a while, this historical transformation of the natural sciences remained incomplete. Though by 1900 historical categories were firmly entrenched in geology, zoology and the human sciences, they had as yet made no foothold in physics and chemistry. Indeed, just in those years when the idea of Fixed Species was finally losing its grip on zoological thought, it was taking an equally firm hold on chemical theory. By the work of Cannizzaro, the correctness of the atomic theory appeared at last to have been put beyond doubt; and, certainly, no-one has ever believed more confidently than the chemists of the 1860's and 1870's that the world consisted ultimately of fixed building-blocks (the atoms of the ninety-two different chemical elements) each with a definite shape, conferred on it at the beginning of things and destined to remain unchanged until their dissolution. Right up to the 1890's, two last immutable features kept their place on the natural scene—the fixed atomic species, and the unchanging laws of nature.

During the twentieth century, historical categories have at last made inroads into fundamental physico-chemical theory. Even before 1900, the discovery of radioactivity had created some embarrassment: evidently not *all* atoms were *absolutely* immutable. In more recent years, we have at last learned to recognise the chemical elements themselves as products of a continuing process, by which an initial supply of hydrogen is transformed in the interior of the stars into heavier and more complex atomic systems. In this sense, we now believe in " the Evolution of the Chemical Elements ".

Physics alone appears, at first glance, to have remained *a-historical*. This shows itself in two ways. In the first place, the forms of theory and argument around which the current debate in physical cosmology has concentrated, themselves hint at a certain historical immaturity. In each historical science in turn, so long as insufficient evidence was yet available for the past course of events to be reconstructed in a satisfactory way, speculation has repeatedly fallen into certain basic patterns—Plato's Big Bang, Aristotle's Steady State, and the Stoics' Cyclic Cosmos being the most typical. Is it a pure coincidence that these three forms of theory have, once again, been playing so large a part in cosmological speculation? Will it turn out in cosmology, as in geology earlier, that our present reliance on these intellectual forms is only a transitory feature of the subject—one which can be outgrown, when enough evidence comes to hand for a detailed empirical reconstruction of the actual course of astronomical development?

The other striking survival from the older, more static view of nature is the very conception of " laws of nature " itself. In physics to-day, this conception plays almost as central a part as it has ever done. Yet, looking back at the way in which the broader panorama of scientific thought has developed during the last two centuries, I am tempted to ask: "Are these supposedly-fixed Laws of Nature also, perhaps, the temporary products of a continuing sequence of cosmic changes? " In the 1930's, Professor E. A. Milne of Oxford used to argue in favour of embracing physics also within an evolutionary view of the natural order. The apparent stability of the laws of nature during the epochs we are in a position to study may (he suggested) only mask a slower, longer-term variability. Indeed, by supposing that the physical laws governing the behaviour of large-scale (macrophysical) objects are slowly changing as compared with those governing sub-atomic (microphysical) processes, Milne even produced a striking explanation of the " red-shift " in the spectrum lines from distant nebulae. (This, of course, is the main observational fact on which cosmological speculation is based, and Milne's theory accounted for it without bringing in the hypothetical " recession of the nebulae " which most other cosmological theories are compelled to assume.)

Milne's theory must, of course, take its chance along with rival cosmological speculations. Yet it is worth noticing that other physicists in the last few years have begun to explore similar possibilities: P. A. M. Dirac, for instance, has been asking whether the fundamental constants of gravitation-theory may not vary from one cosmic epoch to another. Whatever the merits or defects of these particular theories, the very fact that they are now being put forward seriously does something to keep my

112

closing question alive: " Will the Fixed Laws of Nature eventually go the same way as the Fixed Elements of Chemistry, the Fixed Planetary System, the Fixed Frame of the Earth, the Fixed Species of Organic Creatures, the Fixed Laws of Human Nature and Social Organisation . . . ?   Is physics, too, beginning to acquire a historical dimension, a sense (so to speak) of ' cosmic period ' ?   And will variations from epoch to epoch in the fundamental laws and principles of physical theory soon come to appear normal expressions of the evolutionary unfolding of History and Nature, quite as much as—say—the substitution of a modern business suit for Julius Caesar's toga ? "

2

ABSTRACT

OF A

DISSERTATION

READ IN THE

ROYAL SOCIETY OF EDINBURGH,

UPON THE

SEVENTH of MARCH, and FOURTH of APRIL,

M,DCC,LXXXV,

CONCERNING THE

SYSTEM OF THE EARTH,

ITS DURATION, AND STABILITY.

JAMES HUTTON

ABSTRACT, &c.

THE purpose of this Differ-
tation is to form some
estimate with regard to the time
the globe of this Earth has existed, as a world maintaining plants
and animals; to reason with regard to the changes which the
earth has undergone; and to see
how far an end or termination to
this system of things may be perceived, from the consideration of
that

(3)

that which has already come to
pafs.

As it is not in human record,
but in natural hiftory, that we are
to look for the means of afcer-
taining what has already been, it.
is here propofed to examine the
appearances of the earth, in order
to be informed of operations
which have been tranfacted in time
paft. It is thus that, from principles of natural philofophy, we
may arrive at fome knowledge of
order and fyftem in the oeconomy
of this globe, and may form a rational opinion with regard to the
courfe of nature, or to events
which are in time to happen.

The

(4)

The folid parts of the prefent land appear, in general, to have been compofed of the productions of the fea, and of other materials fimilar to thofe now found upon the fhores. Hence we find reafon to conclude,

1*ſt*, That the land on which we reft is not fimple and original, but that it is a compofition, and had been formed by the operation of fecond caufes.

2*dly*, That, before the prefent land was made, there had fubfifted a world compofed of fea and land, in which were tides and currents, with fuch operations at

the

the bottom of the fea as now take place. And,

Laftly, That, while the prefent land was forming at the bottom of the ocean, the former land maintained plants and animals ; at leaft, the fea was then inhabited by animals, in a fimilar manner as it is at prefent.

Hence we are led to conclude, that the greater part of our land, if not the whole, had been produced by operations natural to this globe ; but that, in order to make this land a permanent body, refifting the operations of the waters, two things had been required ;

(6)

red ; 1*ſt*, The * confolidation of maſſes formed by collections of loofe or incoherent materials ; 2*dly*, The elevation of thofe confolidated maſſes from the bottom of the fea, the place where they were collected, to the ſtations in which they now remain above the level of the ocean.

Here are two different changes, which may ſerve mutually to throw

---

* There are two fenfes in which the term *folidity* is ufed ; one of thefe is in oppofition to *fluidity*, the other to *vacuity*. When the change from a fluid ſtate to that of folidity, in the firſt fenfe, is to be expreffed, we fhall employ the term *concretion* ; confequently, the confolidation of a maſs is only to be underſtood as in oppofition to its vacuity, or poroufnefs.

(7)

throw ſome light upon each other; for, as the ſame ſubject has been made to undergo both theſe chan- ges, and as it is from the exami- nation of this ſubject that we are to learn the nature of thoſe events, the knowledge of the one may lead us to ſome underſtanding of the other.

Thus the ſubject is conſidered as naturally divided into two branches, to be ſeparately exami- ned : *Firſt*, by what natural ope- ration ſtrata of looſe materials had been formed into ſolid maf- ſes; *ſecondly*, By what power of nature the conſolidated ſtrata at the

(8)

the bottom of the sea had been transformed into land.

With regard to the *firſt* of theſe, the conſolidation of ſtrata, there are two ways in which this operation may be conceived to have been performed; firſt, by means of the ſolution of bodies in water, and the after concretion of theſe diſſolved ſubſtances, when ſeparated from their ſolvent; *ſecondly*, the fuſion of bodies by means of heat, and the ſubſequent congelation of thoſe conſolidating ſubſtances.

With regard to the operation of water, it is *firſt* conſidered, how

B   far

(9)

31

far the power of this folvent, acting in the natural fituation of
thofe ftrata, might be fufficient to
produce the effect; and here it is
found, that water alone, without
any other agent, cannot be fuppofed capable of inducing folidity
among the materials of ftrata in
that fituation. It is, *2dly*, confidered, how far, fuppofing water
capable of confolidating the ftrata
in that fituation, it might be concluded, from examining natural
appearances, that this had been
actually the cafe? Here again,
having proceeded upon this principle, that water could only con-
folidate ftrata with fuch fubftances as it has the power to diffolve,
and

(10)

32

and having found ftrata confolidated with every fpecies of fubftance, it is concluded, that ftrata
in general have not been confolidated by means of aqueous folution.

With regard to the other probable means, heat and fufion, thefe
are found to be perfectly competent for producing the end in
view, as every kind of fubftance
may by heat be rendered foft, or
brought into fufion, and as ftrata
are actually found confolidated
with every different fpecies of fubftance.

A

(11)

82276

A more particular difcuffion is then entered into: Here, confolidating fubftances are confidered as being claffed under two different heads, viz. Siliceous and fulphureous bodies, with a view to prove, that it could not be by means of aqueous folution that ftrata had been confolidated with thofe particular fubftances, but that their confolidation had been accomplifhed by means of heat and fufion.

Sal Gem, as a fubftance foluble in water, is next confidered, in order to fhow that this body had been laft in a melted ftate; and this example is confirmed by one

of

(12)

of foffile alkali. The cafe of particular feptaria of iron-ftone, as
well as certain cryftallized cavities
in mineral bodies, are then given
as examples of a fimilar fact; and
as containing, in themfelves, a de-
monftration, that all the various
mineral fubftances had been con-
creted and cryftallized immediately from a ftate of fufion.

Having thus proved the actual
fufion of the fubftances with which
ftrata had been confolidated, in
having fuch fluid bodies introduced among their interftices, the
cafe of ftrata, confolidated by
means of the fimple fufion of their
proper materials, is next confidered ;

(13)

35

ed ; and examples are taken from the moſt general ſtrata of the globe, viz. ſiliceous and calcareous. Here alſo demonſtration is given, that this conſolidating operation had been performed by means of fuſion.

Having come to this general concluſion, that heat and fuſion, not aqueous ſolution, had preceded the conſolidation of the looſe materials collected at the bottom of the ſea, thoſe conſolidated ſtrata, in general, are next examined, in order to diſcover other appearances, by which the doctrine may be either confirmed or refuted. Here the changes of ſtrata, from their

(14)

their natural ftate of continuity, by veins and fiffures, are confider-ed ; and the cleareft evidence is hence deduced, that the ftrata have been confolidated by means of fufion, and not by aqueous fo-lution ; for, not only are ftrata in general found interfected with veins and cutters, an appearance inconfiftent with their having been confolidated fimply by previous folution ; but, in proportion as ftrata are more or lefs confolida-ted, they are found with the pro-per correfponding appearances of veins and fiffures.

With regard to the fecond branch, in confidering by what power

(15)

power the confolidated ftrata had been transformed into land, or raifed above the level of the fea, it is fuppofed that the fame power of extreme heat, by which every different mineral fubftance had been brought into a melted ftate, might be capable of producing an expanfive force, fufficient for elevating the land, from the bottom of the ocean, to the place it now occupies above the furface of the fea. Here we are again referred to nature, in examining how far the ftrata, formed by fucceffive fediments or accumulations depofited at the bottom of the fea, are to be found in that regular ftate, which would neceffarily take place

in

(16)

38

in their original production; or if, on the other hand, they are actually changed in their natural situation, broken, twisted, and confounded, as might be expected, from the operation of subterranean heat, and violent expansion. But, as strata are actually found in every degree of fracture, flexure, and contortion, consistent with this supposition, and with no other, we are led to conclude, that our land had been raised above the surface of the sea, in order to become a habitable world; as well as that it had been consolidated by means of the same power of subterranean heat, in order to remain above the level of the sea,

C   and

(17)

and to refift the violent efforts of the ocean.

This theory is next confirmed by the examination of mineral veins, thofe great fiffures of the earth, which contain matter perfectly foreign to the ftrata they traverfe; matter evidently derived from the mineral region, that is, from the place where the active power of fire, and the expanfive force of heat, refide.

Such being confidered as the operations of the mineral region, we are hence directed to look for the manifeftation of this power and force, in the appearances of nature.

(18)

nature. It is here we find eruptions of ignited matter from the
fcattered volcano's of the globe;
and thefe we conclude to be the
effects of fuch a power precifely
as that about which we now inquire. Volcano's are thus confidered as the proper difcharges of
a fuperfluous or redundant power; not as things accidental in the
courfe of nature, but as ufeful for
the fafety of mankind, and as
forming a natural ingredient in
the conftitution of the globe.

The doctrine is then confirmed, by examining this earth, and
by finding every where, befide
the many marks of ancient volcano's,

(19)

41

cano's, abundance of fubterra-
neous or unerupted lava, in the
bafaltic rocks, the Swedifh trap,
the toadftone, ragftone, and whinftone of Britain and Ireland, of
which particular examples are cited; and a defcription given of
the three different fhapes in which
that unerupted lava is found.

The peculiar nature of this fubterraneous lava is then examined;
and a clear diftinction is formed
between this bafaltic rock and the
common volcanic lavas.

Laftly, The extenfion of this
theory, refpecting mineral ftrata,
to all parts of the globe, is made,
by

(20)

42

by finding a perfect fimilarity in the folid land through all the earth, although, in particular places, it is attended with peculiar productions, with which the prefent inquiry is not concerned.

A theory is thus formed, with regard to a mineral fyftem. In this fyftem, hard and folid bodies are to be formed from foft bodies, from loofe or incoherent materials, collected together at the bottom of the fea; and the bottom of the ocean is to be made to change its place with relation to the centre of the earth, to be formed into land above the level of

(21)

of the ſea, and to become a country fertile and inhabited.

That there is nothing viſionary
in this theory, appears from its
having been rationally deduced
from natural events, from things
which have already happened ;
things which have left, in the particular conſtitutions of bodies,
proper traces of the manner of
their production ; and things
which may be examined with all
the accuracy, or reaſoned upon
with all the light, that ſcience can
afford. As it is only by employing ſcience in this manner, that
philoſophy enlightens man with
the knowledge of that wiſdom or

deſign

(22)

44

defign which is to be found in nature, the fyftem now propofed,
from unqueftionable principles,
will claim the attention of fcien-
tific men, and may be admitted in
our fpeculations with regard to
the works of nature, notwithftanding many fteps in the progrefs may remain unknown.

By thus proceeding upon in-
veftigated principles, we are led
to conclude, that, if this part of
the earth which we now inhabit
had been produced, in the courfe
of time, from the materials of a
former earth, we fhould, in the
examination of our land, find data from which to reafon, with regard

(23)

gard to the nature of that world, which had exifted during the period of time in which the prefent earth was forming; and thus we might be brought to underftand the nature of that earth which had preceded this; how far it had been fimilar to the prefent, in producing plants and nourifhing animals. But this interefting point is perfectly afcertained, by finding abundance of every manner of vegetable production, as well as the feveral fpecies of marine bodies, in the ftrata of our earth.

Having thus afcertained a regular fyftem, in which the prefent land of the globe had been firft formed

(24)

46

formed at the bottom of the ocean, and then raifed above the furface of the fea, a queſtion naturally occurs with regard to time; what had been the fpace of time necef-fary for accomplifhing this great work?

In order to form a judgment concerning this fubject, our atten-tion is directed to another pro-grefs in the fyſtem of the globe, namely, the deſtruction of the land which had preceded that on which we dwell. Now, for this purpofe, we have the actual de-cay of the prefent land, a thing conftantly tranfacting in our view, by which to form an eſtimate.

D　　　This

(25)

This decay is the gradual ablution of our foil, by the floods of rain ; and the attrition of the fhores, by the agitation of the waves.

If we could meafure the progrefs of the prefent land, towards its diffolution by attrition, and its fubmerfion in the ocean, we might difcover the actual duration of a former earth ; an earth which had fupported plants and animals, and had fupplied the ocean with thofe materials which the conftruction of the prefent earth required ; confequently, we fhould have the meafure of a correfponding fpace of time, viz. that which had been required in the production of the prefent

(26)

48

prefent land. If, on the contrary, no period can be fixed for the duration or deftruction of the prefent earth, from our obfervations of thofe natural operations, which, though unmeafurable, admit of no dubiety, we fhall be warranted in drawing the following conclufions; 1*ft,* That it had required an indefinite fpace of time to have produced the land which now appears; 2*dly,* That an equal fpace had been employed upon the conftruction of that former land from whence the materials of the prefent came; *laftly,* That there is prefently laying at the bottom of the ocean the foundation of future land, which is to

appear

(27)

appear after an indefinite fpace of
time.

But, as there is not in human
obfervation proper means for mea-
furing the wafte of land upon the
globe, it is hence inferred, that
we cannot eftimate the duration
of what we fee at prefent, nor calculate the period at which it had
begun; fo that, with refpect to
human obfervation, this world has
neither a beginning nor an end.

An endeavour is then made to
fupport the theory by an argument of a moral nature, drawn
from the confideration of a final
caufe. Here a comparifon is formed

(28)

ed between the prefent theory, and thofe by which there is neceffarily implied either evil or diforder in natural things ; and an argument is formed, upon the fuppofed wifdom of nature, for the juftnefs of a theory in which perfect order is to be perceived. For,

According to the theory, a foil, adapted to the growth of plants, is neceffarily prepared, and carefully preferved; and, in the neceffary wafte of land which is inhabited, the foundation is laid for future continents, in order to fupport the fyftem of this living world.

Thus,

(29)

Thus, either in fuppofing Nature wife and good, an argument is formed in confirmation of the theory, or, in fuppofing the theory to be juft, an argument may be eftablifhed for wifdom and benevolence to be perceived in nature. In this manner, there is opened to our view a fubjeſt interefting to man who thinks; a fubjeſt on which to reafon with relation to the fyftem of nature; and one which may afford the human mind both information and entertainment.

(30)

3

ILLUSTRATIONS

OF THE

HUTTONIAN THEORY

OF THE EARTH.

———◆———

By JOHN PLAYFAIR,

F. R. S. EDIN. AND PROFESSOR OF MATHEMATICS
IN THE UNIVERSITY OF EDINBURGH.

Nunc naturalem caufam quærimus et affiduam, non raram et fortuitam.

SENECA.

EDINBURGH:

PRINTED FOR CADELL AND DAVIES, LONDON, AND
WILLIAM CREECH, EDINBURGH.

1802.

ILLUSTRATIONS, &c.

A Very little attention to the phenomena of the mineral kingdom, is fufficient to con-vince us, that the condition of the earth's fur-face has not been the fame at all times that it is at the prefent moment. When we obferve the impreffions of plants in the heart of the hardeft rocks ; when we difcover trees converted into flint, and entire beds of limeftone or of marble compofed of fhells and corals ; we fee the fame individual in two ftates, the moft widely differ-ent from one another ; and, in the latter in-ftance, have a clear proof, that the prefent land was once deep immerfed under the waters of the ocean. If to this we add, that many maffes of rock, the moft folid and compact, confift of no other materials but fand and gravel ; that, on the other hand, loofe gravel, fuch as is form-ed only in beds of rivers, or on the fea-fhore, now abounds in places remote from both : if we reflect, at the fame time, on the irregular

A   and

Editor's Note: A row of asterisks indicates that material has been omitted from the original article.

and broken figure of our continents, and the identity of the mineral ſtrata on oppoſite ſides of the ſame valley, or the ſame inlet of the ſea ; we ſhall ſee abundant reaſon to conclude, that the earth has been the theatre of many great revolutions, and that nothing on its ſurface has been exempted from their effects.

To trace the ſeries of theſe revolutions, to explain their cauſes, and thus to connect toge-ther all the indications of change that are found in the mineral kingdom, is the proper object of a THEORY OF THE EARTH.

But, though the attention of men may be turned to the theory of the earth by a very ſu-perficial acquaintance with the phenomena of geology, the formation of ſuch a theory requires an accurate and extenſive examination of thoſe phenomena, and is inconſiſtent with any but a very advanced ſtate of the phyſical ſciences. There is, perhaps, in thoſe ſciences, no reſearch more arduous than this ; none certainly where the ſubject is ſo complex ; where the appearan-ces are ſo extremely diverſified, or ſo widely ſcattered, and where the cauſes that have ope-rated are ſo remote from the ſphere of ordinary obſervation. Hence the attempts to form a theory of the earth are of very modern origin, and as, from the ſimplicity of its ſubject, aſtro-nomy is the eldeſt, ſo, on account of the com-

plexneſs

plexnefs of its fubject, geology is the youngeft of the fciences.

It is foreign from the prefent purpofe, to enter on any hiftory of the fyftems that, fince the rife of this branch of fcience, have been invented to explain the phenomena of the mineral kingdom. It is fufficient to remark, that thefe fyftems are ufually reduced to two claffes, according as they refer the origin of terreftrial bodies to FIRE or to WATER; and that, conformably to this divifion, their followers have of late been diftinguifhed by the fanciful names of *Vulcanifts* and *Neptunifts*. To the former of thefe Dr HUTTON belongs much more than to the latter; though, as he employs the agency both of fire and of water in his fyftem, he cannot, in ftrict propriety, be arranged with either.

In the fuccinct account which I am now about to give of this fyftem, I fhall confider the mineral kingdom as divided into two parts, namely, ftratified and unftratified fubftances. I fhall treat, firft, of the phenomena peculiar to the ftratified; next, of thofe peculiar to the unftratified; and, laftly, of the phenomena common to both. Beginning, then, with the firft, the fubject naturally divides itfelf into three branches; viz. the *materials*, the *confolidation*, and the *pofition* of the ftrata.

A 2 SECT.

SECTION I.

OF THE PHENOMENA PECULIAR TO STRATIFIED BODIES.

1. *Materials of the Strata.*

1. IT is well known that, on removing the loofe earth which forms the immediate furface of the land, we come to the folid rock, of which a great proportion is found to be regularly difpofed in ftrata, or beds of determinate thicknefs, inclined at different angles to the horizon, but feparated from one another by equidiftant fuperficies, that often maintain their parallelifm to a great extent. Thefe ftrata bear fuch evident marks of being depofited by water, that they are univerfally acknowledged to have had their origin at the bottom of the fea ; and it is alfo admitted, that the materials which they confift of, were then either foft, or in fuch a ftate of comminution and feparation, as rendered them capable of arrangement by the action of the water in which they were immerfed. Thus far moft of the theories of the earth agree ;

but

57

but from this point they begin to diverge, and each to affume a character and direction peculiar to itfelf. Dr Hutton's does fo, by laying down this fundamental propofition, That in all the ftrata we difcover proofs of the materials having exifted as elements of bodies, which muft have been deftroyed before the formation of thofe of which thefe materials now actually make a part *.

2. The calcareous ftrata are the portion of the mineral kingdom that gives the cleareft teftimony to the truth of this affertion. They often contain fhells, corals, and other exuviæ of marine animals in fo great abundance, that they appear to be compofed of no other materials. Though thefe remains of organized bodies are now converted into ftone or into fpar, their fhape and interior ftructure are often fo well preferved, that the fpecies of animal or plant of which they once made a part, can ftill be diftinguifhed and pointed out among the living inhabitants of the ocean.

Others of the calcareous ftrata appear to be compofed of fragments of fome ancient rocks, which, after having been broken, have been again united into a compact ftone. In thefe we find pieces clearly marked as having been once continuous, but now placed at a diftance from

A 3   one

* Hutton's Theory, vol. i. p. 20. &c.

one another, and exhibiting exactly the fame appearances as if they floated in a fluid of the fame fpecific gravity with themfelves.

From thefe, therefore, and a variety of fimilar appearances, Dr Hutton concludes, that the materials of all the calcareous ftrata have been furnifhed, either from the diffolution of former ftrata, or from the remains of organized bodies. But, though this conclufion is meant to be extended to all the calcareous ftrata, it is not afferted that every cubic inch of marble or of limeftone contains in it the characters of its former condition, and of the changes through which it has paffed. It may, however, be fafely affirmed, that there is fcarce any entire ftratum where fuch characters are not to be found. Thefe muft be held as decifive with refpect to the whole fyftem of ftrata to which they belong; they prove the exiftence of calcareous rocks before the formation of the prefent; and, as the deftruction of thofe is evidently adequate to the fupply of the materials of thefe that we now fee, to look for any other fupply were fuperfluous, and could only embarrafs our reafonings by the introduction of unneceffary hypothefes *.

3. The fame conclufions refult from an examination of the filiceous ftrata; under which we may comprehend the common fand-ftone,
and

* Note i.

and alfo thofe pudding-ftones or breccias where the gravel confifts of quartz. In all thefe inftances, it is plain, that the fand or gravel exifted in a ftate quite loofe and unconnected, at the bottom of the fea, previous to its confolidation into ftone. But fuch bodies of gravel or fand could only be formed from the attrition of large maffes of quartz, or from the diffolution of fuch fand-ftone ftrata as exift at prefent; for it will hardly be alleged, that fand is a cryftallization of quartz, formed from that fubftance, when it paffes from a fluid to a folid ftate.

Thofe pudding-ftones in which the gravel is round and polifhed, carry the conclufion ftill farther, as fuch gravel can only be formed in the beds of rivers or on the fhores of the fea; for, in the depths of the ocean, though currents are known to exift, yet there can be no motion of the water fufficiently rapid to produce the attrition required to give a round figure and fmooth furface to hard and irregular pieces of ftone. There muft have exifted, therefore, not only a fea, but continents, previoufly to the formation of the prefent ftrata.

The fame thing is clearly fhewn by thofe petrifactions of wood, where, though the vegetable ftructure is perfectly preferved, the whole mafs is filiceous, and has, perhaps, been found

A 4   in

in the heart of fome mountain, deep imbedded in the folid rock.

4. Characters of the fame import are alfo found among the argillaceous ftrata, though perhaps more rarely than among the calcareous or filiceous. Such are the impreffions of the leaves and ftems of vegetables ; alfo the bodies of fifh and amphibious animals, found very often in the different kinds of argillaceous fchiftus, and in moft inftances having the figure accurately preferved, but the fubftance of the animal replaced by clay or pyrites. Thefe are all remains of ancient feas or continents ; the latter of which have long fince difappeared from the furface of the earth, but have ftill their memory preferved in thofe archives, where nature has recorded the revolutions of the globe.

* * * * * * *

3. *Pofition of the Strata* *.

36. We have feen of what materials the
ftrata are compofed, and by what power they
have been confolidated ; we are next to inquire, from what caufe it proceeds, that they
are now fo far removed from the region which
they originally occupied, and wherefore, from
being all covered by the ocean, they are at
prefent raifed in many places fifteen thoufand
feet above its furface. Whether this great
change of relative place can be beft accounted
for by the depreffion of the fea, or the elevation
of the ftrata themfelves, remains to be confidered.

Of

---

* Theory of the Earth, vol. i. p. 120.

Of thefe two fuppofitions, the former, at firft fight, feems undoubtedly the moft probable, and we feel lefs reluctance to fuppofe, that a fluid, fo unftable as the ocean, has undergone the great revolution here referred to, than that the folid foundations of the land have moved a fingle fathom from their place. This, however is a mere illufion. Such a depreffion of the level of the fea as is here fuppofed, could not happen without a change proportionally great in the folid part of the globe; and, though admitted as true, will be found very inadequate to explain the prefent condition of the ftrata.

37. Suppofing the appearances which clearly indicate fubmerfion under water to reach no higher than ten thoufand feet above the prefent level of the fea, and of courfe the furface of the fea to have been formerly higher by that quantity than it is now; it neceffarily follows, that a bulk of water has difappeared, equal to more than a feven-hundredth part of the whole magnitude of the globe *. The exiftence of empty caverns, of extent fufficient to contain this vaft body of water, and of fuch a convulfion as to lay them open, and give room to the retreat of the fea, are fuppofitions which a philofopher could only be juftified in admitting, if they promifed to furnifh a very complete explanation of appearances.

* NOTE X.

appearances. But this juſtification is entirely wanting in the preſent caſe ; for the retreat of the ocean to a lower level, furniſhes a very partial and imperfect explanation of the phenomena of geology. It will not explain the numberleſs remains of ancient continents that are involved, as we have ſeen, in the preſent, unleſs it be ſuppoſed that the ancient ocean, though it roſe to ſo great a height, had nevertheleſs its ſhores, and was the boundary of land ſtill higher than itſelf. And, as to that which is now more immediately the object of inquiry, the poſition of the ſtrata, though the above hypotheſis would account in ſome ſort for the change of their place, relatively to the level of the ſea ; yet, if it ſhall be proved, that the ſtrata have changed their place relatively to each other, and relatively to the plane of the horizon, ſo as to have had an angular motion impreſſed on them, it is evident that, for theſe facts, the retreat of the ſea does not afford even the ſhadow of a theory.

38. Now, it is certain, that many of the ſtrata have been moved angularly, becauſe that, in their original poſition they muſt have been all nearly horizontal. Looſe materials, ſuch as ſand and gravel ſubſiding at the bottom of the ſea, and having their interſtices filled with water, poſſeſs a kind of fluidity : they are diſpoſed to yield

yield on the fide oppofite to that where the preffure is greateft, and are therefore, in fome degree, fubject to the laws of hydroftatics. On this account they will arrange themfelves in horizontal layers ; and the vibrations of the incumbent fluid, by impreffing a flight motion backward, and forward, on the materials of thefe layers, will very much affift the accuracy of their level.

It is not, however, meant to deny, that the form of the bottom might influence, in a certain degree, the ftratification of the fubftances depofited on it. The figure of the lower beds depofited on an uneven furface, would neceffarily be affected by two caufes ; the inclination of that furface, on the one hand, and the tendency to horizontality, on the other ; but, as the former caufe would grow lefs powerful as the diftance from the bottom increafed, the latter caufe would finally prevail, fo that the upper beds would approach to horizontality, and the lower would neither be exactly parallel to them, nor to one another. Whenever, therefore, we meet with rocks, difpofed in layers quite parallel to one another, we may reft affured, that the inequalities of the bottom have had no effect, and that no caufe has interrupted the ftatical tendency above explained.

Now,

Now, rocks having their layers exactly parallel, are very common, and prove their original horizontality to have been more precife than we could venture to conclude from analogy alone. In beds of fand-ftone, for inftance, nothing is more frequent than to fee the thin layers of fand, feparated from one another by layers ftill finer of coaly, or micaceous matter, that are almoft exactly parallel, and continue fo to a great extent without any fenfible deviation. Thefe planes can have acquired their parallelifm only in confequence of the property of water juft ftated, by which it renders the furfaces of the layers, which it depofites, parallel to its own furface, and therefore parallel to one another. Though fuch ftrata, therefore, may not now be horizontal, they muft have been fo originally ; otherwife it is impoffible to difcover any caufe for their parallelifm, or any rule by which it can have been produced.

39. This argument for the original horizontality of the ftrata, is applicable to thofe that are now fartheft removed from that pofition. Among fuch, for inftance, that are highly inclined, or even quite vertical, and among thofe that are bent and incurvated in the moft fantaftical manner, as happens more efpecially in the

the primary fchifti, we obferve, through all their finuofities and inflections, an equality of thicknefs and of diftance among their component laminæ. This equality could only be produced by thofe laminæ having been originally fpread out on a flat and level furface, from which fituation, therefore, they muft afterwards have been lifted up by the action of fome powerful caufe, and muft have fuffered this difturbance while they were yet in a certain degree flexible and ductile. Though the primary direction of the force which thus elevated them muft have been from below upwards, yet it has been fo combined with the gravity and refiftance of the mafs to which it was applied, as to create a lateral and oblique thruft, and to produce thofe contortions of the ftrata, which, when on the great fcale, are among the moft ftriking and inftructive phenomena of geology.

*   *   *   *   *   *

45. On the whole, therefore, by comparing the actual polition of the ſtrata, their erectneſs, their curvature, the interruptions of their continuity, and the tranſverſe ſtratification of the ſecondary in reſpect of the primary, with the regular and level ſituation which the ſame ſtrata muſt have originally poſſeſſed, we have a complete demonſtration of their having been diſturbed, torn aſunder, and moved angularly, by a force that has, in general, been directed from below upwards. In eſtabliſhing this concluſion, we have reaſoned more from the facts which relate to the *angular elevation* of the ſtrata, than from thoſe which relate to their *abſolute elevation*, or their tranſlation to a greater

diſtance

diftance from the centre of the earth. This
has been done, becaufe the appearances, which
refpect the abfolute lifting up of the ftrata
are more ambiguous than thofe, which refpect
the change of their angular pofition. The
former might be accounted for, could they be
feparated from the latter, in two ways, viz. either by the retreat of the fea, or the raifing up
of the land ; but the latter can be explained
only in one way, and force us of neceffity to acknowledge the exiftence of an expanding power,
which has acted on the ftrata with incredible
energy, and has been directed from the centre
toward the circumference.

46. When we are affured of the exiftence of
fuch a power as this in the mineral regions, we
fhould argue with fingular inconfiftency if we
did not afcribe to it all the other appearances
of motion in thofe regions, which it is adequate
to produce. If nature in her fubterraneous a-
bodes is provided with a force that could burft
afunder the maffy pavement of the globe, and
place the fragments upright upon their edges,
could fhe not, by the fame effort, raife them
from the greateft depths of the fea, to the higheft elevation of the land ? The caufe that is adequate to one of thefe effects, is adequate to them
both together ; for it is a principle well known
in mechanical philofophy, that the force which

D 3   produces

produces a parallel motion, may, according to the way in which it is applied, produce alfo an angular motion, without any diminution of the former effect. It would, therefore, be extremely unphilofophical to fuppofe, that any other caufe has changed the relative level of the ftrata, and the furface of the fea, than that which has, in fo many cafes, raifed the ftrata from a horizontal to a highly inclined, or even vertical fituation : it would be to introduce the action of more caufes than the phenomena require, and to forget, that nature, whofe operations we are endeavouring to trace, combines the poffeffion of infinite refources with the moft economical application of them.

47. From all, therefore, that relates to the pofition of the ftrata, I think I am juftified in affirming, that their difturbance and removal from the place of their original formation, by a force directed from below upwards, is a fact in the natural hiftory of the earth, as perfectly afcertained as any thing which is not the fubject of immediate obfervation. As to the power by which this great effect has been produced, we cannot expect to decide with equal evidence, but muft be contented to pafs from what is certain to what is probable. We may, then, remark, that of the forces in nature to which our experience does in any degree extend,

70

tend, none feems fo capable of the effect we would afcribe to it, as the expanfive power of heat ; a power to which no limits can be fet, and one, which, on grounds quite independent of the elevation of the ftrata, has been already conclu-ded to act with great energy in the fubterra-neous regions. We have, indeed, no other al-ternative, but either to adopt this explanation, or to afcribe the facts in queftion to fome fecret and unknown caufe, though we are ignorant of its nature, and have no evidence of its exift-ence.

* * * * * * *

3. *Granite.*

77. The term Granite is ufed by Dr Hutton to fignify an aggregate ftone, in which quartz, feltfpar and mica are found diftinct from one another, and not difpofed in layers. The addi-tion of hornblend, fchorl, or garnet, to the three ingredients juft mentioned, is not underftood to alter the *genus* of the ftone, but only to confti-tute a fpecific difference, which it is the bufinefs of lithology to mark by fome appropriate cha-racter, annexed to the generic name of granite.

The

The foffil now defined exifts, like whinftone and porphyry, both in maffes and in veins, though moft frequently in the former. It is like them unftratified in its texture, and is regarded here, as being alfo unftratified in its outward ftructure *. One ingredient which is effential to granite, namely, quartz, is not contained in whinftone; and this circumftance ferves to diftinguifh thefe *genera* from one another, though, in other refpects, they feem to be united by a chain of infenfible gradations, from the

F 2   moft

---

* Thofe rocks that confift of the ingredients here enumerated, if they have at the fame time a fchiftofe texture, or a difpofition into layers, are properly diftinguifhed from granite, and called Gneifs, or Granitic Schiftus. But it has been queftioned whether a ftone does not exift compofed of thefe ingredients, and deftitute of a fchiftofe texture, but yet divided into large beds, vifible in its external form. Dr Hutton fuppofes fuch a ftone not to exift, or at leaft not to conftitute any fuch proportion of the mineral kingdom, as to entitle it to particular confideration, in the general fpeculations of geology.

Whether this fuppofition is perfectly correct, may require to be farther confidered: this, however, is certain, that a rock, in all refpects conformable to it, compofes a great proportion of what are ufually called the granite mountains. See NOTE xv.

moft homogeneous bafaltes, to granite the moft highly cryftallized.

78. Granite, it has been juft faid, exifts moft commonly in maffes ; and thefe maffes are rarely, if ever, incumbent on any other rock : they are the bafis on which others reft, and feem, for the moft part, to rife up from under the ancient, or primary ftrata. The granite, therefore, wherever it is found, is inferior to every other rock ; and as it alfo compofes many of the greateft mountains, it has the peculiarity of being elevated the higheft into the atmofphere, and funk the deepeft under the furface, of all the mineral fubftances with which we are acquainted.

Notwithftanding the circumftance of not being alternated with ftratified bodies, which conftitutes a remarkable difference between granite and whinftone, the affinity of thefe foffils is fuch as to make the fimilarity of their origin by no means improbable. Accordingly, in Dr Hutton's theory, granite is regarded as a ftone of more recent formation than the ftrata incumbent on it ; as a fubftance which has been melted by heat, and which, when forced up from the mineral regions, has elevated the ftrata at the fame time.

79. That granite has undergone a change from a fluid to a folid ftate, is evinced from the cryftallized ftructure in which fome of its component

nent

nent parts are ufually found. This cryftallization is particularly to be remarked of the feltfpar, and alfo of the fchorl, where there is any
admixture of that fubftance, whether in flender
fpiculæ, or in larger maffes. The quartz itfelf
is in fome cafes cryftallized, and is fo, perhaps,
more frequently than is generally fuppofed.
The fluidity of granite, in fome former period
of its exiftence, is fo evident from this, as to
make it appear fingular that it fhould ever have
been confidered as a foffil that had remained always the fame, and one, into the origin of which
it was needlefs to inquire. If the regular forms
of cryftallization are not to be received as proofs
of the fubftance to which they belong having
paffed from a fluid to a folid ftate, neither are
the figures of fhells and of other fuppofed petrifactions, to be taken as indications of a paffage
from the animal to the mineral kingdom ; fo
that there is an end of all geological theories,
and of all reafonings concerning the ancient
condition of the globe. To an argument which
ftrikes equally at the root of all theories, it belongs not to this, in particular, to make any reply.

80. We fhall, therefore, confider it as admitted, that the materials of the granite were originally fluid ; and, in addition to this, we think
it can eafily be proved, that this fluidity was

F 3   not

not that of the elements taken feparately, but of the entire mafs. This laft conclufion follows, from the ftructure of thofe fpecimens, where one of the fubftances is impreffed by the forms which are peculiar to another. Thus, in the Portfoy granite *, which Dr Hutton has fo minutely defcribed, the quartz is impreffed by the rhomboidal cryftals of the feltfpar, and the ftone thus formed is compact and highly confolidated. Hence, this granite is not a congeries of parts, which, after being feparately formed, were fomehow brought together and agglutinated ; but it is certain that the quartz, at leaft, was fluid when it was moulded on the feltfpar. In other granites, the impreffions of the fubftances on one another are obferved in a different order, and the quartz gives its form to the feltfpar. This, however, is more unufual; the quartz is commonly the fubftance which has received the impreffions of all the reft ; and the fpiculæ of fchorl often fhoot both acrofs it and the feltfpar.

The ingredients of granite were therefore fluid when mixed, or at leaft when in contact with one another. Now, this fluidity was not the effect of folution in a menftruum ; for, in that cafe, one kind of cryftal ought not to imprefs another, but each of them fhould have its own peculiar fhape.

81. The

---

* Theory of the Earth, vol. i. p. 104.

81. The perfect confolidation of many granites, furnifhes an argument to the fame effect. For, agreeably to what was already obferved, in treating of the ftrata, a fubftance, when cryftallizing, or paffing from a fluid to a folid ftate, cannot be free from porofity, much lefs fill up completely a fpace of a given form, if, at the fame time, any folvent is feparated from it; becaufe the folvent fo feparated would ftill occupy a certain fpace, and when removed by evaporation or otherwife, would leave that fpace empty. The perfect adjuftment, therefore, of the fhape of one fet of cryftallizing bodies, to the fhape of another fet, as in the Portfoy granite, and their confolidation into one mafs, is as ftrong a proof as could be defired, that they cryftallized from a ftate of fimple fluidity, fuch as, of all known caufes, heat alone is able to produce.

82. This conclufion, however, does not reft on a fingle clafs of facts. It has been obferved in many inftances, that where granite and ftratified rocks, fuch as primary fchiftus, are in contact, the latter are penetrated by veins of the former, which traverfe them in various directions. Thefe veins are of different dimenfions, fome being of the breadth of feveral yards, others of a few inches, or even tenths of an inch; they diminifh as they recede from the main bo-

F 4 dy

dy of the granite, to which they are always firmly united, conftituting, indeed, a part of the fame continued rock.

Thefe phenomena, which were firft diftinctly obferved by Dr Hutton, are of great importance in geology, and afford a clear folution of the two chief queftions concerning the relation between granite and fchiftus. As every vein muft be of a date pofterior to the body in which it is contained, it follows, that the fchiftus was not fuper-impofed on the granite, after the formation of this laft. If it be argued, that thefe veins, though pofterior to the fchifti, are alfo pofterior to the granite, and were form-ed by the infiltration of water in which the granite was diffolved or fufpended ; it may be replied, 1mo, That the power of water to diffolve granite, is a poftulatum of the fame kind that we have fo often, and for fuch good reafon, refufed to concede ; and, 2do, That in many inftances the veins proceed from the main body of the granite *upwards* into the fchiftus ; fo that they are in planes much elevated in refpect of the horizon, and have a direction quite oppofite to that which the hypothefis of infiltration requires. It remains certain, therefore, that the whole mafs of granite, and the veins proceeding from it, are coeval, and both of later formation than the ftrata.

Now,

Now, this being eſtabliſhed, and the fluidity
of the veins, when they penetrated into the ſchi-
ſtus, being obvious, it neceſſarily follows, that
the whole granite maſs was alſo fluid at the
ſame time. But this can have been brought
about only by ſubterraneous heat, which alſo
impelled the melted matter againſt the ſuper-
incumbent ſtrata, with ſuch force as to raiſe them
from their place, and to give them that highly inclined poſition in which they are ſtill ſupported
by the granite, after its fluidity has ceaſed. Thus
a concluſion, rendered probable by the cryſtalli-
zation of granite, is eſtabliſhed beyond all contradiction by the phenomena of granitic veins *.

*   *   *   *   *   *   *

* Note xv.

SECTION III.

OF THE PHENOMENA COMMON TO STRATIFIED AND UNSTRATIFIED BODIES.

92. THE feries of changes which foffil bodies are deftined to undergo, does not ceafe with their elevation above the level of the fea ; it affumes, however, a new direction, and from the moment that they are raifed up to the furface, is conftantly exerted in reducing them again under the dominion of the ocean. The folidity is now deftroyed which was acquired in the bowels of the earth ; and as the bottom of the fea is the great laboratory, where loofe materials are mineralized and formed into ftone, the atmofphere is the region where ftones are decompofed, and again refolved into earth.

This decompofition of all mineral fubftances, expofed to the air, is continual, and is brought about by a multitude of agents, both chemical and mechanical, of which fome are known to us, and many, no doubt, remain to be difcovered. Among the various aëriform fluids which compofe our atmofphere, one is already diftinguifhed as the grand principle of mineral decompofition ; the others are not inactive, and to them we muft

G   add

add moiſture, heat, and perhaps light ; ſubſtances which, from their affinities to the elements of mineral bodies, have a power of entering into combination with them, and of thus di-
miniſhing the forces by which they are united
to one another. By the action of air and moiſture, the metallic particles, particularly the
iron, which enters in ſuch abundance into the
compoſition of almoſt all foſſils, becomes oxy-
dated in ſuch a degree as to loſe its tenacity ; ſo
that the texture of the ſurface is deſtroyed, and
a part of the body reſolved into earth.

93. Some earths, again, ſuch as the calcareous, are immediately diſſolved by water ; and
though the quantity ſo diſſolved be extremely
ſmall, the operation, by being continually renewed, produces a flow but perpetual corroſion,
by which the greateſt rocks muſt in time be ſub-
dued. The action of water in deſtroying hard
bodies into which it has obtained entrance, is
much aſſiſted by the viciſſitudes of heat and
cold, eſpecially when the latter extends as far as
the point of congelation ; for the water, when
frozen, occupies a greater ſpace than before, and
if the body is compact enough to refuſe room
for this expanſion, its parts are torn aſunder by
a repulſive force acting in every direction.

94. Beſides theſe cauſes of mineral decompo-
ſition, the action of which we can in ſome mea-
ſure

fure trace, there are others known to us only by their effects.

We fee, for inftance, the pureft rock cryftal affected by expofure to the weather, its luftre tarnifhed, and the polifh of its furface impaired, but we know nothing of the power by which thefe operations are performed. Thus alfo, in the precautions which the mineralogift takes to preferve the frefh fracture of his fpecimens, we have a proof how indifcriminately all the productions of the foffil kingdom are expofed to the attacks of their unknown enemies, and we perceive how difficult it is to delay the beginnings of a procefs which no power whatever can finally counteract.

95. The mechanical forces employed in the difintegration of mineral fubftances, are more eafily marked than the chemical. Here again water appears as the moft active enemy of hard and folid bodies; and, in every ftate, from tranfparent vapour to folid ice, from the fmalleft rill to the greateft river, it attacks whatever has emerged above the level of the fea, and labours inceffantly to reftore it to the deep. The parts loofened and difengaged by the chemical agents, are carried down by the rains, and, in their defcent, rub and grind the fuperficies of other bodies. Thus water, though incapable of acting on hard fubftances by direct attrition, is the

G 2   caufe

caufe of their being fo acted on ; and, when it
defcends in torrents, carrying with it fand, gravel, and fragments of rock, it may be truly faid
to turn the forces of the mineral kingdom againft
itfelf.   Every feparation which it makes is neceffarily permanent, and the parts once detached can never be united, fave at the bottom of
the ocean.

96. But it would far exceed the limits of this
fketch, to purfue the caufes of mineral decompofition through all their forms.   It is fufficient
to remark, that the confequence of fo many mi-
nute, but indefatigable agents, all working together, and having *gravity* in their favour, is a fyftem of univerfal decay and degradation, which
may be traced over the whole furface of the land,
from the mountain top to the fea fhore.   That we
may perceive the full evidence of this truth, one
of the moft important in the natural hiftory of the
globe, we will begin our furvey from the latter
of thefe ftations, and retire gradually toward the
former.

97. If the coaft is bold and rocky, it fpeaks
a language eafy to be interpreted.   Its broken
and abrupt contour, the deep gulphs and falient
promontories by which it is indented, and the proportion which thefe irregularities bear to the force
of the waves, combined with the inequality of
hardnefs in the rocks, prove, that the prefent
                                                line

line of the fhore has been determined by the
action of the fea.   The naked and precipitous
cliffs which overhang the deep, the rocks hollowed, perforated, as they are farther advanced in
the fea, and at laft infulated, lead to the fame
conclufion, and mark very clearly fo many different ftages of decay.   It is true, we do not fee
the fucceffive fteps of this progrefs exemplified
in the ftates of the fame individual rock, but
we fee them clearly in different individuals; and
the conviction thus produced, when the phenomena are fufficiently multiplied and varied, is
as irrefiftible, as if we faw the changes actually
effected in the moment of obfervation.

On fuch fhores, the fragments of rock once
detached, become inftruments of further deftruction, and make a part of the powerful
artillery with which the ocean affails the bul-
warks of the land : they are impelled againft
the rocks, from which they break off other fragments, and the whole are thus ground againft
one another ; whatever be their hardnefs, they
are reduced to gravel, the fmooth furface and
round figure of which, are the moft certain
proofs of a *detritus* which nothing can refift.

98. Again, where the fea-coaft is flat, we have
abundant evidence of the degradation of the
land in the beaches of fand and fmall gravel;
the fand banks and fhoals that are continually

G 3   changing ;

changing ; the alluvial land at the mouths of the rivers ; the bars that feem to oppofe their difcharge into the fea, and the fhallownefs of the fea itfelf. On fuch coafts, the land ufually feems to gain upon the fea, whereas, on fhores of a bolder afpect, it is the fea that generally appears to gain upon the land. What the land acquires in extent, however, it lofes in elevation ; and, whether its furface increafe or diminifh, the depredations made on it are in both cafes evinced with equal certainty.

99. If we proceed in our furvey from the fhores, inland, we meet at every ftep with the fulleft evidence of the fame truths, and particularly in the nature and economy of rivers. Every river appears to confift of a main trunk, fed from a variety of branches, each running in a valley proportioned to its fize, and all of them together forming a fyftem of vallies, communicating with one another, and having fuch a nice adjuftment of their declivities, that none of them join the principal valley, either on too high or too low a level ; a circumftance which would be infinitely improbable, if each of thefe vallies were not the work of the ftream that flows in it.

If indeed a river confifted of a fingle ftream, without branches, running in a ftraight valley, it might be fuppofed that fome great concuffion,

cuffion, or fome powerful torrent, had opened at once the channel by which its waters
are conducted to the ocean; but, when the
ufual form of a river is confidered, the trunk
divided into many branches, which rife at a
great diftance from one another, and thefe again
fubdivided into an infinity of fmaller ramifica-
tions, it becomes ftrongly impreffed upon the
mind, that all thefe channels have been cut by
the waters themfelves; that they have been
flowly dug out by the wafhing and erofion of
the land; and that it is by the repeated touches of the fame inftrument, that this curious
affemblage of lines has been engraved fo deeply
on the furface of the globe.

100. The changes which have taken place in
the courfes of rivers, are alfo to be traced, in many inftances, by fucceffive platforms of flat alluvial land, rifing one above another, and marking the different levels on which the river has
run at different periods of time. Of thefe, the
number to be diftinguifhed, in fome inftances,
is not lefs than four, or even five; and this neceffarily carries us back, like all the operations
we are now treating of, to an antiquity extremely remote: for, if it be confidered, that
each change which the river makes in its bed,
obliterates at leaft a part of the monuments of
former changes, we fhall be convinced, that

G 4   only

only a fmall part of the progreffion can leave
any diftinct memorial behind it, and that there
is no reafon to think, that, in the part which
we fee, the beginning is included *.

101. In the fame manner, when a river under-
mines its banks, it often difcovers depofites of
fand and gravel, that have been made when it
ran on a higher level than it does at prefent.
In other inftances, the fame ftrata are feen on
both the banks, though the bed of the river is
now funk deep between them, and perhaps
holds as winding a courfe through the folid
rock, as if it flowed along the furface; a proof
that it muft have begun to fink its bed, when it
ran through fuch loofe materials as oppofed but
a very inconfiderable refiftance to its ftream.
A river, of which the courfe is both ferpentine
and deeply excavated in the rock, is among the
phenomena, by which the flow wafte of the
land, and alfo the caufe of that wafte, are moft
directly pointed out.

102. It is, however, where rivers iffue through
narrow defiles among mountains, that the identity of the ftrata on both fides is moft eafily re-
cognifed, and remarked at the fame time with
the greateft wonder.   On obferving the Pa-
towmack, where it penetrates the ridge of the
Allegany mountains, or the Irtifh, as it iffues
from the defiles of Altai, there is no man, however

* Note xvi.

ever little addicted to geological fpeculations, who does not immediately acknowledge, that the mountain was once continued quite acrofs the fpace in which the river now flows ; and, if he ventures to reafon concerning the caufe of fo wonderful a change, he afcribes it to fome great convulfion of nature, which has torn the mountain afunder, and opened a paffage for the waters. It is only the philofopher, who has deeply meditated on the effects which action long continued is able to produce, and on the fimplicity of the means which nature employs in all her operations, who fees in this nothing but the gradual working of a ftream, that once flowed as high as the top of the ridge which it now fo deeply interfects, and has cut its courfe through the rock, in the fame way, and almoft with the fame inftrument, by which the lapi-dary divides a block of marble or granite.

* * * * * * *

114. Such, according to Dr Hutton's theory, are the changes which the daily operations of waſte have produced on the ſurface of the globe. Theſe operations, inconſiderable if taken ſeparately, become great, by conſpiring all to the ſame end, never counteracting one another, but proceeding, through a period of indefinite extent, continually in the ſame direction. Thus every thing deſcends, nothing returns upward ; the hard and ſolid bodies every where diſſolve, and the looſe and ſoft no where conſolidate. The powers which tend to preſerve, and thoſe which tend to change the condition of the earth's ſurface, are never *in equilibrio ;* the latter are, in all caſes, the moſt powerful, and, in reſpect of the former, are like *living* in compariſon of *dead* forces. Hence the law of decay is one which ſuffers no exception : The elements of all bodies were once looſe and unconnected, and to the

the fame ftate nature has appointed that they fhould all return.

115. It affords no prefumption againft the reality of this progrefs, that, in refpect of man, it is too flow to be immediately perceived : The utmoft portion of it to which our experience can extend, is evanefcent, in comparifon with the whole, and muft be regarded as the momentary increment of a vaft progreffion, circumfcribed by no other limits than the duration of the world. TIME performs the office of *integrating* the infinitefimal parts of which this progreffion is made up; it collects them into one fum, and produces from them an amount greater than any that can be affigned.

* * * * * * *

117. We are not, however, to imagine, that there is no where any means of repairing this waſte ; for, on comparing the concluſion at which we are now arrived, viz. that the preſent continents are all going to decay, and their materials deſcending into the ocean, with the propoſition firſt laid down, that theſe ſame continents are compoſed of materials which muſt have been collected from the decay of former rocks, it is impoſſible not to recogniſe two correſponding ſteps of the ſame progreſs ; of a progreſs, by which mineral ſubſtances are ſubjected to the ſame ſeries of changes, and alternately waſted away and renovated. In the ſame manner, as the preſent mineral ſubſtances derive their origin from ſubſtances ſimilar to themſelves ; ſo, from the land now going to decay, the ſand and gravel forming on the ſea-ſhore, or in the beds of rivers ; from the ſhells and corals which in ſuch enormous quantities are every day accumulated in the boſom of the ſea ; from the drift wood, and the multitude of vegetable and animal remains continually depoſited in the ocean : from all theſe we cannot doubt, that ſtrata are now forming in thoſe regions, to
which

which nature feems to have confined the powers of mineral reproduction ; from which, after being confolidated, they are again deftined to emerge, and to exhibit a feries of changes fimilar to the paft *.

118. How often thefe viciffitudes of decay and renovation have been repeated, is not for us to determine : they conftitute a feries, of which, as the author of this theory has remarked, we neither fee the beginning nor the end ; a circumftance that accords well with what is known concerning other parts of the economy of the world. In the continuation of the different fpecies of animals and vegetables that inhabit the earth, we difcern neither a beginning nor an end ; and, in the planetary motions, where geometry has carried the eye fo far both into the future and the paft, we difcover no mark, either of the commencement or the termination of the prefent order †. It is unreafonable, indeed, to fuppofe, that fuch marks fhould any where exift. The Author of nature has not given laws to the univerfe, which, like the inftitutions of men, carry in themfelves the elements of their own deftruction. He has not permitted, in his works, any fymptom of infancy or of old age, or any fign by which we may eftimate either their future or their paft duration. He may put an end, as he no doubt gave a beginning,

H 4

* NOTE xix.   † NOTE xx,

ning, to the prefent fyftem, at fome determinate period ; but we may fafely conclude, that this great *cataftrophe* will not be brought about by any of the laws now exifting, and that it is not indicated by any thing which we perceive.

* * * * * * *

126. It is impoffible to look back on the fyftem which we have thus endeavoured to illuftrate, without being ftruck with the novelty and beauty of the views which it fets before us. The very plan and fcope of it diftinguifh it from all other theories of the earth, and point it out as a work of great and original invention. The fole object of fuch theories has hitherto been, to explain the manner in which the prefent laws of the mineral kingdom were firft eftablifhed, or began to exift, without treating of the manner in which they now proceed, and by which their continuance is provided for. The authors of thefe theories have accordingly gone back to a ftate of things altogether unlike the prefent, and have confined their reafonings, or

their

their fictions, to a crisis which never has existed but once, and which never can return. Dr Hutton, on the other hand, has guided his investigation by the philosophical maxim, *Causam naturalem et assiduam quærimus, non raram et fortuitam.* His theory, accordingly, presents us with a system of wise and provident economy, where the same instruments are continually employed, and where the decay and renovation of fossils being carried on at the same time in the different regions allotted to them, preserve in the earth the conditions essential for the support of animal and vegetable life. We have been long accustomed to admire that beautiful contrivance in nature, by which the water of the ocean, drawn up in vapour by the atmosphere, imparts, in its descent, fertility to the earth, and becomes the great cause of vegetation and of life ; but now we find, that this vapour not only fertilizes, but creates the soil; prepares it from the solid rock, and, after employing it in the great operations of the surface, carries it back into the regions where all its mineral characters are renewed. Thus, the circulation of moisture through the air, is a prime mover, not only in the annual succession of the seasons, but in the great geological cycle, by which the waste and reproduction of entire continents is circumscribed. Perhaps a more striking view than this, of the wisdom

dom that prefides over nature, was never pre-
fented by any philofophical fyftem, nor a greater addition ever made to our knowledge of final
caufes. It is an addition which gives confiftency
to the reft, by proving, that equal forefight is exerted in providing for the whole and for the
parts, and that no lefs care is taken to maintain
the conftitution of the earth, than to preferve
the tribes of animals and vegetables which dwell
on its furface. In a word, it is the peculiar excellence of this theory, that it afcribes to the
phenomena of geology an order fimilar to that
which exifts in the provinces of nature with
which we are beft acquainted ; that it produces feas and continents, not by accident, but by
the operation of regular and uniform caufes ;
that it makes the decay of one part fubfervient
to the reftoration of another, and gives ftability
to the whole, not by perpetuating individuals,
but by reproducing them in fucceffion.

* * * * * * *

As another excellence of this theory, I may, perhaps, be allowed to remark, that it extends its confequences beyond thofe to which the author of it has himfelf adverted, and that it affords, which no geological theory has yet done, a fatisfactory explanation of the fpheroidal figure of the earth *.

133. Yet, with all thefe circumftances of originality, grandeur, and fimplicity in its favour, with the addition of evidence as demonftrative as the nature of the fubject will admit, this theory has probably many obftacles to overcome, before it meet the general approbation. The greatnefs of the objects which it fets before us, alarms the imagination; the powers which it fuppofes to be lodged in the fubterraneous regions,

gions; a heat which has fubdued the moft refractory rocks, and has melted beds of marble
and quartz ; an expanfive force, which has folded up, or broken the ftrata, and raifed whole
continents from the bottom of the fea ; thefe
are things with which, however certainly they
may be proved, the mind cannot foon be fami-
liarifed.   The change and movement alfo, which
this theory afcribes to all that the fenfes declare
to be moft unalterable, raife up againft it the
fame prejudices which formerly oppofed the belief in the true fyftem of the world ; and it affords a curious proof, how little fuch prejudices are fubject to vary, that as ARISTAR-
CHUS, an ancient follower of that fyftem, was
charged with impiety for moving the everlafting
VESTA from her place, fo Dr Hutton, nearly on
the fame ground, has been fubjected to the very
fame accufation.   Even the length of time
which this theory regards as neceffary to the revolutions of the globe, is looked on as belonging to the marvellous; and man, who finds
himfelf conftrained by the want of time, or of
fpace in almoft all his undertakings, forgets,
that in thefe, if in any thing, the riches of nature reject all limitation *.

The evidence which muft be oppofed to all
thefe caufes of incredulity, cannot be fully underftood without much ftudy and attention.
                                                    It

* NOTE XXVI.

It requires not only a careful examination of particular inftances, but comprehenfive views of the whole phenomena of geology ; the comparifon of things very remote with one another ; the interpretation of the *obfcure* by the *luminous*, and of the *doubtful* by the *decifive* appearances. The geologift muft not content himfelf with examining the infulated fpecimens of his cabinet, or with purfuing the nice fubtleties of mineralogical arrangement ; he muft ftudy the relations of foffils, as they actually exift ; he muft follow nature into her wildeft and moft inacceffible abodes ; and muft felect, for the places of his obfervations, thofe points, from which the variety and gradation of her works can be moft extenfively and accurately explored. Without fuch an exact and comprehenfive furvey, his mind will hardly be prepared to relifh the true theory of the earth. " *Naturæ enim vis atque majeftas omnibus momentis fide caret, fi quis modo partes atque non totam complectatur animo* *."

*   *   *   *   *   *   *

---

* PLIN. Hift. Nat. lib. vii. cap. i.

Reprinted from *Principles of Geology, or the Modern Changes of the Earth and Its Inhabitants Considered as Illustrative of Geology*, Vol. 1, John Murray (Publishers) Ltd., London, 1872, pp. 88–102

Prejudices Which Have Retarded the Progress of Geology

CHARLES LYELL

CHAPTER V.

PREJUDICES WHICH HAVE RETARDED THE PROGRESS OF GEOLOGY.

PREPOSSESSIONS IN REGARD TO THE DURATION OF PAST TIME—PREJUDICES ARISING FROM OUR PECULIAR POSITION AS INHABITANTS OF THE LAND—OTHERS OCCASIONED BY OUR NOT SEEING SUBTERRANEAN CHANGES NOW IN PROGRESS—ALL THESE CAUSES COMBINE TO MAKE THE FORMER COURSE OF NATURE APPEAR DIFFERENT FROM THE PRESENT—OBJECTIONS TO THE DOCTRINE THAT CAUSES SIMILAR IN KIND AND ENERGY TO THOSE NOW ACTING, HAVE PRODUCED THE FORMER CHANGES OF THE EARTH'S SURFACE, CONSIDERED.

IF we reflect on the history of the progress of geology, as explained in the preceding chapters, we perceive that there have been great fluctuations of opinion respecting the nature of the causes to which all former changes of the earth's surface are referable. The first observers conceived the monuments which the geologist endeavours to decipher to relate to an original state of the earth, or to a period when there where causes in activity, distinct, in kind and degree, from those now constituting the economy of nature. These views were gradually modified, and some of them entirely abandoned, in proportion as observations were multiplied, and the signs of former mutations were skilfully interpreted. Many appearances, which had for a long time been regarded as indicating mysterious and extraordinary agency, were finally recognised as the necessary result of the laws now governing the material world; and the discovery of this unlooked-for conformity has at length induced some philosophers to infer, that, during the ages contemplated in geology, there has never been any interruption to the agency of the same uniform laws of change. The same assemblage of general causes, they conceive, may have been sufficient to produce, by their various combinations, the

endless diversity of effects, of which the shell of the earth
has preserved the memorials; and, consistently with these
principles, the recurrence of analogous changes is expected
by them in time to come.

Whether we coincide or not in this doctrine, we must
admit that the gradual progress of opinion concerning the
succession of phenomena in very remote eras, resembles, in
a singular manner, that which has accompanied the growing
intelligence of every people, in regard to the economy of
nature in their own times.   In an early state of advancement,
when a greater number of natural appearances are unintelligible, an eclipse, an earthquake, a flood, or the approach
of a comet, with many other occurrences afterwards found
to belong to the regular course of events, are regarded as
prodigies.   The same delusion prevails as to moral phenomena, and many of these are ascribed to the intervention
of demons, ghosts, witches, and other immaterial and
supernatural agents.   By degrees, many of the enigmas of
the moral and physical world are explained, and, instead of
being due to extrinsic and irregular causes, they are found
to depend on fixed and invariable laws.   The philosopher at
last becomes convinced of the undeviating uniformity of
secondary causes; and, guided by his faith in this principle,
he determines the probability of accounts transmitted to him
of former occurrences, and often rejects the fabulous tales of
former times, on the ground of their being irreconcilable
with the experience of more enlightened ages.

Prepossessions in regard to the duration of past time.—As a
belief in the want of conformity in the causes by which the
earth's crust has been modified in ancient and modern periods
was, for a long time, universally prevalent, and that, too,
amongst men who were convinced that the order of nature
had been uniform for the last several thousand years, every
circumstance which could have influenced their minds and
given an undue bias to their opinions deserves particular
attention.   Now the reader may easily satisfy himself, that,
however undeviating the course of nature may have been
from the earliest epochs, it was impossible for the first cultivators of geology to come to such a conclusion, so long

as they were under a delusion as to the age of the world, and the date of the first creation of animate beings. However fantastical some theories of the sixteenth century may now appear to us,—however unworthy of men of great talent and sound judgment,—we may rest assured that, if the same misconception now prevailed in regard to the memorials of human transactions, it would give rise to a similar train of absurdities.   Let us imagine, for example, that Champollion, and the French and Tuscan literati when engaged in exploring the antiquities of Egypt, had visited that country with a firm belief that the banks of the Nile were never peopled by the human race before the beginning of the nineteenth century, and that their faith in this dogma was as difficult to shake as the opinion of our ancestors, that the earth was never the abode of living beings until the creation of the present continents, and of the species now existing,—it is easy to perceive what extravagant systems they would frame, while under the influence of this delusion, to account for the monuments discovered in Egypt.   The sight of the pyramids, obelisks, colossal statues, and ruined temples, would fill them with such astonishment, that for a time they would be as men spell-bound—wholly incapable of reasoning with sobriety.   They might incline at first to refer the construction of such stupendous works to some superhuman powers of a primeval world.   A system might be invented resembling that so gravely advanced by Manetho, who relates that a dynasty of gods originally ruled in Egypt, of whom Vulcan, the first monarch, reigned nine thousand years; after whom came Hercules and other demigods, who were at last succeeded by human kings.

When some fanciful speculations of this kind had amused their imaginations for a time, some vast repository of mummies would be discovered, and would immediately undeceive those antiquaries who enjoyed an opportunity of personally examining them ; but the prejudices of others at a distance, who were not eye-witnesses of the whole phenomena, would not be so easily overcome.   The concurrent report of many travellers would, indeed, render it necessary for them to accommodate ancient theories to some of the new facts. and much wit and ingenuity would be required to modify and defend their old positions. Each new invention would violate a greater number of known analogies; for if a theory be required to embrace some false principle, it becomes more visionary in proportion as facts are multiplied, as would be the case if geometers were now required to form an astronomical system on the assumption of the immobility of the earth.

Amongst other fanciful conjectures concerning the history of Egypt, we may suppose some of the following to be started. 'As the banks of the Nile have been so recently colonised for the first time, the curious substances called mummies could never in reality have belonged to men. They may have been generated by some *plastic virtue* residing in the interior of the earth, or they may be abortions of Nature produced by her incipient efforts in the work of creation. For if deformed beings are sometimes born even now, when the scheme of the universe is fully developed, many more may have been " sent before their time, scarce half made up," when the planet itself was in the embryo state. But if these notions appear to derogate from the perfection of the Divine attributes, and if these mummies be in all their parts true representations of the human form, may we not refer them to the future rather than the past? May we not be looking into the womb of Nature, and not her grave? May not these images be like the shades of the unborn in Virgil's Elysium—the archetypes of men not yet called into existence?'

These speculations, if advocated by eloquent writers, would not fail to attract many zealous votaries, for they would relieve men from the painful necessity of renouncing preconceived opinions. Incredible as such scepticism may appear, it has been rivalled by many systems of the sixteenth and seventeenth centuries, and among others by that of the learned Falloppio, who, as we have seen (p. 33), regarded the tusks of fossil elephants as earthy concretions, and the pottery or fragments of vases in the Monte Testaceo, near Rome, as works of nature, and not of art. But when one generation had passed away, and another, not compromised to the support of antiquated dogmas, had succeeded, they would review the evidence afforded by mummies more impartially, and would no longer controvert the preliminary question, that human beings had lived in Egypt before the nineteenth century: so that when a hundred years perhaps had been lost, the industry and talents of the philosopher would be at last directed to the elucidation of points of real historical importance.

But the above arguments are aimed against one only of many prejudices with which the earlier geologists had to contend. Even when they conceded that the earth had been peopled with animate beings at an earlier period than was at first supposed, they had no conception that the quantity of time bore so great a proportion to the historical era as is now generally conceded. How fatal every error as to the quantity of time must prove to the introduction of rational views concerning the state of things in former ages, may be conceived by supposing the annals of the civil and military transactions of a great nation to be perused under the impression that they occurred in a period of one hundred instead of two thousand years. Such a portion of history would immediately assume the air of a romance; the events would seem devoid of credibility, and inconsistent with the present course of human affairs. A crowd of incidents would follow each other in thick succession. Armies and fleets would appear to be assembled only to be destroyed, and cities built merely to fall in ruins. There would be the most violent transitions from foreign or intestine war to periods of profound peace, and the works effected during the years of disorder or tranquillity would appear alike superhuman in magnitude.

He who should study the monuments of the natural world under the influence of a similar infatuation, must draw a no less exaggerated picture of the energy and violence of causes, and must experience the same insurmountable difficulty in reconciling the former and present state of nature. If we could behold in one view all the volcanic cones thrown up in Iceland, Italy, Sicily, and other parts of Europe, during the last five thousand years, and could see the lavas which have flowed during the same period; the dislocations, subsidences, and elevations caused during earthquakes; the lands added to various deltas, or devoured by the sea, together with the effects of devastation by flodds, and imagine that all these events had happened in one year, we must form most exalted ideas of the activity of the agents, and the suddenness of the revolutions. If geologists, therefore, have misinterpreted the signs of a succession of events, so as to conclude that centuries were implied where the characters indicated thousands of years, and thousands of years where the language of Nature signified millions, they could not, if . they reasoned logically from such false premises, come to any other conclusion than that the system of the natural world had undergone a complete revolution.

We should be warranted in ascribing the erection of the great pyramid to superhuman power, if we were convinced that it was raised in one day; and if we imagine, in the same manner, a continent or mountain-chain to have been elevated during an equally small fraction of the time which was really occupied in upheaving it, we might then be justified in inferring, that the subterranean movements were once far more energetic than in our own times. We know that during one earthquake the coast of Chili may be raised for a hundred miles to the average height of about three feet. A repetition of two thousand shocks, of equal violence, might produce a mountain-chain one hundred miles long, and six thousand feet high. Now, should one or two only of these convulsions happen in a century, it would be consistent with the order of events experienced by the Chilians from the earliest times: but if the whole of them were to occur in the next hundred years, the entire district must be depopulated, scarcely any animals or plants could survive, and the surface would be one confused heap of ruin and desolation.

One consequence of undervaluing greatly the quantity of past time, is the apparent coincidence which it occasions of events necessarily disconnected, or which are so unusual, that it would be inconsistent with all calculation of chances to suppose them to happen at one and the same time. When the unlooked-for association of such rare phenomena is witnessed in the present course of nature, it scarcely ever fails to excite a suspicion of the preternatural in those minds which are not firmly convinced of the uniform agency of secondary causes;—as if the death of some individual in whose fate they are interested happens to be accompanied by the appearance of a luminous meteor, or a comet, or the shock of an earthquake. It would be only necessary to multiply such coincidences indefinitely, and the mind of every philosopher would be disturbed. Now it would be difficult to exaggerate the number of physical events, many of them most rare and unconnected in their nature, which were imagined by the Woodwardian hypothesis to have happened in the course of a few months: and numerous other examples might be found of popular geological theories, which require us to imagine that a long succession of events happened in a brief and almost momentary period.

Another liability to error, very nearly allied to the former, arises from the frequent contact of geological monuments referring to very distant periods of time. We often behold, at one glance, the effects of causes which have acted at times incalculably remote, and yet there may be no striking circumstances to mark the occurrence of a great chasm in the chronological series of Nature's archives. In the vast interval of time which may really have elapsed between the results of operations thus compared, the physical condition of the earth may, by slow and insensible modifications, have become entirely altered; one or more races of organic beings may have passed away, and yet have left behind, in the particular region under contemplation, no trace of their existence.

To a mind unconscious of these intermediate events, the passage from one state of things to another must appear so violent, that the idea of revolutions in the system inevitably suggests itself. The imagination is as much perplexed by the deception, as it might be if two distant points in space were suddenly brought into immediate proximity. Let us suppose, for a moment, that a philosopher should lie down to sleep in some arctic wilderness, and then be transferred by a power, such as we read of in tales of enchantment, to a valley in a tropical country, where, on awaking, he might find himself surrounded by birds of brilliant plumage, and all the luxuriance of animal and vegetable forms of which Nature is so prodigal in those regions. The most reasonable supposition, perhaps, which he could make, if by the necromancer's art he were placed in such a situation, would be, that he was dreaming; and if a geologist form theories under a similar delusion, we cannot expect him to preserve more consistency in his speculations than in the train of ideas in an ordinary dream.

It may afford, perhaps, a more lively illustration of the principle here insisted upon, if I recall to the reader's recollection the legend of the Seven Sleepers. The scene of that popular fable was placed in the two centuries which elapsed between the reign of the emperor Decius and the death of Theodosius the younger. In that interval of time (between the years 249 and 450 of our era) the union of the Roman empire had been dissolved, and some of its fairest provinces overrun by the barbarians of the north. The seat of government had passed from Rome to Constantinople, and the throne from a pagan persecutor to a succession of Christian and orthodox princes. The genius of the empire had been humbled in the dust, and the altars of Diana and Hercules were on the point of being transferred to Catholic saints and martyrs. The legend relates, 'that when Decius was still persecuting the Christians, seven noble youths of Ephesus concealed themselves in a spacious cavern in the side of an adjacent mountain, where they were doomed to perish by the tyrant, who gave orders that the entrance should be firmly secured with a pile of huge stones. They immediately fell into a deep slumber, which was miraculously prolonged, without injuring the powers of life, during a period of 187 years. At the end of that time the slaves of Adolius, to whom the inheritance of the mountain had descended, removed the stones to supply materials for some rustic edifice: the light of the sun darted into the cavern, and the seven sleepers were permitted to awake. After a slumber, as they thought, of a few hours, they were pressed by the calls of hunger, and resolved that Jamblichus, one of their number, should secretly return to the city to purchase bread for the use of his companions. The youth could no longer recognise the once familiar aspect of his native country, and his surprise was increased by the appearance of a large cross triumphantly erected over the principal gate of Ephesus. His singular dress and obsolete language confounded the baker, to whom he offered an ancient medal of Decius as the current coin of the empire; and Jamblichus, on the suspicion of a secret treasure, was dragged before the judge. Their mutual enquiries produced the amazing discovery, that two centuries were almost elapsed since Jamblichus and his friends had escaped from the rage of a pagan tyrant.'

This legend was received as authentic throughout the Christian world before the end of the sixth century, and was afterwards introduced by Mahomet as a divine revelation into the Koran, and from hence was adopted and adorned by all the nations from Bengal to Africa who professed the Mahometan faith. Some vestiges even of a similar tradition have been discovered in Scandinavia. 'This easy and universal belief,' observes the philosophical historian of the Decline and Fall, ' so expressive of the sense of mankind, may be ascribed to the genuine merit of the fable itself. We imperceptibly advance from youth to age, without observing the gradual, but incessant, change of human affairs; and even, in our larger experience of history, the imagination is accustomed, by a perpetual series of causes and effects, to unite the most distant revolutions. But if the interval between two memorable eras could be instantly annihilated; if it were possible, after a momentary slumber of two hundred years, to display the new world to the eyes of a spectator who still retained a lively and recent impression of the old, his surprise and his reflections would furnish the pleasing subject of a philosophical romance.'*

Prejudices arising from our peculiar position as inhabitants of the land.—The sources of prejudice hitherto considered may be deemed peculiar for the most part to the infancy of the science, but others are common to the first cultivators of

---

* Gibbon, Decline and Fall, chap. xxxiii.

geology and to ourselves, and are all singularly calculated to produce the same deception, and to strengthen our belief that the course of nature in the earlier ages differed widely from that now established. Although these circumstances cannot be fully explained without assuming some things as proved, which it has been my object elsewhere to demonstrate,* it may be well to allude to them briefly in this place.

The first and greatest difficulty, then, consists in an habitual unconsciousness that our position as observers is essentially unfavourable, when we endeavour to estimate the nature and magnitude of the changes now in progress. In consequence of our inattention to this subject, we are liable to serious mistakes in contrasting the present with former states of the globe. As dwellers on the land, we inhabit about a fourth part of the surface; and that portion is almost exclusively a theatre of decay, and not of reproduction. We know, indeed, that new deposits are annually formed in seas and lakes; and that every year some new igneous rocks are produced in the bowels of the earth, but we cannot watch the progress of their formation; and as they are only present to our minds by the aid of reflection, it requires an effort both of the reason and the imagination to appreciate duly their importance. It is, therefore, not surprising that we estimate very imperfectly the result of operations thus unseen by us; and that, when analogous results of former epochs are presented to our inspection, we cannot immediately recognise the analogy. He who has observed the quarrying of stone from a rock, and has seen it shipped for some distant port, and then endeavours to conceive what kind of edifice will be raised by the materials, is in the same predicament as a geologist, who, while he is confined to the land, sees the decomposition of rocks, and the transportation of matter by rivers to the sea, and then endeavours to picture to himself the new strata which Nature is building beneath the waters.

Prejudices arising from our not seeing subterranean changes. —Nor is his position less unfavourable when, beholding a volcanic eruption, he tries to conceive what changes the

---

* Elements of Geology, 6th edit., 1865 ; and Student's Elements, 1871.

VOL. I.　　　　　　　　　　　　H

column of lava has produced, in its passage upwards, on the intersected strata ; or what form the melted matter may assume at great depths on cooling; or what may be the extent of the subterranean rivers and reservoirs of liquid matter far beneath the surface. It should, therefore, be remembered, that the task imposed on those who study the earth's history requires no ordinary share of discretion ; for we are precluded from collating the corresponding parts of the system of things as it exists now, and as it existed at former periods. If we were inhabitants of another element—if the great ocean were our domain, instead of the narrow limits of the land, our difficulties would be considerably lessened ; while, on the other hand, there can be little doubt, although the reader may, perhaps, smile at the bare suggestion of such an idea, that an amphibious being, who should possess our faculties, would still more easily arrive at sound theoretical opinions in geology, since he might behold, on the one hand, the decomposition of rocks in the atmosphere, or the transportation of matter by running water ; and, on the other, examine the deposition of sediment in the sea, and the imbedding of animal and vegetable remains in new strata. He might ascertain, by direct observation, the action of a mountain torrent, as well as of a marine current ; might compare the products of volcanos poured out upon the land with those ejected beneath the waters ; and might mark, on the one hand, the growth of the forest, and, on the other, that of the coral reef. Yet, even with these advantages, he would be liable to fall into the greatest errors, when endeavouring to reason on rocks of subterranean origin. He would seek in vain, within the sphere of his observation, for any direct analogy to the process of their formation, and would therefore be in danger of attributing them, wherever they are upraised to view, to some ' primeval state of nature.'

But if we may be allowed so far to indulge the imagination, as to suppose a being entirely confined to the nether world—some ' dusky melancholy sprite,' like Umbriel, who could ' flit on sooty pinions to the central earth,' but who was never permitted to 'sully the fair face of light,' and emerge into the regions of water and of air ; and if this being should busy himself in investigating the structure of the globe, he might frame theories the exact converse of those usually adopted by human philosophers. He might infer that the stratified rocks, containing shells and other organic remains, were the oldest of created things, belonging to some original and nascent state of the planet. 'Of these masses,' he might say, ' whether they consist of loose incoherent sand, soft clay, or solid stone, none have been formed in modern times. Every year some of them are broken and shattered by earthquakes, or melted by volcanic fire; and when they cool down slowly from a state of fusion, they assume a new and more crystalline form, no longer exhibiting that stratified disposition and those curious impressions and fantastic markings, by which they were previously characterised. This process cannot have been carried on for an indefinite time, for in that case all the stratified rocks would long ere this have been fused and crystallised. It is therefore probable that the whole planet once consisted of these mysterious and curiously bedded formations at a time when the volcanic fire had not yet been brought into activity. Since that period there seems to have been a gradual development of heat; and this augmentation we may expect to continue till the whole globe shall be in a state of fluidity, or shall consist, in those parts which are not melted, of volcanic and crystalline rocks.'

Such might be the system of the Gnome at the very time that the followers of Leibnitz, reasoning on what they saw on the outer surface, might be teaching the opposite doctrine of gradual refrigeration, and averring that the earth had begun its career as a fiery comet, and might be destined hereafter to become a frozen mass. The tenets of the schools of the nether and of the upper world would be directly opposed to each other, for both would partake of the prejudices inevitably resulting from the continual contemplation of one class of phenomena to the exclusion of another. Man observes the annual decomposition of crystalline and igneous rocks, and may sometimes see their conversion into stratified deposits; but he cannot witness the reconversion of the

H 2

sedimentary into the crystalline by subterranean heat. He is in the habit of regarding all the sedimentary rocks as more recent than the unstratified, for the same reason that we may suppose him to fall into the opposite error if he saw the origin of the igneous class only.

For more than two centuries the shelly strata of the Subapennine hills afforded matter of speculation to the early geologists of Italy, and few of them had any suspicion that similar deposits were then forming in the neighbouring sea. Some imagined that the strata, so rich in organic remains, instead of being due to secondary agents, had been so created in the beginning of things by the fiat of the Almighty. Others, as we have seen, ascribed the imbedded fossil bodies to some plastic power which resided in the earth in the early ages of the world. In what manner were these dogmas at length exploded? The fossil relics were carefully compared with their living analogues, and all doubts as to their organic origin were eventually dispelled. So, also, in regard to the nature of the containing beds of mud, sand, and limestone: those parts of the bottom of the sea were examined where shells are now becoming annually entombed in new deposits. Donati explored the bed of the Adriatic, and found the closest resemblance between the strata there forming, and those which constituted hills above a thousand feet high in various parts of the Italian peninsula. He ascertained by dredging that living testacea were there grouped together in precisely the same manner as were their fossil analogues in the inland strata; and while some of the recent shells of the Adriatic were becoming incrusted with calcareous rock, he observed that others had been newly buried in sand and clay, precisely as fossil shells occur in the Subapennine hills.

In like manner, the volcanic rocks of the Vicentin had been studied in the beginning of the last century; but no geologist suspected, before the time of Arduino, that these were composed of ancient submarine lavas. During many years of controversy, the popular opinion inclined to a belief that basalt and rocks of the same class had been precipitated from a chaotic fluid, or an ocean which rose at successive periods over the continents, charged with the component elements of the rocks in question. Few will now dispute that it would have been difficult to invent a theory more distant from the truth; yet we must cease to wonder that it gained so many proselytes, when we remember that its claims to probability arose partly from the very circumstance of its confirming the assumed want of analogy between geological causes and those now in action. By what train of investigations were geologists induced at length to reject these views, and to assent to the igneous origin of the trappean formations? By an examination of volcanos now active, and by comparing their structure and the composition of their lavas with the ancient trap rocks.

The establishment, from time to time, of numerous points of identification, drew at length from geologists a reluctant admission, that there was more correspondence between the condition of the globe at remote eras and now, and more uniformity in the laws which have regulated the changes of its surface, than they at first imagined. If, in this state of the science, they still despaired of reconciling every class of geological phenomena to the operations of ordinary causes, even by straining analogy to the utmost limits of credibility, we might have expected, at least, that the balance of probability would now have been presumed to incline towards the close analogy of the ancient and modern causes. But, after repeated experience of the failure of attempts to speculate on geological monuments, as belonging to a distinct order of things, new sects continued to persevere in the principles adopted by their predecessors. They still began, as each new problem presented itself, whether relating to the animate or inanimate world, to assume an original and dissimilar order of nature; and when at length they approximated, or entirely came round to an opposite opinion, it was always with the feeling, that they were conceding what they had been justified à *priori* in deeming improbable. In a word, the same men who, as natural philosophers, would have been most incredulous respecting any extraordinary deviations from the known course of nature, if reported to have happened *in their own time*, were equally disposed, as geologists, to expect the proofs of such deviations at every period of the past.

I shall proceed in the following chapters to enumerate some of the principal difficulties still opposed to the theory of the uniform nature and energy of the causes which have worked successive changes in the crust of the earth, and in the condition of its living inhabitants.   The discussion of so important a question on the present occasion may appear premature, but it is one which naturally arises out of a review of the former history of the science.   It is, of course, impossible to enter into such speculative topics, without occasionally carrying the novice beyond his depth, and appealing to facts and conclusions with which he will be unacquainted, until he has studied some elementary work on geology, but it may be useful to excite his curiosity, and lead him to study such works by calling his attention at once to some of the principal points of controversy.*

* In the earlier editions of this work, a fourth book was added on Geology Proper, or Systematic Geology, containing an account of the former changes of the animate and inanimate creation, brought to light by an examination of the crust of the earth. This I after- wards (in 1838) expanded into a separate publication called the Elements or Manual of Geology, of which a sixth edition appeared, January 1865, and the greater part of which is embodied in the Student's Elements published in 1871.

Reprinted from *History of the Inductive Sciences from the Earliest to the Present Time*, Vol. 2, Appleton-Century-Crofts, New York, 1872, pp. 586–598

The Two Antagonist Doctrines of Geology

WILLIAM WHEWELL

CHAPTER VIII.

THE TWO ANTAGONIST DOCTRINES OF GEOLOGY.

Sect. 1.—*Of the Doctrine of Geological Catastrophes.*

THAT great changes, of a kind and intensity quite different from the common course of events, and which may therefore properly be called *catastrophes*, have taken place upon the earth's surface, was an opinion which appeared to be forced upon men by obvious facts. Rejecting, as a mere play of fancy, the notions of the destruction of the earth by cataclysms or conflagrations, of which we have already spoken, we find that the first really scientific examination of the materials of the earth, that of the Sub-Apennine hills, led men to draw this inference. Leonardo da Vinci, whom we have already noticed for his early and strenuous assertion of the real marine origin of fossil impressions of shells, also maintained that the bottom of the sea had become the top of the mountain ; yet his mode of explaining this may perhaps be claimed by the modern advocates of uniform causes as more allied to their opinion, than to the doctrine of catastrophes.[1] But Steno, in 1669, approached nearer to this doctrine; for he asserted that Tuscany must have changed its face at intervals, so as to acquire six different configurations, by the successive breaking down of the older strata into inclined positions, and the horizontal deposit of new ones upon them. Strabo, indeed, at an earlier period had recourse to earthquakes, to explain the occurrence of shells in mountains; and Hooke published the same opinion later. But the Italian geologists prosecuted their researches under the advantage of having, close at hand, large collections of conspicuous and consistent phenomena. Lazzaro Moro, in 1740, attempted to apply the theory of earthquakes to the Italian strata; but both he and his expositor, Cirillo Generelli, inclined rather to reduce the violence of these operations within the ordinary course of nature,[2] and thus leant to the doctrine of uniformity, of which we have afterwards to speak. Moro was encouraged in this line of speculation by the extraordinary occurrence, as it was deemed by most persons, of the rise of a new volcanic island from a deep part of the Mediterranean, near Santorino, in 1707.[3] But in other countries, as the geological facts were studied, the doctrine of catastrophes appeared to gain ground. Thus in England, where, through a large part of the country, the coal-measures are extremely inclined and contorted, and covered over by more horizontal fragmentary beds, the opinion that some violent catastrophe had occurred to dislocate them, before the superincumbent strata were deposited, was strongly held. It was conceived that a period of violent and destructive action must have succeeded to one of repose; and that, for a time, some unusual and paroxysmal forces must have been employed in elevating and breaking the pre-existing strata, and wearing their fragments into smooth pebbles, before nature subsided into a new age of tranquillity and vitality. In like manner Cuvier, from the alternations of fresh-water and salt-water species in the strata of Paris, collected the opinion of a series of great revolutions, in which " the thread of induction was broken." Deluc and others, to whom we owe the first steps in geological dynamics, attempted carefully to distinguish between causes now in action, and those which have ceased to act; in which latter class they reckoned the causes which have

---

[1] "Here is a part of the earth which has become more light, and which rises, while the opposite part approaches nearer to the centre, and what was the bottom of the sea is become the top of the mountain."—Venturi's *Léonardo au Vinci.*

[2] Lyell, i. 3. p. 64. (4th ed.)   [3] Ib. p. 60.

elevated the existing continents. This distinction was assented to by many succeeding geologists. The forces which have raised into the clouds the vast chains of the Pyrenees, the Alps, the Andes, must have been, it was deemed, something very different from any agencies now operating.

This opinion was further confirmed by the appearance of a complete change in the forms of animal and vegetable life, in passing from one formation to another. The species of which the remains occurred, were entirely different, it was said, in two successive epochs : a new creation appears to have intervened ; and it was readily believed that a transition, so entirely out of the common course of the world, might be accompanied by paroxysms of mechanical energy. Such views prevail extensively among geologists up to the present time : for instance, in the comprehensive theoretical generalizations of Elie de Beaumont and others, respecting mountain-chains, it is supposed that, at certain vast intervals, systems of mountains, which may be recognized by the parallelism of course of their inclined beds, have been disturbed and elevated, lifting up with them the aqueous strata which had been deposited among them in the intervening periods of tranquillity, and which are recognized and identified by means of their organic remains : and according to the adherents of this hypothesis, these sudden elevations of mountain-chains have been followed, again and again, by mighty waves, desolating whole regions of the earth.

The peculiar bearing of such opinions upon the progress of physical geology will be better understood by attending to the *doctrine of uniformity*, which is opposed to them, and with the consideration of which we shall close our survey of this science, the last branch of our present task.

Sect. 2.—*Of the Doctrine of Geological Uniformity.*

THE opinion that the history of the earth had involved a series of catastrophes, confirmed by the two great classes of facts, the symptoms of mechanical violence on a very large scale, and of complete changes in the living things by which the earth had been tenanted, took strong hold of the geologists of England, France, and Germany. Hutton though he denied that there was evidence of a beginning of the present state of things, and referred many processes in the formation of strata to existing causes, did not assert that the elevatory forces which raise continents from the bottom of the ocean, were of the same order, as well as of the same kind, with the volcanoes and earthquakes which now shake the surface.   His doctrine of uniformity was founded rather on the supposed analogy of other lines of speculation, than on the examination of the amount of changes now going on.   "The Author of nature," it was said, "has not permitted in His works any symptom of infancy or of old age, or any sign by which we may estimate either their future or their past duration :" and the example of the planetary system was referred to in illustration of this.[4]   And a general persuasion that the champions of this theory were not disposed to accept the usual opinions on the subject of creation, was allowed, perhaps very unjustly, to weigh strongly against them in the public opinion.

While the rest of Europe had a decided bias towards the doctrine of geological catastrophes, the phenomena of Italy, which, as we have seen, had already tended to soften the rigor of that doctrine, in the progress of speculation from Steno to Generelli, were destined to mitigate it still more, by converting to the belief of uniformity transalpine geologists who had been bred up in the catastrophist creed.   This effect was, indeed, gradual.   For a time the distinction of the *recent* and the *tertiary* period was held to be marked and strong.   Brocchi asserted that a large portion of the Sub-Apennine fossil shells belonged to a living species of the Mediterranean Sea : but the geologists of the rest of Europe turned an incredulous ear to this Italian tenet ; and the persuasion of the distinction of the tertiary and the recent period was deeply impressed on most geologists by the memorable labors of Cuvier and Brongniart on the Paris basin.   Still, as other tertiary deposits were examined, it was found that they could by no means be considered as contemporaneous, but that they formed a chain of posts, advancing nearer and nearer to the recent period.   Above the strata of the basins of London and Paris,[5] lie the newer strata of Touraine, of Bourdeaux, of the valley of the Bormida and the Superga near Turin, and of the basin of Vienna, explored by M. Constant Prevost Newer and higher still than these, are found the Sub-Apennine formations of Northern Italy, and probably of the same period, the English "crag" of Norfolk and Suffolk.   And most of these marine formations are associated with volcanic products and fresh-water deposits, so as to imply apparently a long train of alternations of corresponding processes.   It may easily be supposed that, when the subject had assumed this form, the boundary of the present and past condition of the earth

---

[4] Lyell, i. 4, p. 94.   [5] Lyell, 1st ed. vol. iii. p. 61.

was in some measure obscured.   But it was not long before a very able attempt was made to obliterate it altogether.   In 1828, Mr. Lyell set out on a geological tour through France and Italy.[*]   He had already conceived the idea of classing the tertiary groups by reference to the number of recent species which were found in a fossil state. But as he passed from the north to the south of Italy, he found, by communication with the best fossil conchologists, Borelli at Turin, Guidotti at Parma, Costa at Naples, that the number of extinct species decreased; so that the last-mentioned naturalist, from an examination of the fossil shells of Otranto and Calabria, and of the neighboring seas, was of opinion that few of the tertiary shells were of extinct species.   To complete the series of proof, Mr. Lyell himself explored the strata of Ischia, and found, 2000 feet above the level of the sea, shells, which were all pronounced to be of species now inhabiting the Mediterranean; and soon after, he made collections of a similar description on the flanks of Etna, in the Val di Noto, and in other places.

The impression produced by these researches is described by himself.[']   "In the course of my tour I had been frequently led to reflect on the precept of Descartes, that a philosopher should once in his life doubt everything he had been taught; but I still retained so much faith in my early geological creed as to feel the most lively surprize on visiting Sortino, Pentalica, Syracuse, and other parts of the Val di Noto, at beholding a limestone of enormous thickness, filled with recent shells, or sometimes with mere casts of shells, resting on marl in which shells of Mediterranean species were imbedded in a high state of preservation.   All idea of [necessarily] attaching a high antiquity to a regularly-stratified limestone, in which the casts and impressions of shells alone were visible, vanished at once from my mind.   At the same time, I was struck with the identity of the associated igneous rocks of the Val di Noto with well-known varieties of 'trap' in Scotland and other parts of Europe ; varieties which I had also seen entering largely into the structure of Etna.

"I occasionally amused myself," Mr. Lyell adds, "with speculating on the different rate of progress which geology might have made, had it been first cultivated with success at Catania, where the phenomena above alluded to, and the great elevation of the modern tertiary beds in the Val di Noto, and the changes produced in the historical era by the Calabrian earthquakes, would have been familiarly known."

---

[*] 1st ed. vol. iii. Pref.   ['] Lyell, 1st ed. Pref. x.

Before Mr. Lyell entered upon his journey, he had put into the hands of the printer the first volume of his "Principles of Geology, being an attempt to explain the former Changes of the Earth's Surface *by reference to the Causes now in Operation.*" And after viewing such phenomena as we have spoken of, he, no doubt, judged that the doctrine of catastrophes of a kind entirely different from the existing course of events, would never have been generally received, if geologists had at first formed their opinions upon the Sicilian strata. The boundary separating the present from the anterior state of things crumbled away; the difference of fossil and recent species had disappeared, and, at the same time, the changes of position which marine strata had undergone, although not inferior to those of earlier geological periods, might be ascribed, it was thought, to the same kind of earthquakes as those which still agitate that region. Both the supposed proofs of catastrophic transition, the organical and the mechanical changes, failed at the same time ; the one by the removal of the fact, the other by the exhibition of the cause. The powers of earthquakes, even such as they now exist, were, it was supposed, if allowed to operate for an illimitable time, adequate to produce all the mechanical effects which the strata of all ages display. And it was declared that all evidence of a beginning of the present state of the earth, or of any material alteration in the energy of the forces by which it has been modified at various epochs, was entirely wanting.

Other circumstances in the progress of geology tended the same way. Thus, in cases where there had appeared in one country a sudden and violent transition from one stratum to the next, it was found, that by tracing the formations into other countries, the chasm between them was filled up by intermediate strata ; so that the passage became as gradual and gentle as any other step in the series. For example, though the conglomerates, which in some parts of England overlie the coal-measures, appear to have been produced by a complete discontinuity in the series of changes; yet in the coal-fields of Yorkshire, Durham, and Cumberland, the transition is smoothed down in such a way that the two formations pass into each other. A similar passage is observed in Central-Germany, and in Thuringia is so complete, that the coal-measures have sometimes been considered as subordinate to the *todtliegendes.*[8]

Upon such evidence and such arguments, the doctrine of catastro-

---

[8] De la Beche, p. 414, *Manual.*

phes was rejected with some contempt and ridicule; and it was main
tained, that the operation of the causes of geological change may properly and philosophically be held to have been uniform through all
ages and periods.   On this opinion, and the grounds on which it has
been urged, we shall make a few concluding remarks.

It must be granted at once, to the advocates of this geologica'
uniformity, that we are not arbitrarily to assume the existence of
catastrophes.   The degree of uniformity and continuity with which
terremotive forces have acted, must be collected, not from any gratuitous hypothesis, but from the facts of the case.   We must suppose
the causes which have produced geological phenomena, to have been
as similar to existing causes, and as dissimilar, as the effects teach us.
We are to avoid all bias in favor of powers deviating in kind and
degree from those which act at present; a bias which, Mr. Lyell asserts,
has extensively prevailed among geologists.

But when Mr. Lyell goes further, and considers it a merit in a course
of geological speculation that it *rejects* any difference between the intensity of existing and of past causes, we conceive that he errs no less
than those whom he censures.   "An *earnest and patient endeavor to
reconcile* the former indication of change,"[9] with *any* restricted class
of causes,—a habit which he enjoins,—is not, we may suggest, the
temper in which science ought to be pursued.   The effects must themselves teach us the nature and intensity of the causes which have
operated; and we are in danger of error, if we seek for slow and shun
violent agencies further than the facts naturally direct us, no less than
if we were parsimonious of time and prodigal of violence.   *Time*, in-
exhaustible and ever accumulating his efficacy, can undoubtedly do
much for the theorist in geology; but *Force*, whose limits we cannot
measure, and whose nature we cannot fathom, is also a power never
to be slighted: and to call in the one to protect us from the other, is
equally presumptuous, to whichever of the two our superstition leans.
To invoke Time, with ten thousand earthquakes, to overturn and set
on edge a mountain-chain, should the phenomena indicate the change
to have been sudden and not successive, would be ill excused by pleading the obligation of first appealing to known causes.[10]

---

[9] Lyell, B. iv. c. i. p. 328, 4th ed.

[10] [2nd Ed.] [I have, in the text, quoted the fourth edition of Mr. Lyell's
Principles, in which he recommends " an earnest and patient endeavor to reconcile the former indications of change with the evidence of gradual mutatior

In truth, we know causes only by their effects; and in order to .earn the nature of the causes which modify the earth, we must study them through all ages of their action, and not select arbitrarily the period in which we live as the standard for all other epochs. The forces which have produced the Alps and Andes are known to us by experience, no less than the forces which have raised Etna to its present height; for we learn their amount in both cases by their results. Why, then, do we make a merit of using the latter case as a measure for the former? Or how can we know the true scale of such force, except by comprehending in our view all the facts which we can bring together?

In reality when we speak of the *uniformity* of nature, are we not obliged to use the term in a very large sense, in order to make the doctrine at all tenable? It includes catastrophes and convulsions of a very extensive and intense kind; what is the limit to the violence which we must allow to these changes? In order to enable ourselves to represent geological causes as operating with uniform energy through all time, we must measure our time by long cycles, in which repose and violence alternate; how long may we extend this cycle of change, the repetition of which we express by the word *uniformity*?

And why must we suppose that all our experience, geological as well as historical, includes more than *one* such cycle? Why must we insist upon it, that man has been long enough an observer to obtain the *average* of forces which are changing through immeasurable time?

---

now in progress." In the sixth edition, in that which is, I presume, the corresponding passage, although it is transferred from the fourth to the first Book (B. i. c. xiii. p. 325) he recommends, instead, "an earnest and patient inquiry how far geological appearances are reconcileable with the effect of changes now in progress." But while Mr. Lyell has thus softened the advocate's character in his language in this passage, the transposition which I have noticed appears to me to have an opposite tendency. For in the former edition, the causes now in action were first described in the second and third Books, and the great problem of Geology, stated in the first Book, was attempted to be solved in the fourth. But by incorporating this fourth Book with the first, and thus prefixing to the study of existing causes arguments against the belief of their geological insufficiency, there is an appearance as if the author wished his reader to be prepared by a previous pleading against the doctrine of catastrophes, before he went to the study of existing causes. The Doctrines of Catastrophes and of Uniformity, and the other leading questions of the Palætiological Sciences, are further discussed in the *Philosophy of the Inductive Sciences* Book x.]

Vol. II.—38.

The analogy of other sciences has been referred to, as sanctioning this attempt to refer the whole train of facts to known causes. To have done this, it has been said, is the glory of Astronomy : she seeks no hidden virtues, but explains all by the force of gravitation, which we witness operating at every moment. But let us ask, whether it would really have been a merit in the founders of Physical Astronomy, to assume that the celestial revolutions resulted from any selected class of known causes? When Newton first attempted to explain the motions of the moon by the force of gravity, and failed because the measures to which he referred were erroneous, would it have been philosophical in him, to insist that the difference which he found ought to be overlooked, since otherwise we should be compelled to go to causes other than those which we usually witness in action? Or was there any praise due to those who assumed the celestial forces to be the same with gravity, rather than to those who assimilated them with any other known force, as magnetism, till the calculation of the laws and amount of these forces, from the celestial phenomena, had clearly sanctioned such an identification? We are not to select a conclusion now well proved, to persuade ourselves that it would have been wise to assume it anterior to proof, and to attempt to philosophize in the method thus recommended.

Again, the analogy of Astronomy has been referred to, as confirming the assumption of perpetual uniformity. The analysis of the heavenly motions, it has been said, supplies no trace of a beginning, no promise of an end. But here, also, this analogy is erroneously applied. Astronomy, as the science of cyclical motions, has nothing in common with Geology. But look at Astronomy where she has an analogy with Geology ; consider our knowledge of the heavens as a palætiological science ;—as the study of a past condition, from which the present is derived by causes acting in time. Is there then no evidence of a beginning, or of a progress? What is the import of the Nebular Hypothesis? A luminous matter is condensing, solid bodies are forming, are arranging themselves into systems of cyclical motion ; in short, we have exactly what we are told, on this analogy, we ought not to have ;—the beginning of a world. I will not, to justify this argument, maintain the truth of the nebular hypothesis ; but if geologists wish to borrow maxims of philosophizing from astronomy, such speculations as have led to that hypothesis must be their model.

Or, let them look at any of the other provinces of palætiological speculation ; at the history of states, of civilization, of languages. We may assume some *resemblance* or connexion between the principles which determined the progress of government, or of society, or of literature, in the earliest ages, and those which now operate; but who has speculated successfully, assuming an *identity* of such causes? Where do we now find a language in the process of formation, unfolding itself in inflexions, terminations, changes of vowels by grammatical relations, such as characterize the oldest known languages? Where do we see a nation, by its natural faculties, inventing writing, or the arts of life, as we find them in the most ancient civilized nations? We may assume hypothetically, that man's faculties develop themselves in these ways; but we see no such effects produced by these faculties, in our own time, and now in progress, without the influence of foreigners.

Is it not clear, in all these cases, that history does not exhibit a series of cycles, the aggregate of which may be represented as a uniform state, without indication of origin or termination? Does it not rather seem evident that, in reality, the whole course of the world, from the earliest to the present times, is but *one* cycle, yet unfinished;—offering, indeed, no clear evidence of the mode of its beginning; but still less entitling us to consider it as a repetition or series of repetitions of what had gone before?

Thus we find, in the analogy of the sciences, no confirmation of the doctrine of uniformity, as it has been maintained in Geology. Yet we discern, in this analogy, no ground for resigning our hope, that future researches, both in Geology and in other palætiological sciences, may throw much additional light on the question of the uniform or catastrophic progress of things, and on the earliest history of the earth and of man. But when we see how wide and complex is the range of speculation to which our analogy has referred us, we may well be disposed to pause in our review of science;—to survey from our present position the ground that we have passed over;—and thus to collect. so far as we may, guidance and encouragement to enable us to advance in the track which lies before us.

Before we quit the subject now under consideration, we may, however, observe, that what the analogy of science really teaches us, as the most promising means of promoting this science, is the strenuous cultivation of the two subordinate sciences, Geological Knowledge of Facts, and Geological Dynamics. These are the two provinces of knowledge—corresponding to Phenomenal Astronomy, and Mathematical Mechanics—which may lead on to the epoch of the Newton of geology. We may, indeed, readily believe that we have much to do in both these departments. While so large a portion of the globe is geologically unexplored;—while all the general views which are to extend our classifications satisfactorily from one hemisphere to another, from one zone to another, are still unformed; while the organic fossils of the tropics are almost unknown, and their general relation to the existing state of things has not even been conjectured;—how can we expect to speculate rightly and securely, respecting the history of the whole of our globe? And if Geological Classification and Description are thus imperfect, the knowledge of Geological Causes is still more so. As we have seen, the necessity and the method of constructing a science of such causes, are only just beginning to be perceived. Here, then, is the point where the labors of geologists may be usefully applied; and not in premature attempts to decide the widest and abstrusest questions which the human mind can propose to itself.

It has been stated,[11] that when the Geological Society of London was formed, their professed object was to multiply and record observations, and patiently to await the result at some future time; and their favorite maxim was, it is added, that the time was not yet come for a General System of Geology. This was a wise and philosophical temper, and a due appreciation of their position. And even now, their task is not yet finished; their mission is not yet accomplished. They have still much to do, in the way of collecting Facts; and in entering upon the exact estimation of Causes, they have only just thrown open the door of a vast Labyrinth, which it may employ many generations to traverse, but which they must needs explore, before they can penetrate to the Oracular Chamber of Truth.

I REJOICE, on many accounts, to find myself arriving at the termination of the task which I have attempted. One reason why I am glad to close my history is, that in it I have been compelled, especially in the latter part of my labors, to speak as a judge respecting eminent philosophers whom I reverence as my Teachers in those very sciences on which I have had to pronounce a judgment;—if, indeed, even the appellation of Pupil be not too presumptuous. But I doubt not that such men are as full of candor and tolerance, as they are of knowledge and thought. And if they deem, as I did, that such a history of

---

[11] Lyell, B. i. c. iv. p. 103.

science ought to be attempted, they will know that it was not only the historian's privilege, but his duty, to estimate the import and amount of the advances which he had to narrate; and if they judge, as I trust they will, that the attempt has been made with full integrity of intention and no want of labor, they will look upon the inevitable imperfections of the execution of my work with indulgence and hope.

There is another source of satisfaction in arriving at this point of my labors. If, after our long wandering through the region of physical science, we were left with minds unsatisfied and unraised, to ask, "Whether this be all?"—our employment might well be deemed weary and idle. If it appeared that all the vast labor and intense thought which has passed under our review had produced nothing but a barren Knowledge of the external world, or a few Arts ministering merely to our gratification; or if it seemed that the methods of arriving at truth, so successfully applied in these cases, aid us not when we come to the higher aims and prospects of our being;—this History might well be estimated as no less melancholy and unprofitable than those which narrate the wars of states and the wiles of statesmen. But such, I trust, is not the impression which our survey has tended to produce. At various points, the researches which we have followed out, have offered to lead us from matter to mind, from the external to the internal world; and it was not because the thread of investigation snapped in our hands, but rather because we were resolved to confine ourselves, for the present, to the material sciences, that we did not proceed onwards to subjects of a closer interest. It will appear, also, I trust, that the most perfect method of obtaining speculative truth,—that of which I have had to relate the result,—is by no means confined to the least worthy subjects; but that the Methods of learning what is really true, though they must assume different aspects in cases where a mere contemplation of external objects is concerned, and where our own internal world of thought, feeling, and will, supplies the matter of our speculations, have yet a unity and harmony throughout all the possible employments of our minds. To be able to trace such connexions as this, is the proper sequel, and would be the high reward, of the labor which has been bestowed on the present work. And if a persuasion of the reality of such connexions, and a preparation for studying them, have been conveyed to the reader's mind while he has been accompanying me through our long survey, his time may not have been employed on

124

these pages in vain. However vague and hesitating and obscure may be such a persuasion, it belongs, I doubt not, to the dawning of a better Philosophy, which it may be my lot, perhaps, to develop more fully hereafter, if permitted by that Superior Power to whom all sound philosophy directs our thoughts.

6

Copyright © 1965 by the American Association for the Advancement of Science

Reprinted from *Science*, **148**, 754–759 (1965)

The Method of Multiple Working Hypotheses

With this method the dangers of parental
affection for a favorite theory can be circumvented.

T. C. Chamberlin

As methods of study constitute the leading theme of our session, I have chosen as a subject in measurable consonance the method of multiple working hypotheses in its application to investigation, instruction, and citizenship.

There are two fundamental classes of study. The one consists in attempting to follow by close imitation the processes of previous thinkers, or to acquire by memorizing the results of their investigations. It is merely secondary, imitative, or acquisitive study. The other class is primary or creative study. In it the effort is to think independently, or at least individually, in the endeavor to discover new truth, or to make new combinations of truth, or at least to develop an individualized aggregation of truth. The endeavor is to think for one's self, whether the thinking lies wholly in the fields of previous thought or not. It is not necessary to this habit of study that the subject-material should be new; but the process of thought and its results must be individual and independent, not the mere following of previous lines of thought ending in predetermined results. The demonstration of a problem in Euclid precisely as laid down is an illustration of the former; the demonstration of the same proposition by a method of one's own or in a manner distinctively individual is an illustration of the latter; both lying entirely within the realm of the known and the old.

Creative study, however, finds its largest application in those subjects in which, while much is known, more remains to be known. Such are the fields which we, as naturalists, cultivate; and

we are gathered for the purpose of developing improved methods lying largely in the creative phase of study, though not wholly so.

Intellectual methods have taken three phases in the history of progress thus far. What may be the evolutions of the future it may not be prudent to forecast. Naturally the methods we now urge seem the highest attainable. These three methods may be designated, first, the method of the ruling theory; second, the method of the working hypothesis; and, third, the method of multiple working hypotheses.

In the earlier days of intellectual development the sphere of knowledge was limited, and was more nearly within the compass of a single individual; and those who assumed to be wise men, or aspired to be thought so, felt the need of knowing, or at least seeming to know, all that was known as a justification of their claims. So, also, there grew up an expectancy on the part of the multitude that the wise and the learned would explain whatever new thing presented itself. Thus pride and ambition on the one hand, and expectancy on the other, developed the putative wise man whose knowledge boxed the compass, and whose acumen found an explanation for every new puzzle which presented itself. This disposition has propagated itself, and has come down to our time as an intellectual predilection, though the compassing of the entire horizon of knowledge has long since been an abandoned affectation. As in the earlier days, so still, it is the habit of some to hastily conjure up an explanation for every new phenomenon that presents itself.

Interpretation rushes to the forefront as the chief obligation pressing upon the putative wise man. Laudable as the effort at explanation is in itself, it is to be condemned when it runs before a serious inquiry into the phenomenon itself. A dominant disposition to find out what is, should precede and crowd aside the question, commendable at a later stage, "How came this so?" First full facts, then interpretations.

Premature Theories

The habit of precipitate explanation leads rapidly on to the development of tentative theories. The explanation offered for a given phenomenon is naturally, under the impulse of self-consistency, offered for like phenomena as they present themselves, and there is soon developed a general theory explanatory of a large class of phenomena similar to the original one. This general theory may not be supported by any further considerations than those which were involved in the first hasty inspection. For a time it is likely to be held in a tentative way with a measure of candor. With this tentative spirit and measurable candor, the mind satisfies its moral sense, and deceives itself with the thought that it is proceeding cautiously and impartially toward the goal of ultimate truth. It fails to recognize that no amount of provisional holding of a theory, so long as the view is limited and the investigation partial, justifies an ultimate conviction. It is not the slowness with which conclusions are arrived at that should give satisfaction to the moral sense, but the thoroughness, the completeness, the all-sidedness, the impartiality, of the investigation.

It is in this tentative stage that the affections enter with their blinding influence. Love was long since represented as blind, and what is true in the personal realm is measurably true in the intellectual realm. Important as

Thomas C. Chamberlin (1843–1928), a geologist, was president of the University of Wisconsin at the time this lecture was written. Later he was professor and director of the Walker Museum of the University of Chicago. In 1893 he founded the *Journal of Geology*, which he edited until his death. In 1908 he was president of the AAAS. The article is reprinted from *Science* (old series), **15**, 92 (1890).

the intellectual affections are as stimuli and as rewards, they are nevertheless dangerous factors, which menace the integrity of the intellectual processes. The moment one has offered an original explanation for a phenomenon which seems satisfactory, that moment affection for his intellectual child springs into existence; and as the explanation grows into a definite theory, his parental affections cluster about his intellectual offspring, and it grows more and more dear to him, so that, while he holds it seemingly tentative, it is still lovingly tentative, and not impartially tentative. So soon as this parental affection takes possession of the mind, there is a rapid passage to the adoption of the theory. There is an unconscious selection and magnifying of the phenomena that fall into harmony with the theory and support it, and an unconscious neglect of those that fail of coincidence. The mind lingers with pleasure upon the facts that fall happily into the embrace of the theory, and feels a natural coldness toward those that seem refractory. Instinctively there is a special searching-out of phenomena that support it, for the mind is led by its desires. There springs up, also, an unconscious pressing of the theory to make it fit the facts, and a pressing of the facts to make them fit the theory. When these biasing tendencies set in, the mind rapidly degenerates into the partiality of paternalism. The search for facts, the observation of phenomena and their interpretation, are all dominated by affection for the favored theory until it appears to its author or its advocate to have been overwhelmingly established. The theory then rapidly rises to the ruling position, and investigation, observation, and interpretation are controlled and directed by it. From an unduly favored child, it readily becomes master, and leads its author whithersoever it will. The subsequent history of that mind in respect to that theme is but the progressive dominance of a ruling idea.

Briefly summed up, the evolution is this: a premature explanation passes into a tentative theory, then into an adopted theory, and then into a ruling theory.

When the last stage has been reached, unless the theory happens, perchance, to be the true one, all hope of the best results is gone. To be sure, truth may be brought forth by an in-

Thomas Chrowder Chamberlin was noted for his contributions to glaciology and for his part in formulating the Chamberlin-Moulton (planetesimal) hypothesis of the origin of the earth.

vestigator dominated by a false ruling idea. His very errors may indeed stimulate investigation on the part of others. But the condition is an unfortunate one. Dust and chaff are mingled with the grain in what should be a winnowing process.

Ruling Theories Linger

As previously implied, the method of the ruling theory occupied a chief place during the infancy of investigation. It is an expression of the natural infantile tendencies of the mind, though in this case applied to its higher activities, for in the earlier stages of development the feelings are relatively greater than in later stages.

Unfortunately it did not wholly pass away with the infancy of investigation, but has lingered along in individual instances to the present day, and finds illustration in universally learned men and pseudo-scientists of our time.

The defects of the method are obvious, and its errors great. If I were to name the central psychological fault, I should say that it was the admission of intellectual affection to the place that should be dominated by impartial intellectual rectitude.

So long as intellectual interest dealt chiefly with the intangible, so long it was possible for this habit of thought

to survive, and to maintain its dominance, because the phenomena themselves, being largely subjective, were plastic in the hands of the ruling idea; but so soon as investigation turned itself earnestly to an inquiry into natural phenomena, whose manifestations are tangible, whose properties are rigid, whose laws are rigorous, the defects of the method became manifest, and an effort at reformation ensued. The first great endeavor was repressive. The advocates of reform insisted that theorizing should be restrained, and efforts directed to the simple determination of facts. The effort was to make scientific study factitious instead of causal. Because theorizing in narrow lines had led to manifest evils, theorizing was to be condemned. The reformation urged was not the proper control and utilization of theoretical effort, but its suppression. We do not need to go backward more than twenty years to find ourselves in the midst of this attempted reformation. Its weakness lay in its narrowness and its restrictiveness. There is no nobler aspiration of the human intellect than desire to compass the cause of things. The disposition to find explanations and to develop theories is laudable in itself. It is only its ill use that is reprehensible. The vitality of study quickly disappears when the object sought is a mere collocation of dead unmeaning facts.

The inefficiency of this simply repressive reformation becoming apparent, improvement was sought in the method of the working hypothesis. This is affirmed to be *the* scientific method of the day, but to this I take exception. The working hypothesis differs from the ruling theory in that it is used as a means of determining facts, and has for its chief function the suggestion of lines of inquiry; the inquiry being made, not for the sake of the hypothesis, but for the sake of facts. Under the method of the ruling theory, the stimulus was directed to the finding of facts for the support of the theory. Under the working hypothesis, the facts are sought for the purpose of ultimate induction and demonstration, the hypothesis being but a means for the more ready development of facts and of their relations, and the arrangement and preservation of material for the final induction.

It will be observed that the distinction is not a sharp one, and that a working hypothesis may with the utmost ease degenerate into a ruling theory. Affection may as easily cling about an hypothesis as about a theory, and the demonstration of the one may become a ruling passion as much as of the other.

A Family of Hypotheses

Conscientiously followed, the method of the working hypothesis is a marked improvement upon the method of the ruling theory; but it has its defects—defects which are perhaps best expressed by the ease with which the hypothesis becomes a controlling idea. To guard against this, the method of multiple working hypotheses is urged. It differs from the former method in the multiple character of its genetic conceptions and of its tentative interpretations. It is directed against the radical defect of the two other methods; namely, the partiality of intellectual parentage. The effort is to bring up into view every rational explanation of new phenomena, and to develop every tenable hypothesis respecting their cause and history. The investigator thus becomes the parent of a family of hypotheses: and, by his parental relation to all, he is forbidden to fasten his affections unduly upon any one. In the nature of the case, the danger that springs from affection is counteracted, and therein is a radical difference between this method and the two preceding. The investigator at the outset puts himself in cordial sympathy and in parental relations (of adoption, if not of authorship) with every hypothesis that is at all applicable to the case under investigation. Having thus neutralized the partialities of his emotional nature, he proceeds with a certain natural and enforced erectness of mental attitude to the investigation, knowing well that some of his intellectual children will die before maturity, yet feeling that several of them may survive the results of final investigation, since it is often the outcome of inquiry that several causes are found to be involved instead of a single one. In following a single hypothesis, the mind is presumably led to a single explanatory conception. But an adequate explanation often involves the co-ordination of several agencies, which enter into the combined result in varying proportions. The true explanation is therefore necessarily complex. Such complex explanations of phenomena are specially encouraged by the method of multiple hypotheses, and constitute one of its chief merits. We are so prone to attribute a phenomenon to a single cause, that, when we find an agency present, we are liable to rest satisfied therewith, and fail to recognize that it is but one factor, and perchance a minor factor, in the accomplishment of the total result. Take for illustration the mooted question of the origin of the Great Lake basins. We have this, that, and the other hypothesis urged by different students as the cause of these great excavations; and all of these are urged with force and with fact, urged justly to a certain degree. It is practically demonstrable that these basins were river-valleys antecedent to the glacial incursion, and that they owe their origin in part to the pre-existence of those valleys and to the blocking-up of their outlets. And so this view of their origin is urged with a certain truthfulness. So, again, it is demonstrable that they were occupied by great lobes of ice, which excavated them to a marked degree, and therefore the theory of glacial excavation finds support in fact. I think it is furthermore demonstrable that the earth's crust beneath these basins was flexed downward, and that they owe a part of their origin to crust deformation. But to my judgment neither the one nor the other, nor the third, constitutes an adequate explanation of the phenomena. All these must be taken together, and possibly they must be supplemented by other agencies. The problem, therefore, is the determination not only of the participation, but of the measure and the extent, of each of these agencies in the production of the complex result. This is not likely to be accomplished by one whose working hypothesis is pre-glacial erosion, or glacial erosion, or crust deformation, but by one whose staff of working hypotheses embraces all of these and any other agency which can be rationally conceived to have taken part in the phenomena.

A special merit of the method is, that by its very nature it promotes thoroughness. The value of a working hypothesis lies largely in its suggestiveness of lines of inquiry that might otherwise be overlooked. Facts that are trivial in themselves are brought into significance by their bearings upon the hypothesis, and by their causal indications. As an illustration, it is only necessary to cite the phenomenal influence which the Darwinian hypothesis has exerted upon the investigations of the past two decades. But a single working hypothesis may lead investigation along a given line to the neglect of others equally important; and thus, while inquiry is promoted in certain quarters, the investigation lacks in completeness. But if all rational hypotheses relating to a subject are worked co-equally, thoroughness is the presumptive result, in the very nature of the case.

In the use of the multiple method, the re-action of one hypothesis upon another tends to amplify the recognized scope of each, and their mutual conflicts whet the discriminative edge of each. The analytic process, the development and demonstration of criteria, and the sharpening of discrimination, receive powerful impulse from the co-ordinate working of several hypotheses.

Fertility in processes is also the natural outcome of the method. Each hypothesis suggests its own criteria, its own means of proof, its own methods of developing the truth; and if a group of hypotheses encompass the subject on all sides, the total outcome of means and of methods is full and rich.

The use of the method leads to certain peculiar habits of mind which deserve passing notice, since as a factor of education its disciplinary value is one of importance. When faithfully pursued for a period of years, it develops a habit of thought analogous to the method itself, which may be designated a habit of parallel or complex thought. Instead of a simple succession of thoughts in linear order, the procedure is complex, and the mind appears to become possessed of the power of simultaneous vision from different standpoints. Phenomena appear to become capable of being viewed analytically and synthetically at once. It is not altogether unlike the study of a landscape, from which there comes into the mind myriads of lines of intelligence, which are received and co-ordinated simultaneously, producing a complex impression which is recorded and studied directly in its complexity. My description of this process

T. C. Chamberlin published two papers under the title of "The method of multiple working hypotheses." One of these papers, first published in the *Journal of Geology* in 1897, was quoted by John R. Platt in his recent article "Strong inference" (*Science,* 16 Oct. 1964). Platt wrote: "This charming paper deserves to be reprinted." Several readers, having had difficulty obtaining copies of Chamberlin's paper, expressed agreement with Platt. One wrote that the article had been reprinted in the *Journal of Geology* in 1931 and in the *Scientific Monthly* in November 1944. Another sent us a photocopy. Several months later still another wrote that the Institute for Humane Studies (Stanford, Calif.) had reprinted the article in pamphlet form this year. On consulting the 1897 version, we found a footnote in which Chamberlin had written: "A paper on this subject was read before the Society of Western Naturalists in 1892, and was published in a scientific periodical." Library research revealed that "a scientific periodical" was *Science* itself, for 7 February 1890, and that Chamberlin had actually read the paper before the Society of Western Naturalists on 25 October 1889. The chief difference between the 1890 text and the 1897 text is that, as Chamberlin wrote in 1897: "The article has been freely altered and abbreviated so as to limit it to aspects related to geological study." The 1890 text, which seems to be the first and most general version of "The method of multiple working hypotheses," is reprinted here. Typographical errors have been corrected, and subheadings have been added.

is confessedly inadequate, and the affirmation of it as a fact would doubtless challenge dispute at the hands of psychologists of the old school; but I address myself to naturalists who I think can respond to its verity from their own experience.

Drawbacks of the Method

The method has, however, its disadvantages. No good thing is without its drawbacks; and this very habit of mind, while an invaluable acquisition for purposes of investigation, introduces difficulties in expression. It is obvious, upon consideration, that this method of thought is impossible of verbal expression. We cannot put into words more than a single line of thought at the same time; and even in that the order of expression must be conformed to the idiosyncrasies of the language, and the rate must be relatively slow. When the habit of complex thought is not highly developed, there is usually a leading line to which others are subordinate, and the difficulty of expression does not rise to serious proportions; but when the method of simultaneous vision along different lines is developed so that the thoughts running in different channels are nearly equivalent, there is an obvious embarrassment in selection and a disinclination to make the attempt. Furthermore, the impossibility of expressing the mental operation in words leads to their disuse in the silent process of

thought, and hence words and thoughts lose that close association which they are accustomed to maintain with those whose silent as well as spoken thoughts run in linear verbal courses. There is therefore a certain predisposition on the part of the practitioner of this method to taciturnity.

We encounter an analogous difficulty in the use of the method with young students. It is far easier, and I think in general more interesting, for them to argue a theory or accept a simple interpretation than to recognize and evaluate the several factors which the true elucidation may require. To illustrate: it is more to their taste to be taught that the Great Lake basins were scooped out by glaciers than to be urged to conceive of three or more great agencies working successively or simultaneously, and to estimate how much was accomplished by each of these agencies. The complex and the quantitative do not fascinate the young student as they do the veteran investigator.

Multiple Hypotheses and
Practical Affairs

It has not been our custom to think of the method of working hypotheses as applicable to instruction or to the practical affairs of life. We have usually regarded it as but a method of science. But I believe its application to practical affairs has a value coordinate with the importance of the

affairs themselves. I refer especially to those inquiries and inspections that precede the coming-out of an enterprise rather than to its actual execution. The methods that are superior in scientific investigation should likewise be superior in those investigations that are the necessary antecedents to an intelligent conduct of affairs. But I can dwell only briefly on this phase of the subject.

In education, as in investigation, it has been much the practice to work a theory. The search for instructional methods has often proceeded on the presumption that there is a definite patent process through which all students might be put and come out with results of maximum excellence; and hence pedagogical inquiry in the past has very largely concerned itself with the inquiry, "What is the best method?" rather than with the inquiry, "What are the special values of different methods, and what are their several advantageous applicabilities in the varied work of instruction?" The past doctrine has been largely the doctrine of pedagogical uniformitarianism. But the faculties and functions of the mind are almost, if not quite, as varied as the properties and functions of matter: and it is perhaps not less absurd to assume that any specific method of instructional procedure is more effective than all others, under any and all circumstances, than to assume that one principle of interpretation is equally applicable to all the phenomena of nature. As there is an endless variety of mental processes and combinations and an indefinite number of orders of procedure, the advantage of different methods under different conditions is almost axiomatic. This being granted, there is presented to the teacher the problem of selection and of adaptation to meet the needs of any specific issue that may present itself. It is important, therefore, that the teacher shall have in mind a full array of possible conditions and states of mind which may be presented, in order that, when any one of these shall become an actual case, he may recognize it, and be ready for the emergency.

Just as the investigator armed with many working hypotheses is more likely to see the true nature and significance of phenomena when they present themselves, so the instructor equipped with a full panoply of hypotheses ready for application more readily recognizes the actuality of the situation, more accurately measures its significance, and more appropriately applies the methods which the case calls for.

The application of the method of multiple hypotheses to the varied affairs of life is almost as protean as the phases of that life itself, but certain general aspects may be taken as typical of the whole. What I have just said respecting the application of the method to instruction may apply, with a simple change of terms, to almost any other endeavor which we are called upon to undertake. We enter upon an enterprise in most cases without full knowledge of all the factors that will enter into it, or all of the possible phases which it may develop. It is therefore of the utmost importance to be prepared to rightly comprehend the nature, bearings, and influence of such unforeseen elements when they shall definitely present themselves as actualities. If our vision is narrowed by a preconceived theory as to what will happen, we are almost certain to misinterpret the facts and to misjudge the issue. If, on the other hand, we have in mind hypothetical forecasts of the various contingencies that may arise, we shall be the more likely to recognize the true facts when they do present themselves. Instead of being biased by the anticipation of a given phase, the mind is rendered open and alert by the anticipation of any one of many phases, and is free not only, but is predisposed,

to recognize correctly the one which does appear. The method has a further good effect. The mind, having anticipated the possible phases which may arise, has prepared itself for action under any one that may come up, and it is therefore ready-armed, and is predisposed to act in the line appropriate to the event. It has not set itself rigidly in a fixed purpose, which it is predisposed to follow without regard to contingencies. It has not nailed down the helm and predetermined to run a specific course, whether rocks lie in the path or not; but, with the helm in hand, it is ready to veer the ship according as danger or advantage discovers itself.

It is true, there are often advantages in pursuing a fixed predetermined course without regard to obstacles or adverse conditions. Simple dogged resolution is sometimes the salvation of an enterprise; but, while glorious successes have been thus snatched from the very brink of disaster, overwhelming calamity has in other cases followed upon this course, when a reasonable regard for the unanticipated elements would have led to success. So there is to be set over against the great achievements that follow on dogged adherence great disasters which are equally its result.

Danger of Vacillation

The tendency of the mind, accustomed to work through multiple hypotheses, is to sway to one line of policy or another, according as the balance of evidence shall incline. This is the soul and essence of the method. It is in general the true method. Nevertheless there is a danger that this yielding to evidence may degenerate into unwarranted vacillation. It is not always possible for the mind to balance evidence with exact equipoise, and to determine, in the midst of the execution of an enterprise, what is the measure of probability on the one side or the other; and as difficulties present themselves, there is a danger of being biased by them and of swerving from the course that was really the true one. Certain limitations are therefore to be placed upon the application of the method, for it must be remembered that a poorer line of policy consistently adhered to may bring better results than a vacillation between better policies.

There is another and closely allied danger in the application of the method. In its highest development it presumes a mind supremely sensitive to every grain of evidence. Like a pair of delicately poised scales, every added particle on the one side or the other produces its effect in oscillation. But such a pair of scales may be altogether too sensitive to be of practical value in the rough affairs of life. The balances of the exact chemist are too delicate for the weighing-out of coarse commodities. Despatch may be more important than accuracy. So it is possible for the mind to be too much concerned with the nice balancings of evidence, and to oscillate too much and too long in the endeavor to reach exact results. It may be better, in the gross affairs of life, to be less precise and more prompt. Quick decisions, though they may contain a grain of error, are oftentimes better than precise decisions at the expense of time.

The method has a special beneficent application to our social and civic relations. Into these relations there enter, as great factors, our judgment of others, our discernment of the nature of their acts, and our interpretation of their motives and purposes. The method of multiple hypotheses, in its application here, stands in decided contrast to the method of the ruling theory or of the simple working hypothesis. The primitive habit is to interpret the acts of others on the basis of a theory. Childhood's unconscious theory is that the good are good, and the bad are bad. From the good the child expects nothing but good; from the bad, nothing but bad. To expect a good act from the bad, or a bad act from the good, is radically at variance with childhood's mental methods. Unfortunately in our social and civic affairs too many of our fellow-citizens have never outgrown the ruling theory of their childhood.

Many have advanced a step farther, and employ a method analogous to that of the working hypothesis. A certain presumption is made to attach to the acts of their fellow-beings, and that which they see is seen in the light of that presumption, and that which they construe is construed in the light of that presumption. They do not go to the lengths of childhood's method by assuming positively that the good are wholly good, and the bad wholly bad; but there is a strong presumption in their minds that he concerning whom they have an ill opinion will act from corresponding motives. It requires positive evidence to overthrow the influence of the working hypothesis.

The method of multiple hypotheses assumes broadly that the acts of a fellow-being may be diverse in their nature, their moves, their purposes, and hence in their whole moral character; that they may be good though the dominant character be bad; that they may be bad though the dominant character be good; that they may be partly good and partly bad, as is the fact in the greater number of the complex activities of a human being. Under the method of multiple hypotheses, it is the first effort of the mind to see truly what the act is, unbeclouded by the presumption that this or that has been done because it accords with our ruling theory or our working hypothesis. Assuming that acts of similar general aspect may readily take any one of several different phases, the mind is freer to see accurately what has actually been done. So, again, in our interpretations of motives and purposes, the method assumes that these may have been any one of many, and the first duty is to ascertain which of possible motives and purposes actually prompted this individual action. Going with this effort there is a predisposition to balance all evidence

fairly, and to accept that interpretation to which the weight of evidence inclines, not that which simply fits our working hypothesis or our dominant theory. The outcome, therefore, is better and truer observation and juster and more righteous interpretation.

Imperfections of Knowledge

There is a third result of great importance. The imperfections of our knowledge are more likely to be detected, for there will be less confidence in its completeness in proportion as there is a broad comprehension of the possibilities of varied action, under similar circumstances and with similar appearances. So, also, the imperfections of evidence as to the motives and purposes inspiring the action will become more discernible in proportion to the fulness of our conception of what the evidence should be to distinguish between action from the one or the other of possible motives. The necessary result will be a less disposition to reach conclusions upon imperfect grounds. So, also, there will be a less inclination to misapply evidence; for, several constructions being definitely in mind, the indices of the one motive are less liable to be mistaken for the indices of another.

The total outcome is greater care in ascertaining the facts, and greater discrimination and caution in drawing conclusions. I am confident, therefore, that the general application of this method to the affairs of social and civic life would go far to remove those misunderstandings, misjudgments, and misrepresentations which constitute so pervasive an evil in our social and our political atmospheres, the source of immeasurable suffering to the best and most sensitive souls. The misobservations, the misstatements, the misinterpretations, of life may cause less gross suffering than some other evils; but they, being more universal and more subtle, pain. The remedy lies, indeed, partly in charity, but more largely in correct intellectual habits, in a predominant, ever-present disposition to see things as they are, and to judge them in the full light of an unbiased weighing of evidence applied to all possible constructions, accompanied by a withholding of judgment when the evidence is insufficient to justify conclusions.

I believe that one of the greatest moral reforms that lies immediately before us consists in the general introduction into social and civic life of that habit of mental procedure which is known in investigation as the method of multiple working hypotheses.

7

Reprinted from *Science*, **3**(53), 1–13 (1896)

The Origin of Hypotheses, Illustrated by the Discussion of a Topographic Problem*

G. K. GILBERT

AN important part—in some respects the most important part—of the work of science is the explanation of the facts of Nature. The process through which natural phenomena are explained is called the 'method of hypotheses,' and though it is familiar to most of my audience I shall nevertheless describe it briefly for the purpose of directing special attention to one of its factors.

The hypothesis has been called a 'scientific guess,' and unless the title 'guess' carries with it something of disrespect it is not inappropriate. When the investigator, having under consideration a fact or group of facts whose origin or cause is unknown, seeks to discover their origin,

his first step is to make a guess. In other words, he frames a hypothesis or invents a tentative theory. Then he proceeds to test the hypothesis, and in planning a test he reasons in this way: If the phenomenon was really produced in the hypothetic manner, then it should possess, in addition to the features already observed, certain other specific features, and the discovery of these will serve to verify the hypothesis. Resuming

*Annual Address of the President of the Geological Society of Washington ; read December 11, 1895, to the Scientific Societies of Washington. By special arrangement, through the Joint Commission of those societies, this number of SCIENCE is mailed to all members.

its examination, he searches for these particular features. If they are found the theory is supported; and in case the features thus predicted and discovered are numerous and varied, the theory is accepted as satisfactory. But if the reëxamination reveals features inconsistent with the tentative theory, the theory is thereby discredited, and the investigator proceeds to frame and test a new one. Thus, by a series of trials, inadequate explanations are one by one set aside, and eventually an explanation is discovered which satisfies all requirements.

When the subject of study is one of wide interest it usually happens that several investigators coöperate in the invention and testing of hypotheses. Often each investigator will originate a hypothesis, and a series of rigorous tests will be applied through the endeavor of each one to establish his own by overthrowing all others. The different theories are rivals competing for ascendancy, and their authors are also rivals, ambitious for the credit of discovery. The personal factor thus introduced tends to bias the judgment and is to that extent unfavorable to the progress of science; but the conflict of theories, leading, as it eventually must, to the survival of the fittest, is advantageous. Fortunately there is a mode of using hypotheses which regulates the personal factor without restricting the competition of theories, and this has found favor with the greatest investigators. It has recently been formulated and ably advocated by our fellow-member, Prof. T. C. Chamberlin, who calls it the 'method of multiple hypotheses.'*

In the application of this method the student of a group of phenomena, instead of inventing and testing hypotheses one at a time, devises at an early stage as many as possible, and then, treating them as rival claimants, assigns to himself the rôle of

*The Method of Multiple Working Hypotheses, Science (1st series), Vol. XV. (1890) pp. 92–96.

judge. Returning to the study of nature, he seeks for special features which cannot consist with all the hypotheses, and may therefore serve to discriminate among them. Thus by a series of crucial tests he eliminates one after another of the tentative theories until but a single one remains, and he then proceeds to apply such tests as he may to the survivor.

In these methods of work, whether theories are examined successively or simultaneously, there are two steps involving the initiative of the investigator; he invents hypotheses and he invents tests for them. It is to the intellectual character of these inventions that your attention is invited.

The mental process by which hypotheses are suggested is obscure. Ordinarily they flash into consciousness without premonition, and it would be easy to ascribe them to a mysterious intuition or creative faculty; but this would contravene one of the broadest generalizations of modern psychology. Just as in the domain of matter nothing is created from nothing, just as in the domain of life there is no spontaneous generation, so in the domain of mind there are no ideas which do not owe their existence to antecedent ideas which stand in the relation of parent to child. It is only because our mental processes are largely conducted outside the field of consciousness that the lineage of ideas is difficult to trace.

To explain the origin of hypotheses I have a hypothesis to present,—not, indeed, as original, for it has been at least tacitly assumed by various writers on scientific method, but rather as worthy of more general attention and recognition. It is that hypotheses are always suggested through analogy. The unexplained phenomenon on which the student fixes his attention resembles in some of its features another phenomenon of which the explanation is known. Analogic reasoning suggests that the desired explanation is similar in character to the known, and this suggestion constitutes the production of a hypothesis.

To test this hypothesis of hypotheses I have for some years endeavored to analyze the methods employed by myself and some of my associates in geologic research, and this study has proved so interesting in connection with the investigation of a peculiar crater in Arizona, that I shall devote the remainder of my hour to an outline of that investigation.

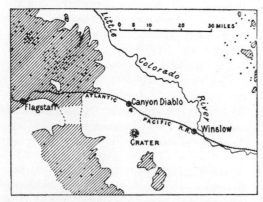

Fig. 1.—Map of part of northern Arizona. The shaded areas are covered by volcanic rocks. Dots mark ancient volcanic vents.

In northeastern Arizona there is an arid plain beneath whose scanty soil are level beds of limestone. At one point the plain is interrupted by a bowl-shaped or saucer-shaped hollow, a few thousand feet broad and a few hundred feet deep ; and about this hollow is an approximately circular rim rising one or two hundred feet above the surface of the plain (Plate 1, Figs. 2 and 3). In other words, there is a crater ; but the crater differs from the ordinary volcanic structure of that name in that it contains no volcanic rock. The circling sides of the bowl show limestone and sandstone, and the rim is wholly composed of these materials. On the slopes of this crater and on the plain round about many pieces of iron have been found, not iron ore, but the metal itself, and this substance is foreign to the limestone of the plain and to all other formations of the region.* The features of the locality thus include three things of unusual character and requiring explanation : First, the crater composed of non-volcanic rock ; second, the scattered iron masses ; third, the association of crater and iron. To account for these phenomena a number of theories have been suggested.

In the year 1886 a company of shepherds encamped on the slopes of the crater and pastured their sheep on the surrounding plain. Mathias Armijo, one of their number, found a piece of iron, and, deceived by its lustrous surface, supposed it to be silver. The mistake was quickly corrected by his fellows, but his discovery excited their interest, and other pieces of iron were soon found. The curiosity of the shepherds was aroused also by the crater, and they invented a theory which is admirable for its simplicity : The crater was produced by an explosion, the material of the rim being thrown out from the central cavity, and the iron was thrown out from the same cavity at the same time. You will observe that this theory is comprehensive. It accounts for the crater, the iron, and the association of the two. As I have never met these first students of the phenomena I have had no opportunity to make inquiry as to the origin of their theory ; but its close relation to the theories of geologic disturbance which are current in mining districts suggests that it also sprung from the familiar process of blasting. As the firing of a blast opens a cavity and heaps dislocated rock masses in an irregular way, the unlearned miner finds in natural blasting an easy explanation of hollows and uplifts.

Four years later a man by the name of Craft saw in the iron a possibility of profit. Setting up a heap of stone to mark the spot,

---

* The crater is locally known as Coon Mountain, or Coon Butte. The iron is known to literature as the Canyon Diablo fall of meteorites.

he located a mining claim; and going to the city of Albuquerque, he announced that he had a vein of pure iron 40 yards wide and two miles long, and offered to sell his property to a railway company. The samples he submitted were examined by an assayer, and the officers of the company gave consideration to his proposal, agreeing to send a representative to examine the property. The negotiation was not concluded, because Mr. Craft, having borrowed money on the strength of his great expectations, mysteriously disappeared, but the incident served to give information of the locality to a scientific observer. The assayer forwarded a piece of the iron to the late Dr. A. E. Foote, the mineralogist, who visited the place, collected a quantity of the iron and examined the crater. In the summer of 1891 he communicated his observations to the American Association for the Advancement of Science,* which that year was the guest of the scientific societies of Washington, and his paper aroused much interest. For the crater of non-volcanic rock he offered no explanation, but the iron he pronounced of celestial origin—a shower of fallen meteors. It has long been known that many of the bodies which reach the earth from outer space are composed of iron, and that such iron is of peculiar character, having a certain crystalline structure, being alloyed with nickel, and including nodules of certain substances which are not found in any other association. So Doctor Foote, in characterizing the iron as meteoric, merely referred it to a well-established class. His explanation was not tentative, but final, and has not been called in question by any subsequent investigator.

In the discussion following the reading of his paper a new hypothesis was proposed,

* A new locality for meteoric iron with a preliminary notice of the discovery of diamonds in the iron. Proc. Am. Ass. Adv. Science, Vol. 40, pp. 279–283.

and as this was offered by myself I can trace its origin with comparative confidence. The crust of the earth is not equally dense at all points, but some parts are heavier than others. Not only are there variations from hill to hill and from formation to formation, but the continents are in general composed of lighter materials than the ocean beds, and one side of the sphere is so much heavier than the other that its attraction pulls most of the water away from the other side. Among the various theories that have been proposed for the origin of the planet there is one which ascribes it to the falling together under mutual attraction of many smaller celestial bodies, and it has been suggested that the variations in the crust may represent original differences of the concurrent masses. Speculating on such lines I had asked myself what would result if another small star should now be added to the earth, and one of the consequences which had occurred to me was the formation of a crater, the suggestion springing from the many familiar instances of craters formed by collision. A raindrop falling on soft ooze produces a miniature crater; so does a pebble thrown into a pool of pasty mud. A larger crater is made when a steel projectile is fired against steel armor plate; and analogy easily bridged the interval from the cannon ball to the asteroid. So when Dr. Foote described a limestone crater in association with iron masses from outer space, it at once occurred to me that the theme of my speculation might here find its realization. The suggested explanation assumes that the shower of falling iron masses included one larger than the rest, and that this greater mass, by the violence of its collision, produced the crater. Here again you will observe that a single theory explains the crater, the iron and their association.

The thought of examining the scar produced on the earth by the collision of a star was so attractive that I desired to visit the crater, but as that was not immediately practicable I arranged to have it visited by one of my colleagues. A few months later Mr. Willard D. Johnson spent several days at the locality, making a sketch map and describing the various features. When he reached the rim of the crater he found it to consist chiefly of limestone strata inclined outward, and his first thought was that the rim might be the remnant of the dome of strata over a laccolite. The laccolite is a peculiar volcanic product. The molten lavas which make volcanoes rise from deep sources through cracks or passages among the rocks and flow out over the surface of the land, but sometimes rising lavas fail to reach the surface, and accumulate at lower levels, opening for themselves bubble-shaped chambers over which the strata are arched. In the dome-like structures thus produced the rocks dip outward in all directions from a central region, and this outward dip was the feature which, through analogy, suggested to Mr. Johnson a laccolitic origin. His first idea, however, was not long retained, for examining the walls and bottom of the crater he found no trace of the igneous rocks of which laccolites are composed, and the theory afforded no aid in accounting for the hollow. He therefore dismissed it and sought another. He may have considered several others, but the only one placed on record is an explosion theory. In some way, probably by volcanic heat, a body of steam was produced at a depth of some hundreds or thousands of feet, and the explosion of this steam produced the crater. The fall of iron was independent, and the association of the two occurrences in the same locality is accidental.* As Mr. Johnson is at once a civil engineer and a student of geology and geography, he had at command

* Mr. Johnson's discussion of the problem was communicated to me in a personal letter. G. K. G.

as basis for analogic reasoning the explosive phenomena associated with the arts and also those which belong to the history of volcanoes, and we may assume that these suggested his theory.

Mr. Johnson's account of the crater was much fuller than Dr. Foote's, but instead of satisfying my curiosity tended rather to whet it, and I availed myself of the first opportunity to make a personal visit. Four hypotheses had now been made, but only two survived. The theory of the shepherds, deriving the iron from the cavity of the crater, was disproved by Dr. Foote's determination of the meteoric character of the iron. The laccolitic theory had been promptly set aside by Mr. Johnson. There remained the theory of a star's collision and the theory of a steam explosion. If my visit was to aid in the determination of the problem of cause it must gather the data which would discriminate between these two theories, and an attempt was accordingly made to devise crucial tests. If the crater was produced by the collision and penetration of a stellar body that body now lay beneath the bowl, but not so if the crater resulted from explosion. Any observation which would determine the presence or absence of a buried star might therefore serve as a crucial test. Direct exploration by means of a shaft or drill hole could not be undertaken on account of the expense, but two indirect methods seemed feasible.

If the crater were produced by explosion the material contained in the rim, being identical with that removed from the hollow, is of equal amount; but if a star entered the hole the hole was partly filled thereby, and the remaining hollow must be less in volume than the rim. The presence or absence of the star might therefore be tested by measuring the cubic contents of the hollow and of the rim and comparing the two. Of the intellectual origin of this test perhaps the most that can be said is that it is a test by quantities, and that the experienced investigator, having previously found relations of quantity the most satisfactory criteria, habitually employs them whenever the circumstances permit.

Again it occurred to me that the stellar body would presumptively be composed, like the smaller masses round about, of iron, and that its presence or absence might, therefore, be determined by means of the magnetic needle. If it were absent the compass would point in the same direction, whatever its position with reference to the crater, whether within or without, on one side or the other, near by or miles away ; but if a mass of iron large enough to produce the crater lay beneath it its attraction would pull the needle one way or the other, producing local variations. Doubtless the suggestion of this test came from knowledge of the methods employed in searching for magnetic iron ore in northern Michigan, where the prospector carries the dip needle to and fro through the forest, and by means of its changes of direction determines the position and extent of bodies of ore.

As an equipment for these measurements I provided myself with the instruments necessary to make an accurate topographic map, and obtained, through the courtesy of the Coast and Geodetic Survey, a full set of instruments for the observation of terrestrial magnetism. I was so fortunate, also, as to secure the coöperation of an expert magnetic observer, Mr. Marcus Baker, of the Geological Survey, and together we set out for Arizona.

At this time it seemed to me that the presumption was in favor of the theory ascribing the crater to a falling star, because that theory explained, while its rival did not, the close association of the crater with the shower of celestial iron. So far as we know, a falling meteor is just as likely to reach any one spot on the earth's surface as any other, and it is, therefore, entirely possible that the coincidence of the meteoric locality with the locality of the crater has no special significance; but if the two phenomena are not connected by a causal relation, it is no more probable that the crater should coincide in place with one of the 165 meteoric falls recorded within the bounds of the United States than that it should occupy any other spot of our broad domain. A rough estimate shows the probability of non-coincidence to be at least 800 times as great as the probability of coincidence. This by no means warrants the conclusion that an explanation ascribing a causal relation is 800 times as probable as one ascribing fortuitous coincidence, but it legitimately inclines the mind toward causality in the absence of more direct and authoritative evidence.

This point is illustrated by the investigation of the peculiar sky colors observed twelve years ago. Considering the phenomenon of coloration in its entirety—character, distribution and duration—it was not merely rare, it was unique. In the same year a tremendous volcanic explosion occurred in the Straits of Sunda, and that also was unique in intensity. The coincidence of the two, which in this case was a matter of time rather than place, led to the belief that the one was caused by the other, and this belief was held by many men of science before an adequate explanation of the mode of causation had been suggested.

So when Mr. Baker and I started for the crater it seemed rather probable than otherwise that we should find a local deflection of the magnetic needle, and that we should find the material of the rim more than sufficient to fill the hollow it surrounds.

Before our journey was ended another explanation suggested itself. Mr. Johnson had described the crater as not truly circular but somewhat oval, the longer diameter lying east and west. He noted also that

FIG. 1.—Lonar Lake, India, occupying an explosion crater. From Newbold's *Summary of the Geology of Southern India*, Jour. Roy. Asiatic Soc., vol. 9, p. 40. London, 1848.

FIG. 2.—The limestone crater of Arizona, Coon Butte, as seen from the south. Photograph of a model by Mr. Victor Mindeleff.

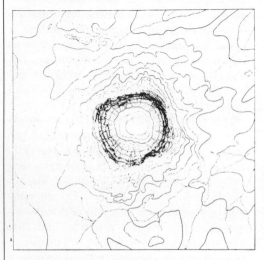

FIG. 3.—Contour map of Coon Butte. The vertical distance from contour to contour is ten feet. Lines of drainage are dotted.

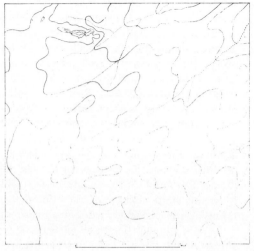

FIG. 4.—Restoration of the site of Coon Butte before the formation of the crater. Contour interval, ten feet; lines of drainage dotted. Compare with Fig. 3.

FIG. 5.—Volcanic cinder-cone, with crater, north of San Francisco Mountain, Arizona. The position of the crater, at top of the hill, is characteristic of most volcanoes. Compare Fig, 2, where the crater lies chiefly below the level of the plain.

FIG. 6.—Craters made by throwing clay balls at a clay target. A ball of the same size is shown. 1 shows the effect of high velocity, 2 of low.

ILLUSTRATIONS TO ARTICLE BY G. K. GILBERT ON HYPOTHESES.

FIG. 1.—Rim of Coon Butte, with part of inner face.

FIG. 2.—Block of limestone on outer slope of Coon Butte, one-half mile from rim.

FIG. 3.—Largest block of limestone on rim of Coon Butte. Diameter, 60 feet.

FIG. 4.—Outer slope of Coon Butte.

FIG. 5.—Interior of Coon Butte, as seen from the talus on one side. The cliff below the rim is of limestone.

FIG. 6.—Exterior of Coon Butte, as seen from the surrounding plain.

the rim was bulkier on the east side than on the west, and that nearly all the iron had been found east of the crater. The new explanation was that a star, falling obliquely from the western sky, struck the earth and bounded off, finally coming to rest at some point farther east. The idea was of course derived from the ricochet of projectiles; I had seen the mark left by a rifle ball where it rebounded from a plowed field. This explanation could be tested by a simple examination of the topographic form; and it may be as well to anticipate here the order of the narrative, and say that the form of the crater was found to be quite inconsistent with the ricochet hypothesis. The difference of the two diameters is quite small; the eastern rim is but little more massive than the western; and the dislocation of the rocks in the western rim is of such character that it could not have been produced by a body descending obliquely toward the east.

Arriving at the crater we spent two

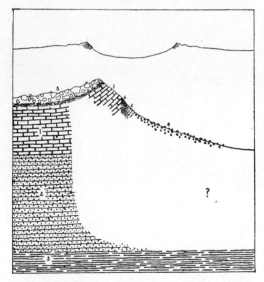

FIG. 2.—The upper diagram is profile across the crater; the lower, a cross-section of the rim. 1, limestone. 2, sandstone. 3, red shale. 4, crushed rock. 5, loose blocks of limestone and sandstone. 6, talus of debris fallen from 1 and 2 above.

weeks in topographic and magnetic surveys and the study of local details. The diameter of the bowl, measured from rim to rim, is about three-fourths of a mile. Its depth below the rim is from 550 to 600 feet; below the plain, 400 feet, the rim being 150 to 200 feet high. The rim is in part composed of limestone strata like those which underlie the plain, but turned up, so as to incline steeply away from the hollow on all sides. On the inclined strata rests a mantle of loose fragments which are in part of limestone and in part of sandstone. The limestone masses are fragments of the formation occurring just beneath, and the sandstone masses are fragments of a formation which underlies the limestone formation. Most of the masses are of moderate size, but others are large, the limestone reaching a diameter of 60 feet, and the sandstone about 100 feet. (Plate 2. Figs. 1–4.). They are irregularly mingled, one material predominating in one tract and the other in another. The limestone is the more conspicious because withstanding better the attacks of the weather. In fact the larger blocks of sandstone have been so far washed away that they do not project above the surface. From the crest of the rim outward this loose material occupies the surface for an average distance of half a mile, being characterized by rolling or hummocky topography. At greater distances it is thinly spread and the constituent blocks are small. At one mile it is represented only by scattered fragments, but these continue with diminishing frequency to a distance of three and a half miles.

Inside the rim the edges of limestone strata occupy the slope for a space of 150 to 250 feet. They are succeeded in several places by sandstone strata, but the sandstone does not hold its original relation to the limestone; it is separated by a vertical zone of crushed rock, and there is other evidence that it has been faulted upward.

The lower slopes are occupied by fragments of limestone and sandstone with an arrangement showing that they have fallen from the cliff above so as to constitute a talus of the ordinary type, and the central tract is composed of fine material of the same kind. Whatever may have been the original shape of the pit, its present form has resulted from subsequent modification under the action of rain and frost.

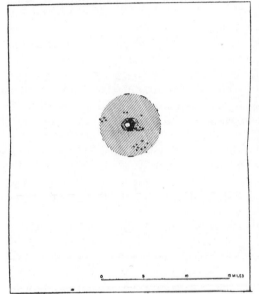

Fig. 3.—Distribution Chart. The inner line is the rim of crater. In the inner shaded area the loose debris has a depth of more than one foot. In the outer shaded area are scattered blocks; where the area is bounded by a line its limit was surveyed. The chief district of small iron masses is shown by dots. Large iron masses are indicated by crosses. The distribution of the iron is chiefly on the authority of Mr. F. W. Volz, of Canyon Diablo, A. T.

No iron has been found within the crater, but a great number of fragments were obtained from the outer slopes where they rested on the mantle of loose blocks. Many others were obtained from the plain within the region of scattered debris, and others, though a smaller number, from the outer

plain. One large piece was discovered eight miles east of the crater, or almost twice as distant as any fragments of the ejected limestone. Another was long ago discovered twenty miles to the southward, but what became of it is not known, and it has not been definitely identified as a member of the same meteoric shower. Most of the masses are small. There have been found more than one thousand, possibly more than two thousand, pieces weighing less than an ounce; others weigh a few ounces or a few pounds; forty or fifty exceed one hundred pounds, and two exceed one thousand pounds. The total weight of all finds is probably ten tons. At the time of my visit I was told that all had been discovered east of a north and south line passing through the middle of the crater, but this may have been an accident of the method of search, for more recently six large ones have been reported from points west of that line.

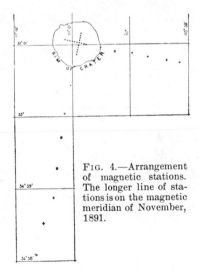

Fig. 4.—Arrangement of magnetic stations. The longer line of stations is on the magnetic meridian of November, 1891.

The magnetic survey by Mr. Baker included the selection and mapping of a system of stations, and the observation at each of the three magnetic elements: the horizontal component of direction, or the compass bearing; the vertical component of direction, or the inclination of the dip needle ; and the intensity of the magnetic force. Two lines of stations, at right angles to each other, were carried across the crater, and one of these lines was extended to a distance of three and a half miles on the plain. When the results were tabulated and compared, the magnetism was found to be constant in direction and intensity at all the stations, the deviations from uniformity being not greater than the unavoidable errors of observation. So if the crater contains a mass of iron its attraction is too feeble to be detected by the instruments employed. That we might learn the precise meaning of this result, the delicacy of the instruments was afterward tested at the Washington Navy Yard, by observing their behavior when placed in certain definite relations to a group of iron cannon whose weight was known, and the following conclusions were reached : If a mass of iron equivalent to a sphere 1500 feet in diameter is buried beneath the crater it must lie at least 50 miles below the surface ; if a mass 500 feet in diameter lies there its depth is not less than 10 miles. So the theory of a great iron meteor is negatived by the magnetic results, unless we may suppose either that the meteor was quite small as compared to the diameter of the crater, or that it penetrated to a very great depth.

The topographic survey was executed with such detail as to warrant the drawing of contour lines for each ten feet of height. (Plate 1, Fig. 3.) During its progress the configuration of the surrounding country was carefully studied, and its general plan was found to be so simple and regular that the original contours before the creation of the crater could be restored without great liability of error. (Plate 1, Fig. 4.) Such restoration was made, and with its aid two quantities were afterwards computed: first, the cubic contents of the rim so far as it projects above the ancient surface ; second,

the cubic contents of the hollow so far as it lies below the ancient surface. The two volumes were compared with each other and also with the volume of a spherical projectile estimated as competent to produce the crater. From experiments with balls of clay fired against a target of the same material it seems probable that a crater 4,000 feet in diameter might be produced by a swift-moving meteor with a diameter of 1,500 feet. (Plate 1, Fig. 6.) It seems possible, though not probable, that it could be made by a mass 750 feet in diameter. The volume of the greater assumed projectile is 60 million cubic yards ; the volume of the lesser, $7\frac{1}{2}$ million yards. The magnitude of the hollow was found to be 82 million yards, and the magnitude of the rim was also found to be 82 million yards. It, therefore, appears that if the rim were to be dug away down to the level of the ancient plain, and the material tightly packed within the hollow of the crater, it would suffice to precisely fill that hollow and restore the ancient plain. The excess of matter required by the theory of a buried star was not found.

Thus each of the two experiments whose testimony had been invoked declared against the theory of a colliding meteor ; and the expectation founded on the high improbability of fortuitous coincidence nevertheless failed of realization.

Attention being now directed to the only surviving theory, that of steam explosion, all the various features discovered in the local study were considered with reference to it. To describe and discuss them on this occasion would lead too far from our subject, and they may be passed by with the remark that, while not all are as yet fully understood, they seem not to oppose the theory.

For the sake of applying another quantitative test, an attempt was made to ascertain whether the energy which could be developed by heating the water contained in the sandstone formation would be sufficient, when the overlying strata gave way, to hurl their fragments out upon the surounding plain. As the data were quite indefinite the computation could result only in a rough approximation, and there is no need to weary you with its details, but it served to show that the assumed cause was of the same order of magnitude as the result accomplished. The idea of applying such a test needs no specific explanation, because quantitative tests of this particular type are among the most familiar resources of investigation. Whenever a tentative theory involves the application of force or the expenditure of energy the investigator (or his critic) habitually asks whether the assumed cause affords a sufficient amount of force or of energy.

Practically the same conclusion was reached in a more satisfactory way by studying the accounts of other natural explosions where steam was the agent. At several epochs in its history the top of Mount Vesuvius has been torn away by a sudden convulsion. In Java the summit of Mount Tomboro was blown away, with the production of a great crater which now contains a lake, and a similar catastrope occurred on the slope of Mount Pepandaján. The great explosion of Krakatoa, in 1883, demolished several volcanic islands and created others, reconstructing the topography of a district in the Straits of Sunda. On July 15, 1888, a great opening was torn in the Japanese mountain Kobandai, the summit and part of one side being removed. The last mentioned instance is the most available for comparison because the agency of steam distinctly appeared, and because the history of the event has been admirably reported by two Japanese geologists, Profs. Sekiya and Kikuchi, of Tokio. *

* The Eruption of Bandai-san. Trans. Seismological Soc. of Japan, Vol. XIII., (1890), pp. 139–222. 9 plates.

There were in this case about twenty explosions, all occurring within the space of one or two minutes. A cloud of rock fragments ascended to a height of 4,000 feet. The greater number, moving obliquely away from the mountain side, fell upon its lower slope, down which they rolled as an avalanche for a distance of five miles, overwhelming several villages and transforming a fertile plain into a rocky desert. In other directions fell showers of stones, and a cloud of dust descending more slowly. The resulting crater, less regular in form than the subject of our study, was nineteen times as capacious, and from its bottom fierce jets of steam issued for weeks and even months. Kobandai is a volcanic peak, and although it had been quiescent for ten centuries there can be little doubt that the steam it evolved was generated by volcanic heat.

The competency of volcanic steam for the production of a crater is thus shown by a parallel instance. and the only conspicuous difference between the Japanese case and the Arizonian lies in the fact that in the one the disrupted rock was volcanic and in the other it was not. This difference seems unessential, for in neither case was there an eruption of liquid rock; the ancient lavas of Kobandai had been cold for ages, and their relation to the catastrophe was wholly passive. Moreover, the manifestation of volcanic energy is no more exceptional on the Arizona plateau than in the Bandai district. The little limestone crater is in the midst of a great volcanic district. (Fig. 1, Page 3, and Plate 1, Fig. 5.) The nearest volcanic crater is but ten miles distant, and within a radius of fifty miles are hundreds of vents from which lava has issued during the later geologic periods.

In following this line of thought I have but reversed the logical route by which Mr. Johnson probably reached his theory, verifying the theory by recomparison with its source.

Yet other verification was afterwards found through the published accounts of certain small craters in Germany, France and India. In the valley of the Rhine are a number of circular basins, for the most part containing lakes and hence called *maars*. They are depressed below the level of the surrounding plain, and some of them are surrounded by raised rims. The descriptions are somewhat conflicting, but it is clear that some of the basins are hollowed chiefly from non-volcanic rocks, limestone, sandstone and slate, and that their rims are composed in part of fragments of similar rocks.* The Indian crater (Plate 1, Fig. 1), which also contains a lake, is hollowed from a volcanic rock, the Deccan trap, and shows no other material ; but in other features it parallels so closely the Arizona crater that I quote from Doctor Blanford's description: "The surrounding country for hundreds of miles consists entirely of Deccan trap ; in this rock, at Lonar, there is a nearly circular hollow about 300 to 400 feet deep, and rather more than a mile in diameter, containing at the bottom a shallow lake of salt water without any outlet. * * * * The sides of the hollow to the north and northeast are absolutely level with the surrounding country, whilst in all other directions there is a raised rim, never exceeding 100 feet in height, and frequently only 40 or 50, composed of blocks of basalt, irregularly piled, and precisely similar to the rock exposed on the sides of the hollow. The dip of the surrounding traps is away from the hollow, but very low.

"It is impossible to ascribe this hollow to any other cause than volcanic explosion."†

* Volcanos. By G. Poulett Scrope. London, 1872. Pp. 369–384. Die Vulkane der Eifel, in ihrer Bildungsweise erlautert. By Dr. Herman Vogelsang. Naturkundige verhandelingen von der hollandische Maalschappij der Wetenschappen te Haarlem. Vol. 21, Part 1. Pp. 41–76.

†A Manual of the Geology of India, by H. B. Medlicott and W. T. Blanford. Part I., pp. 379–380 8vo. Calcutta, 1879.

For the sake of completeness, mention should be made of two other hypotheses, which resemble the laccolitic suggestion in that each was based on a single feature of the crater but failed to find verification in any other feature. The fact that the pit occurs in limestone suggested that it might be what is called a *limestone sink*, a cavern having been made by the solution of the rock and the roof having afterwards fallen in.* The fact that the loose debris of the rim lies in hummocks with intervening hollows, and thus resembles in its topographic character the terminal moraine of a glacier, suggested that ice was concerned in its distribution.

Yet another hypothesis, and the last that need be mentioned, was made by welding together two which had preceded. It is a general fact that causes are complex, and as the explanations which first suggest themselves are apt to be simple, it often occurs that the theory finally adopted combines elements of two or more of the theories tentatively proposed. The expert constructor of theories is therefore prone to suspect that rival explanations embody half truths, and to seek for methods of combination. The combination proposed in this case utilizes the theory of meteoric impact and the theory of volcanic explosion, and its author is Mr. Warren Upham. His suggestion is that, by some volcanic process, heat had been engendered among the rocks of the locality, so that the conditions were ripe for an explosion, and that the mine was actually fired by a falling star, whose collision ruptured a barrier between water and hot rock, or in some other way touched the volcanic button.† It will be noted that this explanation demands a coincidence of what may be called the second order, for the colliding star is supposed not only to have chanced upon the prepared

* This suggestion was made by a correspondent.

† American Geologist, Vol. 13, (1894), p. 116; also a personal letter.

locality, but to have arrived opportunely at the critical epoch.

Still another contribution to the subject, while it does not increase the number of hypotheses, is nevertheless important in that it tends to diminish the weight of the magnetic evidence and thus to reopen the question which Mr. Baker and I supposed we had settled. Our fellow-member, Mr. Edwin E. Howell, through whose hands much of the meteoric iron has passed,

Fig. 5.—Iron meteorite found near the crater. Weighs 161½ pounds. Property of Mr. Edwin E. Howell, of Washington, D. C.

points out that each of the iron masses, great and small, is in itself a complete individual. They have none of the characters that would be found if they had been broken one from another, and yet, as they are all of one type and all reached the earth within a small district, it must be supposed that they were originally connected in some way. Reasoning by analogy from the characters of other meteoric bodies, he infers that the irons were all included in a large mass of some different material, either crystalline rock, such as constitutes the class of meteorites called 'stony,' or else a compound of iron and sulphur, similar to certain nodules discovered inside the iron masses when sawn in two. Neither of these materials is so enduring as the iron, and the fact that they are not now found on the plain does not prove their original

absence. Moreover, the plain is strewn in the vicinity of the crater with bits of limonite, a mineral frequently produced by the action of air and water on iron sulphide, and this material is much more abundant than the iron. If it be true that the iron masses were thus imbedded, like plums, in an astral pudding, the hypothetic buried star might have great size and yet only small power to attract the magnetic needle. Mr. Howell also proposes a qualification of the test by volumes, suggesting that some of the rocks beneath the buried star might have been condensed by the shock so as to occupy less space. * These considerations are eminently pertinent to the study of the crater and will find appropriate place in any comprehensive discussion of its origin ; but the fact which is peculiarly worthy of note at the present time is their ability to unsettle a conclusion that was beginning to feel itself secure. This illustrates the tentative nature, not only of the hypotheses of Science, but of what Science calls its results. The method of hypotheses, and that method is the method of Science, founds its explanations of Nature wholly on observed facts, and its results are ever subject to the limitations imposed by imperfect observation. However grand, however widely accepted, however useful its conclusion, none is so sure that it can not be called in question by a newly discovered fact. In the domain of the world's knowledge there is no infallibility.

And now let us return for a moment from the illustrative investigation to the hypothesis of hypotheses. If my idea is correct —if it be true that tentative explanations are always founded on accepted explanations of similar phenomena—then fertility of invention implies a wide and varied knowledge of the causes of things, and the

* Mr. Howell's suggestions were communicated orally and are here published by permission.

understanding of Nature in many of her varied aspects is an essential part of the intellectual equipment of the investigator. Moreover, mankind, collectively, through the agency of its men of science and inventors, is an investigator, slowly unravelling the complex of Nature and weaving from the disentangled thread the fabric of civilization. Its material, social and intellectual condition advances with the progress of its knowledge of natural laws and is wholly dependent thereon. As an investigator it makes each new conquest by the aid of possessions earlier acquired, and the breadth of its domain each day is the foundation and measure of its daily progress. Knowledge of Nature is an account at bank, where each dividend is added to the principal and the interest is ever compounded; and hence it is that human progress, founded on natural knowledge, advances with ever increasing speed.

8

Reprinted from *Science*, **63**(1636), 463–468 (1926)

The Value of Outrageous Geological Hypotheses[1]

W. M. DAVIS

MEETINGS of geological societies in these modern days are often somewhat prosaic as compared to those of an earlier time when the limits as well as the methods of geological speculation were less defined than now, and when contradictory differences of opinion were commonly expressed even with regard to fundamental ideas concerning the conditions and processes of earth history. That was a time when the scientific imagination, not so much hampered as it is now by standardized principles, was accustomed to roam with little restraint over the unexplored fields of geological investigation; a time when the facts regarding the earth's crust had been gathered from a relatively small part of its surface, when a theory was thought to be established if it explained nothing more than the facts which it had been invented to explain, and when lively discussion as to the merits of rival theories too often degenerated into polemical diatribes between rival theorists.

In those earlier days, attendance at the meetings of Section E of the American Association for the Advancement of Science—the only meetings in which geologists from different parts of the country were then brought together—was likely to be rewarded by a vigorous, not to say vituperative dispute between Marsh and Cope, not merely as to the completed structures and systematic relationships of the fossil vertebrates that they were finding in the fresh-water Tertiary deposits of the west, but also as to mere priority in finding and naming the fossils; and so eager was each of those eminent worthies to secure his prior claim for a new find before the other came upon it that, according to stories then current, one or both of them sometimes, while still in the western field, resorted to the telegraphic announcement of a name for a newly discovered fossil to be published in the eastern newspapers. In the years, half a century ago, when I first attended the meetings of the Boston Society of Natural History, they were occasionally the scene of emphatic contradictions between T. Sterry Hunt and M. E. Wadsworth on matters petrographic, for that recondite branch of geological science was then just taking form among us. Hunt knew exactly how rocks ought, in accordance with his theoretical views of terrestrial chemistry, to be constituted; while

[1] An address delivered before the Leconte Club of the University of California at its annual meeting at Berkeley, February 21, 1925.

Wadsworth, in view of his observational study of thin sections, knew exactly how rocks are constituted; and each of these convinced positivists maintained his view with earnest vehemence.

It is as a result of many verbal battles then fought without asking or giving quarter that geology has come in these modern days to be a relatively well-restrained and orderly science. How much more carefully are facts scrutinized, and how much larger and safer is the inductive base of our generalizations now than formerly. How narrowly limited is the special field, either in subject or locality, upon which a member of the Geological Society of America now ventures to address his colleagues; so narrow that he often has it pretty much all to himself, and so thoroughly does he cover it that when his statement is completed there is little or nothing left for any one else to say. How much more rigorously logical is the guidance of the train of thought by which advance is made from the facts of observation to the conclusions of theory; and if by good fortune a hearer differs from a speaker as to the track along which the train of thought should be directed, how seldom does he intimate his difference of judgment in any but the most courteous manner! How utterly extinct is rudely polemical dissension; so extinct indeed that the younger geologists of to-day must be surprised to learn that it ever flourished. I wonder sometimes if those younger men do not find our meetings rather demure, not to say a trifle dull; and whether they would not enjoy a return to the livelier manners of earlier times.

Yes, our meetings are certainly prosaic to-day as compared to those of the earlier formative period when speculation was freer and when differences of opinion on major principles were almost the rule rather than the exception. Our younger members may perhaps experience a feeling of disappointment, or even of discouragement at the unanimity with which the conclusions of an elder are received by a geological audience; for it must dampen the enthusiasm of beginners if they gain the impression that all the larger generalizations of our science have been established, thus leaving for them to discover only items of localized fact. And a like feeling of discouragement must often be shared by the chairman of a meeting when, after his encouraging invitation, "This interesting paper is now open for discussion," only silence follows. Are we not in danger of reaching a stage of theoretical stagnation, similar to that of physics a generation ago, when its whole realm appeared to have been explored? We shall be indeed fortunate if geology is so marvelously enlarged in the next thirty years as physics has been in the last thirty. But to make such progress, violence must be done to many of our accepted principles; and it is here that the value of outrageous hypotheses, of which I wish to speak, appears. For inasmuch as the great advances of physics in recent years and as the great advances of geology in the past have been made by outraging in one way or another a body of preconceived opinions, we may be pretty sure that the advances yet to be made in geology will be at first regarded as outrages upon the accumulated convictions of to-day, which we are too prone to regard as geologically sacred.

It was outrageous, two centuries ago, to interpret fossils as records of ancient life; for that interpretation did violence to the view then accepted as to the manner in which the earth had been formed and as to the date at which life had come to exist upon it. It was outrageous, little more than a century ago, to discover fossils of marine organisms in the disordered strata of lofty mountains high above sea level; for that discovery did violence to the ideas then obtaining as to the stability of the earth's crust. And it was equally outrageous, half a century ago, to be told that after mountains had been lifted up, they might in time be worn down to lowlands again, for that idea did violence to the views that had then come to be held regarding the instability of the earth's crust. It was an outrage upon the tacitly accepted principles of geological climatology, based on the postulate of a cooling earth, that there already should have been a glacial period in the past; and for that matter, the form in which the glacial theory was first promulgated was truly enough outrageous; nevertheless it now, in a much modified form, holds good as a standardized geological verity.

It was altogether outrageous to think that man had long been an inhabitant of the earth, instead of looking upon him as a new comer; and it was equally outrageous to discover that the sequence of fossils preserved in successive stratified formations indicated such a progression of life as would result from the evolution of later forms from earlier forms, instead of simply an arbitrary succession of independent creations. It is still rather outrageous to think that the earth has long been and possibly is still heating itself up by the slow compression of an originally uncompacted interior under the weight of a heavy exterior, instead of thinking that it has long been and still is cooling by the slow loss of a great original store of heat. And in view of the many evidences of crowding in the outer crust, it may be thought wantonly outrageous to look upon the earth as possessing an expanding interior which, like the caged starling, "wants to get out." Yet I believe it the part of wisdom to view even that outrage, as well as the Wegener outrage of wandering continents and the Joly outrage of periodical subcrustal heating-up and breaking out, calmly, as if they were all possibilities; and it may also come to be the part of wisdom to ask ourselves in what way and how far our present conception of the earth must be modified in order to transform such outraging possibilities into reasonable actualities; for that is precisely the way in which the above-listed outrages and many others have gained an established place in our science. Of course, if we do not approve of the necessary modifications we may reject them, and with them the outrages that they countenance.

Let it be noted in passing that the omission of the original L from the leading word of the preceding paragraph unfortunately results in its being pronounced as if it were derived from "out" and "rage"; its true meaning would be better indicated if its form were ultrageous, as it might well have been had not the L been lost on the way from Latin through French into English; for with the L preserved, the T and R would be joined in the second syllable and properly separated from the first. A word of opposite meaning would then be, not in-rageous, but in-trageous; and our language would be much more symmetrically developed if that and many similar opposites were added to it. However, if we are not allowed to say ultrageous, we might—or at least those of us who pronounce the French-English word, "route," like the English word "root" might—say oo-trageous, and thus reasonably avoid the implication of an erroneous popular etymology. But this is an irrelevant digression.

All that was necessary to make the outrageous occurrence of fossils reasonable and believable was to remodel our conception of the earth from that of a recently-and-ready-made planet into that of a very ancient and slowly changing planet, on which life had existed for ages and ages, always under the influence of environing conditions and in the presence of slow-acting processes very much like those of to-day; and when the ideal counterpart of the actual earth was once conceived in this fashion, the earth was still found to be just as comfortable a planet to live on as it had been in association with the earlier concept of a ready-made earth. All that was needed to explain the occurrence of marine fossils in the disordered strata of mountain tops was to replace the concept of an immovable earth's crust by that of a deformable crust; and although the rate of deformation was at first thought to have been violently rapid, the need of such hurry was later seen to be no need at all, but only a fancy; and thereafter the deformation was conceived to be a slow process. And so it has been with one of these outrages after another: their accommodation is easily accomplished by merely replacing one concept of the earth, under which they are unacceptable, by another under which they are acceptable; and the replacement once made, we are just as happy as we were before. To be sure, the

process of replacement may be mentally uncomfortable, even distressing, while it is going on; but the moral of that is that we must not allow our concepts of the earth, in so far as they transcend the reach of observation, to root themselves so deeply and so firmly in our minds that the process of uprooting them causes mental discomfort; and one of the best aids toward the realization of this moral may be found in frequently making explicit announcement of all the unproved postulates on which our favorite concepts are based; for then we shall not be so likely to forget that they are all preceded by a great big IF.

We shall be aided in following this counsel if we strive to recognize how far most of our concepts of the earth really do transcend the short reach of observation. It is usual for a field observer to record that he has seen, for example, a ridge of sandstone; yet all that he has actually seen is a series of small and disconnected sandstone outcrops, perhaps not occupying more than a twentieth or a hundredth of the ridge surface; and the composition of the rest of the surface and of all the interior of the ridge is only a matter of inference; truly, a good and justifiable inference, but not the less an inference for being good and justifiable. Similarly, it is customary for a field geologist to record the presence of a fault when he detects the repetition of a given sequence of strata, and indeed to believe in the displacement that the term, fault, implies, as if it as well as the recurrence of the sequence of strata were a fact of observation; yet not only are the underground extensions of the strata and their long-past displacement merely matters of inference, but even the fault-fracture itself is usually inferred instead of being seen; or if seen at all, it is seen only in small linear extent, thus leaving all the rest of its superficial trace as well as all of its surface, either lost in the air or buried underground, to the imagination. In thus making distinction between the few facts of actual observation and their large extension in a superstructure of inference, it is not intended to impugn for a moment the validity of well-reasoned superstructures, but only to emphasize the inevitable disproportion that must exist between them and their observed basis; and thus to make clearer the enormously speculative nature of geological science. For let it be noted that, in the case of a fault, we have to do with a double inference; first, the inference as to underground structures from surface outcrops; second, the inference of displacement because of the repetition of the inferred underground structures. Nevertheless, we believe that faults actually exist.

The very foundation of our science is only an inference; for the whole of it rests on the unprovable assumption that, all through the inferred lapse of time which the inferred performance of inferred geological processes involves, they have been going on in a manner consistent with the laws of nature as we know them now. We seldom realize the magnitude of that assumption. A philosopher of the would-be absolute school once said to me, in effect: "You geologists have an easy way of solving difficult questions: you account for the structures of the earth's crust by assuming that time and processes have been going on for millions and millions of years in the past as they go on to-day; but how do you know that time did not begin only a few hundred thousand years ago after the earth had been suddenly created in imitation of what it would have been if it had been slowly constructed in the manner that you assume?" The answer is as easy as the question: We do not *know;* we merely make a pragmatic choice between the concept of such an imitative creation which seems to us absurd, and a long and orderly evolution which seems to us reasonable. We might, to be sure, were we disposed to be disputatious, turn upon the would-be absolutist and ask him what he is going to do about it; but we have better use for our time than that.

The more clearly the immensely speculative nature of geological science is recognized, the easier it becomes to remodel our concepts of any inferred terrestrial conditions and processes in order to make outrages upon them not outrageous. The more definitely it is understood that the concept of a shrinking earth is based upon certain anterior concepts as to the status of its unobservable interior, the more readily can we entertain the concept of an expanding earth, based upon certain other concepts as to the status of its interior; and it is that particular outrage upon our standardized beliefs that I propose we should contemplate, calmly if possible, and patiently at any rate. To encourage our patience, let me recall another outrageous idea of recent introduction, which in itself is only a sort of reaction from an outrage of somewhat earlier invention and a return toward a more primitive view; namely, the recent idea that those topographical features which we call mountains owe their leading feature, namely, their height, not as has been until lately supposed to a vertical movement of escape from the horizontal thrust by which their rocks have been crowded together, but to an uplifting force which acted long after the rocks were crowded together, and in which, as was thought when the view of a mobile earth crust was first promulgated, no component of horizontal thrusting is necessarily involved. A chief difference between that primitive view and its revival in the recent outrage is that the first view took little account of erosion and implied that each individual ridge and peak was the result of an individual or localized uplift; while the second view takes great account of erosion, not only

in ascribing the present intermont valleys to the long and slow action of that patient process during and after recent uplift, but still more in ascribing the destruction of the surface inequalities, that must have been earlier produced when horizontal thrusting forces crowded the mountain rocks together, to a vastly longer action of erosion before the recent uplift of the worn-down mass was begun; for where in the whole world can we find mountains that to-day owe their height to an upward escape from horizontal thrusting; in other words, where in the world can we find any existing mountains that are still in the cycle of erosion which was introduced by an upward escape from the horizontal thrusting that deformed their rocks, and not in a later cycle of erosion which was introduced by uplift alone after the inequality of surface form due to earlier thrusting had been greatly reduced, if not practically obliterated!

The conventional phrase, horizontal compression, has been avoided in the preceding paragraph and the alternative phrase, horizontal thrusting, has been used in its stead, in order to prepare the way for the rather mild idea that the same terrestrial forces which produce great overthrusts may also, if somewhat differently applied, produce rock folds, slaty cleavage, and various other phenomena ordinarily explained under the earlier phrase; and thus to prepare the further way for the altogether outrageous idea that overthrusts do not result from the effort of the outer crust to adjust itself to a cooling and shrinking interior, but from the effort of an in-any-way warming and expanding interior to rearrange the outer crust. Of course, this is "impossible"; that is, it is impossible in an earth of the kind that we ordinarily imagine the earth to be; but it is not at all impossible in an earth of the kind in which it would be possible. Our task therefore is to try to discover, as judicially and as complacently as we may, what sort of an earth that sort of an earth would be; and then to entertain the concept of that sort of an earth as hospitably as we can and to examine the behavior of such an earth at our leisure.

If an earth with an expanding interior had nothing more to do than to stretch its crust, there would be little trouble in our endeavor; but the concept before us compels the expanding earth to do various other things also; and especially to produce great crustal overthrusts, the cross-country advance of which is measured in tens or twenties of miles. Hence the outward radial push of the expanding interior must somehow be turned into an almost tangential thrust; and how that is to be done it is difficult to imagine. However, there is no reason for immediate discouragement on that account; it is very natural that our imagination to-day should fall short of conceiving all the possible behaviors of a warming and expanding

earth, because we are not practised in imagining—
that is, in making an image of—that sort of an earth,
although a good beginning in that direction has been
made in such an essay as that by Boucher on "The
pattern of the earth's mobile belts."[2] But we surely
have yet much to learn as to what may be all the
various reactions of an expanding earth-interior on
the shell that encloses it, even though many possible
reactions may be now conceived.

For example, let the enclosing shell be defined as
that part of the whole sphere which is exterior to
the depth at which the next inner shell is warming
more rapidly than any other. If that depth be great,
the chief thrust of the expanding interior will be exerted on a thick shell; if small, on a thin shell; and
the effects of interior expansion visible on the surface
will surely be different in the two cases. It seems
conceivable that the total thrust of expansion may, in
so far as it produces batholithic movement, be slowly
concentrated at the weakest part of the shell, and
there permit the interior movement to be locally increased by the conversion of cubic expansion into
linear expansion or intrusion; this being the converse
of the process by which an unduly heavy and therefore isostatically subsiding part of the crust may
slowly distribute its local movement through the whole
of the interior and there produce a diminished spherical extension, as Lawson has suggested. It seems
also conceivable that the movement of a localized
batholithic introduction may find advantage in making an outward escape from compelling interior pressures, by changing the direction of its ascent from
vertical to oblique, and thus diminishing the rate at
which it has to raise the overhead crust. Whether an
obliquely ascending mass of this kind could eventually,
as it approaches the surface, drive along a slice of
crust ahead of it and thus produce what we call
on overthrust, is evidently problematical; but if an
overthrust could be produced in that way it would be
gratifying in certain respects.

If an obliquely ascending batholithic intrusion
works its way through a heavy shell toward the surface and there drives ahead of it a crustal slice which
we recognize as an overthrust, the oblique emergence
of such an overthrust slice will cause an underdrag of
the covering rocks in the rear of thrusting advance,
and thus displace them with more or less extensional
jostling so that they will cover a greater breadth of
surface than that which they occupied before being
underdragged. Surely the need of some such under-
dragging ought to have been recognized long ago
when the prevalence of so-called normal faults which
indicate superficial extension, over other faults which
indicate compression, was inductively established; and
the need is still greater to-day when great faults, such

[2] *Journ. Geol.*, xxxii, 1924, 265–290.

as those of the Basin Ranges, have been found to dip
at moderate angles, such as 50° or 40° to the down-
throw; for such was the conclusion reached by Gilbert in his latest season of field work in the Great
Basin a little over ten years ago.[3] It is true that some
geologists maintain the possibility of producing so-
called normal faults as an indirect effect of horizontal
compressional forces; but even if so contradictory an
effect may thus be possibly produced, it by no means
follows that such faults can not also be produced much
more directly by extensional forces; and the possible
cause and working of extensional forces should therefore be investigated; for there is no generally accepted mechanism, like an underdrag, adequate for
the strong extensional dislocation of crustal blocks
with diverse displacements, to be found in the usual
schemes of dynamical geology; and in the lack of
such a mechanism, any process, even a fantastic process, that will cause a strong underdrag seems worthy
of at least an hour's consideration.

But let no one imagine that I here put forth the
idea of an expanding earth interior, with its imagined consequences of an obliquely out-and-over-
thrust mass exerting an underdrag on the superficial
crust in its rear, as an idea to be believed. I do
not believe it myself, and am therefore doubly far
from asking any one else to believe it. The idea is
set forth simply as an outrage, to do violence to certain generally established views about the earth's behavior that perhaps do not deserve to be regarded as
established; and it is set forth chiefly as a means of
encouraging the contemplation of other possible behaviors; not, however, merely a brief contemplation
followed by an off-hand verdict of "impossible" or
"absurd," but a contemplation deliberate enough to

[3] Since giving the address on which this article is based
I have had opportunity of seeing several Basin Ranges,
some in southeastern California, in company with that
most competent of guides, Dr. L. F. Noble, of the U. S.
Geological Survey, and some in Utah in the helpful company of Professors Schneider and Mathew, of the State
University at Salt Lake City, and of Professor M. O.
Hayes, of Brigham Young University at Provo; and the
evidence then found for the occurrence of slanting fault
surfaces seems to me indisputable, not only in the Basin
Ranges themselves but also in the much longer bounding
ranges of the Wasatch mountains on the east and the
Sierra Nevada on the west. Far from the Range blocks
being vertically uplifted without compression, as Gilbert
first proposed in his report on the Wheeler Survey fifty
years ago, still farther from their being the crowded
blocks of a collapsed arch, as others have supposed, the
Basin Range blocks seem to be the irregularly uplifted
and diversely tilted blocks of a former lowland of erosion
which has suffered a pronounced extension of its former
east-west breadth, as I have briefly stated in the Proceedings of the National Academy of Sciences for July,
1925.

seek out just what conditions would make the outrage seem permissible and reasonable.

Let me close this address by explaining to this hospitable and sympathetic conclave why it seems peculiarly appropriate for me, an easterner, to set before the westerners here gathered the particular outrage with which I have detained them. It is because my contacts with the geology of the Pacific slope during the winter of 1924–25—very unconformable contacts, because of my preconceptions—have been outraging the views that I have more or less unconsciously gained on the Atlantic slope as to the demure quietude of the later geological periods. In the east, the Miocene, Pliocene and Pleistocene have witnessed only leisurely processes of degradation, deposition and deformation, all of small relatively measure; but here on the Pacific slope those periods have been characterized by an extraordinary activity; deposits of enormous thickness have been laid down, and those deposits have been deformed and eroded on a scale that is really rather disconcerting. Is it not fair, therefore, that in return for the incredible stories that have been told me here as to what has happened lately in Californian geology, I should take a turn at telling some outrageously impossible stories myself? In any case, there stand the Basin Range fault blocks, just beyond the eastern skyline of California, displaced in such a manner as to extend over a greater breadth of country than that which they previously occupied; and if it is not possible to explain their extension by underdrag, as an indirect reaction of a passive exterior crust on an expanding earth interior, then we must ask by what other outrageous process it is proposed to explain them.

9

Reprinted from *Bull. Geol. Soc. Amer.*, **44**, 461–493 (June 30, 1933)

RÔLE OF ANALYSIS IN SCIENTIFIC INVESTIGATION [1] *

BY DOUGLAS JOHNSON

(Read before Section E of the American Association for the Advancement of Science, December 27, 1932)

CONTENTS

INTRODUCTION

Custom decrees that the Chairman of your Section should, at the interval of a year following his presidency, deliver before you an appropriate address. It has seemed to me that I could best command your interest in some field of discussion where every one of us, geographer and geologist alike, has had experience. So I have selected the broad field connoted by the highly inclusive term, "scientific investigation"; and I would direct your attention, not to any particular results of such investigation, but to a concrete problem of method which I suppose must concern every scientific worker. This problem can briefly be stated as follows: What is the

---

[1] Manuscript received by the Secretary of the Society April 20, 1933.

* Expanded form of address as retiring Vice-President and Chairman of Section E, American Association for the Advancement of Science.

(461)

precise rôle of analysis in a properly conceived and successfully executed scientific investigation?

To answer such a question it is obvious that we must first determine, as best we may, what constitutes "a properly conceived and successfully executed scientific investigation." The early part of my address will be directed to this end. The latter part may then the more effectively discuss the value of analysis as an instrument of precision in research.

For a preface to my remarks I can do no better than quote from the opening paragraphs of Gilbert's presidential address before the American Society of Naturalists, delivered almost half a century ago: "In the statement of these considerations it is impossible to avoid that which is familiar, and even much that is trite. Indeed all expectation of entertaining or edifying you with the original or the new may as well be disclaimed at the outset. I shall merely attempt to outline certain familiar principles, the common property of scientific men, with such accentuations of light and shade as belong to my individual point of view." [2]

Obviously, my particular point of view, like Gilbert's, must be that of the student of earth science. It goes without saying that I am not competent to speak of methods of research in chemistry and physics, where experiment plays a far larger rôle than in geology and geography. So, also, the biologist, the astronomer, and investigators in other fields must speak for themselves. For this reason the title of my address may seem unduly ambitious. Yet I prefer the broader, more inclusive term "scientific investigation" to the more restricted, if more accurate, "geologic and geographic investigation," because it seems to me probable that some of the principles here discussed may find application beyond the limits of my particular field.

One characteristic of earth science deserves special emphasis in this connection. Most problems in geology, including geomorphology, involve the invisible past. We study phenomena produced long ago; and even where the responsible processes continue to operate, it is usually at a rate so slow as to give limited and imperfect information concerning their past performance. Furthermore, we can utilize experiment in solving our problems only in restricted measure. For us, the scientific method involves extended and deeply penetrating mental processes to which special attention must be given in our discussion.

It is a pleasure to express here my indebtedness to several of my Colum-

---

[2] G. K. Gilbert: The Inculcation of Scientific Method by Example, with an Illustration drawn from the Quaternary Geology of Utah. Amer. Jour. Sci., 3rd ser., vol. 31, p. 284, 1886.

bia colleagues, Robert S. Woodworth, Professor of Psychology; Adam
Leroy Jones, Associate Professor of Philosophy; Sam F. Trelease, Professor of Botany; and George B. Pegram, Professor of Physics; who were
good enough to read the address in manuscript form and give me the
benefit of helpful criticisms. To Professor William Morris Davis I am
indebted not only for very full comments on the manuscript which have
enabled me greatly to improve its form and contents, but above all for
penetrating and constructive discussions of method during association
with him for thirty years as student, colleague, and friend. But neither
Professor Davis nor others have any responsibility for the views expressed
in these pages.[3]

THE SCIENTIFIC METHOD

"Nothing has such power to broaden the mind as the ability to investigate
systematically and truly all that comes under thy observation in life."—Meditations of Marcus Aurelius.

BACKGROUND AND DEFINITION

The quest of Truth remains today, as it has remained throughout the
ages, the supreme manifestation of the human intellect. And the orderly
procedure of the mind by which this quest is prosecuted has long provided
a fascinating subject for study. From the days of Zeno, Socrates, Plato,
and Aristotle the "intellectual processes operative in the acquisition and
in the creation of knowledge" have been intensively investigated. Out of
these studies came the deductive, or Aristotelian, logic of the ancient
Greeks. Perverted by the scholastic philosophers of the Middle Ages, the
deductive process had fallen into disrepute when the scientific era opened
with the great discoveries of the fifteenth and sixteenth centuries. William
Gilbert in his "Treatise on Magnetism" and Francis Bacon in his "Novum
Organum" condemned the scholastic method of reasoning, and pointed the
way to a process in which the mind should start with the more common
and obvious facts of nature, then move forward to things more complex
until generalizations could be formulated and hidden causes discovered.
Thus was born modern inductive logic, sometimes called "the scientific
method." [4] But the scientific method of today is as far removed from that

---

[3] Professor Davis, in particular, dissents from my use of "inductive inference"; from
the treatment of "Verification and Elimination" as a single stage in investigation; from
the use of "Interpretation" rather than "Explanation" as the name of the final stage;
and from various other phrases and passages in the text. Nor does he recognize that
analysis is a process properly involved in the several stages of an investigation.

[4] Cf. D. S. Robinson: The Principles of Reasoning, pp. 333-343, 1930.

advocated by Bacon as is modern deductive logic from the legalistic quibbles of the schoolmen.

For purposes of this discussion I shall define "scientific method" as any method which effectively utilizes the several powers of the mind in a systematic, impersonal, non-emotional, unprejudiced effort to discover the origin and relationships of phenomena, and the laws which control their manifestation. The powers of the mind must be *effectively* utilized, for incorrect habits of thought may lead to the obscuring, rather than to the discovering, of truth. Not one capacity of the mind, but its *several capacities,* must be utilized. Observation, memory, comparison, classification, generalization, analysis and synthesis, induction and deduction, invention, experimentation, prediction, verification, revision, confirmation and interpretation—all these may, and a number of them must, be present in any truly scientific investigation. The procedure must be *systematic,*[5] for unguided, haphazard inquiry leads to omission of essentials, and consequent falsifying of results. It must be *impersonal,* since the discovery of things as they are is impossible to him who seeks to find them as he or others think they should be. It must be *non-emotional,* for when emotion claims the throne, reason abdicates. And finally, the procedure must be *unprejudiced,* for biased judgment is a poor instrument with which to seek the truth.

PLACE OF DEDUCTION IN THE SCIENTIFIC METHOD

Of all the factors involved in the scientific method as above defined, only one is likely to be challenged by any considerable number of men. Over three hundred years ago Francis Bacon was declaiming against the employment of hypothesis and deduction in scientific investigation. To-day there are those who echo his criticism. You may call to mind the names of eminent geographers and geologists who view with disfavor Davis's extended employment of the deductive method in geomorphology. Their criticism is born, not so much of failure to recognize that Davis has made skillful and fruitful use of the method, as of a deep conviction that the method itself is dangerous, and apt to give results of artificial, rather than intrinsic, value.

What is this "deduction" which the ancient Greeks developed, the scholastic philosophers perverted, Francis Bacon condemned, and some of our colleagues hesitate to employ? Since we are to hold it an essential of correct scientific method, it behooves us to understand each other when

---

[5] "Systematic" does not necessarily imply unvarying order in the sequence of procedure. If the system exists, one may utilize its different parts in variable order, providing all parts are ultimately utilized.

we use the term, deduction, and to make sure that we all draw the same distinctions between it and the inductive process. Unfortunately, authorities are far from agreement in defining logic and the nature and scope of its parts. This is true of induction to a greater degree than it is of deduction. When the elect disagree, the layman who must use their terms can do little more than specify the sense in which he employs them. This I do, with apologies to the psychologist and logician who find that in these pages "mind," "mental faculties," "induction," "deduction," and "analysis" are employed with less skill and precision than they could wish. To the student of natural science the terms as here employed will, I believe, convey the general conceptions essential to our discussion.·

Both induction and deduction start with something, do something with it, and get something else. In inductive reasoning we start with observed facts, subject them to certain logical processes, and seek thereby to discover general principles, or laws. In deductive reasoning we start with some generalization, subject it to certain other logical processes, and seek thereby to derive specific consequences which should correspond to observed facts. Induction proceeds from the particular to the general: deduction proceeds from the general to the particular.

When employing the inductive method the investigator is deeply concerned with the validity of the observations which serve as the starting point of his mental processes. If the supposed facts are not facts, it matters little how correct may be his mental processes; the results obtained are not trustworthy. But when employing the deductive method, the investigator is not primarily concerned with the truth of the generalization which serves as the starting point of his deductive reasoning. For the time being he takes the general proposition for granted, and concerns himself with the consequences which may logically be deduced from it. If the general proposition be erroneous, he will discover its invalidity when he has derived from it specific consequences, compared these with known or newly observed facts, and found an obvious lack of accordance between deduced expectations and observed realities.

In both types of reasoning, appeal is made to observed facts, although not necessarily to the same facts. But in one case, facts come into the picture first; in the other, last.

It is easy to see why logicians should hold that induction is the reverse of deduction. But it is a fallacy for the scientific investigator to suppose that the two methods of reasoning are antagonistic, or that one of them is good, the other, bad. The wise investigator will use one to supplement

the other ; for he secures great advantage if he first employs inductive
reasoning to derive from observed facts certain general conclusions, then
reverses the process and, using the conclusions as working hypotheses,
deduces their reasonable consequences, checking these last against observed facts as the best proof of the correctness of his reasoning.

NATURE OF THE SCIENTIFIC METHOD

With deduction accorded its rightful place in the scheme of scientific
reasoning, we are ready to inquire more closely into the precise nature
of the scientific method.   Fortunately, distinguished guidance is here
afforded us.   Gilbert's paper on "The Inculcation of Scientific Method
by Example" [6] placed in clear relief the outstanding differences between
the theorist and the investigator, defined the rôle of hypothesis and
deduction in research, and emphasized the superiority of multiple working
hypotheses as an instrument for discovering scientific truth.   Chamberlin's essay on "The Method of Multiple Working Hypotheses" [7] later
demonstrated the dangerous defects of what he terms the method of the
ruling theory and the method of the single working hypothesis, and convincingly advocated the employment of a variety of working hypotheses
as the most effective means of discovering all facts pertinent to a problem
and the best guarantee of arriving at their correct interpretation.   Davis
has in most striking manner illuminated the deductive processes essential to this last method,[8] applying them to a wide range of geomorphic
problems.   An experience of more than thirty years devoted to scientific
study, in the course of which I have given conscious attention to the
relative value of work accomplished by different methods of investigation,
convinces me that I shall make no mistake if I follow the lead of the
masters just cited, and accept the method of multiple working hypotheses
as the best procedure in research.   And I hope it will not seem presumptuous if I venture to lay before you certain considerations not fully discussed by Gilbert, Chamberlin, or Davis, which seem to me essential to
the best understanding and use of the method they regarded as superior
to all others.

---

[6] G. K. Gilbert : The Inculcation of Scientific Method by Example, with an Illustration
drawn from the Quaternary Geology of Utah.   Amer. Jour. Sci., 3rd ser., vol. 31, p. 284,
1886.

[7] T. C. Chamberlin : The Method of Multiple Working Hypotheses.   Jour. Geol., vol. 5,
pp. 837-848, 1897.

[8] See especially his paper on the "Disciplinary Value of Geography, Part I, The Science
of Geographical Investigation."   Pop. Sci. Mo., vol. 78, pp. 105-119, 1911.

STAGES OF INVESTIGATION

In actual use the method of multiple working hypotheses naturally resolves itself into fairly distinct steps, or stages. It is desirable to summarize these, and to state the essential nature of each stage, before proceeding to our inquiry as to the rôle analysis should play in each.

The first stage is that of *Observation,* during which the investigator takes cognizance of certain facts or things, and of the existence of a problem concerning their interpretation. Here, and repeatedly in succeeding stages, memory plays an important rôle, the investigator recalling from past experience, acquired either personally or through the work of others, additional things of similar character, already named and catalogued in his mind.

In the second stage, *Classification* occurs. The observed facts are compared, both among themselves and with others called up from memory; fundamental, as opposed to superficial, resemblances are noted; and like facts are grouped together.

Conditions are then ripe for the third stage, in which *Generalization* takes place. Here, the inductive process comes most prominently into play, the mind inducing from the classified facts broad generalizations respecting them. These generalizations may take the form of a law or principle expressing some significant relation between the observed facts. Such law or principle may, in turn, constitute a partial or complete explanation of the facts, as did Harvey's famous induction regarding circulation of the blood. More often, perhaps, the generalization merely paves the way for an explanation, which comes fully into being in the next, or fourth, stage.

Here, *Invention* further extends the inductive process. If the inductive inference of stage three involves a complete explanation of the facts, the investigator deliberately proceeds to invent additional possible explanations. If the earlier induction provided but a partial explanation, or merely paved the way toward an explanation, invention completes the process, and then presses on to the formulation of additional explanations for the same facts. It is this deliberate effort to invent the greatest possible number of reasonable explanations for a given phenomenon or set of phenomena which distinguishes the method of multiple working hypotheses from other methods of research. For it is these tentative explanations which become the working hypotheses of the next, or fifth, stage. Where others have previously invented explanations of the phenomena in question, these are, by the impartial investigator, tentatively taken into his family of working hypotheses, and accorded the same hospitable consideration given to those of his own creation. The precise nature of the inventive process is a debatable question which need not here concern us.[9] It is sufficient for our purposes to know that it is a real and valuable faculty of the human intellect, and that it is capable of deliberate cultivation.

In the fifth stage, *Verification* and *Elimination* require extended use of deductive reasoning. The investigator now takes each working hypothesis in turn, and, reversing the mental processes previously employed, first deduces its reasonable consequences as fully and as precisely as possible, then confronts the deduced expectations with facts observed in the field or established by experimentation, thus verifying the competence of the hypothesis to explain the facts, or eliminating it as incompetent and hence unworthy of further consideration. As will later appear, the verification of this stage is limited, and requires some measure of independent confirmation.

Confirmation and *Revision* is secured in the sixth stage. Each hypothesis which survives stage four requires further study. Its deduced consequences guide the investigator in predicting the existence of facts not yet discovered, or in devising experiments which will establish facts not previously known. The establishing of these facts, or the failure to establish them, will confirm the validity of the hypothesis, or require its revision, possibly even its ultimate rejection. From this severer test there should emerge one or more hypotheses which alone or jointly offer satisfactory explanation of the observed facts.

The investigator now passes to the seventh and last stage, fully formulating his ultimate *Interpretation* of the phenomena investigated.

No one would pretend that each of the seven stages described above is a sealed compartment, in which the mind performs no function appropriate to the other stages. The human intellect is an unruly member, leaping forward and backward in a most wilful way. One could not, even if he would, carry through an investigation strictly in the orderly sequence just enumerated. Nor is it desirable that he should do so. The development of any stage may throw valuable light on the procedure appropriate to some earlier or some later stage; and the intellect instinctively, inevitably turns backward or rushes forward to profit by this better illumination.

Furthermore, the solution of many problems is not the work of a single

---

[9] "No one has ever given any explanation how the hypotheses arise in the mind." Encyclopedia Britannica, Article on *Induction*, 11th Edition, 1910. For a hypothesis as to the origin of hypotheses, see G. K. Gilbert: "The Origin of Hypotheses," Science, n. s., vol. 3, pp. 1-13, 1896.

mind, but of the collective mind of those who, sometimes over long periods, contribute to the advancement of knowledge in a given field. The individual thus enters upon an investigation when others have more or less completely traversed some or all of the stages enumerated above. If wise, he will retrace their steps sufficiently to verify or revise such of their results as he wishes to utilize in his own study. But in view of the work already done, his mind is likely to push forward here, or turn back there, as expediency rather than system may dictate. There is thus set up a plexus of paths by which the mind traverses every part of the problem under investigation.

The plexus will vary for different problems, and must vary widely for problems in such widely different fields as physics and botany, chemistry and geomorphology. I have merely set forth what seems to me the normal, logical sequence of scientific research, viewed from the standpoint of a student of geomorphology. In this field certainly, possibly in other fields also, any well-ordered investigation will be found to present the seven enumerated stages in more or less distinct form, however complex may be the paths by which the mind traverses these stages.

RÔLE OF ANALYSIS IN SCIENTIFIC RESEARCH

"Look beneath the surface; let not the several quality of a thing nor its worth escape thee."—Meditations of Marcus Aurelius.

DEFINITION OF SCIENTIFIC ANALYSIS

What is analysis? The dictionaries tell us that it is the process of separating a thing or a concept into its constituent parts, in order to arrive at the essential or ultimate elements, causes, or principles; that it is the tracing of things back to their sources; and that it is designed to clarify and test knowledge. The chemist analyzes a complex substance to determine its precise composition. For the purposes of our discussion I would define scientific analysis as "the process of separating observations, arguments, and conclusions into their constituent parts, tracing each part back to its source and testing its validity, for the purpose of clarifying and perfecting knowledge."

What rôle should analysis play in the seven stages of a scientific investigation as previously defined? In proceeding to this inquiry I shall again fall back upon a precedent established by Gilbert. He realized the disadvantage of discussing abstract propositions, when the human mind grasps more readily things presented to it in concrete terms. Accordingly, he illustrated his conceptions of correct scientific method by experience gained in his study of the shores of Lake Bonneville. I am persuaded we can pursue our discussion most easily if I illustrate my observations, so far as feasible, by reference to some concrete example. For sake of simplicity I shall select a common phenomenon found on the rocky shores of many lands, and in tracing the part played by analysis at each stage of a comprehensive scientific investigation, shall illustrate my points freely by referring to the study of this phenomenon, the origin of which is still a subject of dispute. Those who follow this discussion should not attach undue importance to the particular example selected for illustrative purposes, nor to the statements made concerning it. I have deliberately selected a debatable matter, the study of which is not yet completed. Our concern here is not with the origin of the phenomenon, but with the method of investigating its origin.

Along the rocky shores bordering the sea are found nearly horizontal platforms or benches (figure 1), from a few feet to a few hundred feet in width, cut in solid rock. At their seaward margins these benches terminate in relatively abrupt slopes which descend toward or even into the water. From their inner, or landward, margins rise steep slopes or rocky cliffs. In elevation the benches range from one or two feet above ordinary high tide indefinitely upward, often several hundred feet above the same datum plane.

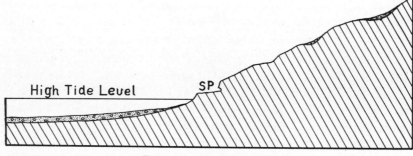

FIGURE 1.—*Marine Benches*

The combination of bench and cliff bordering the sea might arouse no significant reaction in the mind of the ordinary observer. But such a combination instantly excites the brain of the geomorphologist to activity. Memory recalls textbook diagrams, explanatory titles under photographs of similar features, discussions of the origin of such features in geologic and geographic treatises, perhaps earlier field examination of such forms. Rapidly the mind compares the forms, the situations, the mutual relations of the assembled features in the various cases called before it by observation and memory, and before the observer is conscious of the process he has leaped to an inductive inference: that both bench and cliff were cut into the margins of the land by wave erosion. In this case the inductive inference involves an explanation; but it is only a partial explanation. If the bench is wave-carved, why is it exposed to view above the level of the sea? The mind rushes on: Perhaps the land has been raised since the bench was carved; perhaps the sealevel has dropped.

The observer must now be on his guard against a dangerous tendency of the human intellect: the tendency to accept as valid a plausible explanation, and then to look for facts in support of that explanation. He must deliberately repress the tendency toward premature conclusions, and begin the task of gathering all the facts upon which alone can a·satisfactory explanation be based. As already stated, we are not primarily concerned with the course of his particular investigation of these interesting forms, nor with the conclusions he may reach respecting their origin; but we shall draw freely upon this imaginary hypothetical study, to illustrate the uses of analysis in scientific research.

A final word of caution before we enter upon our task of tracing the rôle analysis plays in the several stages of an investigation. One should be careful not to confound the analytical process with the various processes which constitute the successive stages of research. It is our thesis that analysis is not one of these stages; rather, it is something which may be employed in many, indeed in all, of them. Analysis is not Observation; it is not Classification. It is an instrument of precision which may be so employed as to render Observation more complete and more accurate, Classification more effective and significant; and so on through each succeeding stage of a properly conceived and successfully executed scientific investigation.

I. ANALYSIS IN THE STAGE OF OBSERVATION

The initial stage in scientific investigation is normally that of *Observation*. The first employment of this mental process may be made incidentally, perhaps almost passively. But once the investigator realizes that the observed facts present a problem, for the solution of which additional facts are desirable, he becomes an active inquirer, seeking to discover all facts bearing on the problem. He observes as widely and as accurately as possible; the observed facts being automatically recorded in his memory, but also, since the memory is notoriously fallible, deliberately recorded in his notebook if he be a prudent investigator.

Since we are here concerned merely with the recording of external facts observed by the eye, it might appear that analysis has no rôle to play in this initial stage of an investigation. But let us dissect the matter and scrutinize its parts, first turning our attention to the material facts which are the object of observation. Can the analytical powers of the mind be brought to bear upon the facts themselves in such manner as to aid the investigator?

Let our hypothetical student of coastal benches answer. Having observed one such bench, and had his curiosity aroused to investigate, he begins his search for other examples. Very soon he is confronted by the necessity of discriminating coastal benches of the type described above from other forms similar in some respects yet different in others. He needs a name by which to designate the type forming the particular object of inquiry, and tentatively calls them "elevated marine benches" or simply "marine benches," because of their resemblance to submarine benches formed by wave abrasion and because of their apparent marine origin.

The question then arises: What constitutes a marine bench in the sense in which he is using the term? Is it any nearly level surface in the vicinity of the shore, terminated seaward by an abrupt descent, and landward by cliffs which rise steeply to higher levels? This combination of slopes appears repeatedly in Nature, as the product of a variety of causes. The non-critical investigator, who welcomes every shelf and scarp facing the sea as one more link in the chain of facts he seeks to explain, is almost sure to accumulate, without knowing it, an assortment of unrelated observations, many of them irrelevant to his problem. His investigation is seriously compromised from its very beginning. How can he explain the origin and history of elevated marine benches if he unwittingly bases his reasoning on a confused mixture of wave-carved surfaces, benches due to differential weathering, rock terraces cut by a river before drowning brought in the sea, notches cut by a glacial stream flowing between waning continental ice and the sloping hillside, and terraces due to landslides, faulting, or monoclinal warping? As all these features are found in the immediate vicinity of the shore in different places, and as all of them have repeatedly been ascribed to wave erosion, it is clear that the field data must, in the first instance, be subjected to rigid scrutiny.

The facts can not be lumped. They must be analyzed—separated into their constituent parts, and each part tested as to its relevancy to the problem under investigation. Here, the analytical process imposes a very uncomfortable but highly essential rule: that every fact the relevancy of which is in any degree doubtful be excluded from the group of facts for which explanation is being sought. A certain shelf, or terrace, closely resembles other marine benches, but its surface corresponds to the surface of a resistant rock layer. Obviously, this particular form need not necessarily have the same origin as those cut indifferently across various rock structures. Although the observer may feel certain that this feature is identical in origin with the other forms he is investigating, the analytical process demands that it be excluded from the study. And for a thoroughly sufficient reason. If this one terrace be really a product of differential weathering, and yet be included among the facts later to be explained, the explanation then advanced must be competent to account for it just as fully as for the others. This will inevitably exclude from consideration any hypothesis competent to explain the other benches only, and may thus prevent discovery of the true explanation of the forms constituting the main object of inquiry.

I have said that the rule requiring exclusion of all data of doubtful relevancy is an uncomfortable one. The investigator does not willingly push aside some of his most spectacular facts. He is disturbed when analysis points a questioning finger at one after another of his bits of observational data. He may even find that if he rigidly applies the exclusion rule, the quota of competent data ultimately admitted to consideration is pitifully small. It may not even be sufficient to establish any theory on a sound basis. And the investigator craves a theory which will explain the facts.

Let us frankly admit the discomfort rigid analysis imposes; but let us, at the same time, acknowledge that there is no guarantee of safety in scientific investigation save that offered by such analysis. If the facts are all of equivocal character, lending themselves to a variety of interpretations, the formulation of a definite theory must await discovery of new and less equivocal facts. Postponement of judgment may bring discomfort; it does not bring disaster.

Thus far we have discussed the necessity of analyzing the things observed. But the investigator must probe more deeply than that. He must analyze the observational process itself. Does he really see what he thinks he sees? Not unless he is constantly on guard against the well-known dangers to which observation is subject. Chief among these, perhaps, is the tendency to include inference with fact, to confuse theory with observation. The observer thinks he sees a bench cut in granite, when in fact all he really observes is a few scattered outcrops of granite protruding through the soil; from these he infers, perhaps erroneously, that the vastly larger invisible mass is likewise granite. In the case imagined a moment ago the investigator may report that he saw one wave-carved bench which did not bevel the rock layers. All he really saw was a shelf, or bench, parallel to the rock layers. He inferred it was wave-carved, when in fact it may have been produced by differential weathering of weaker beds overlying a more resistant layer.

Another danger to which observation is subject is that of ocular deception. This danger looms large when one is looking for certain specific forms in the landscape. The power of suggestion is greater than many realize, and one looking for marine benches is in danger of finding them in faint undulations or irregularities of the terrain, which under other circumstances would not impress him as significant. If he is looking for benches which he expects to be horizontal, the inclined position of something he mistakes for such a bench may quite escape his eye, even when the inclination is so marked as to be quite obvious to another who is without preconception as to what should be observed.

Incompleteness of observation is another common danger. The eye tends to pick from the landscape that which seems to it for the moment significant, and fails to note much that later stages of the study may show to be vitally important.

Even where the initial observation is fairly complete, the benefits which should be derived from it may be lost through failure to record permanently in memory and in notebook all the facts observed. Many studies of shore terraces have been vitiated because only part of the relevant facts were recorded by the investigator.

No one is wholly free from the dangers of defective observation. But one who acquires the habit of analyzing his observational powers and processes is less exposed to these dangers than are those who give no conscious attention to this initial step of a scientific investigation. In research, as in other things, to be forewarned is to be forearmed; and the investigator who is conscious of the dangers inevitably associated with the observational process will be on his guard against those dangers. He will deliberately exclude all inferences from his group of supposedly observed facts, will avoid the possibility of ocular deception by pushing each observation far enough to establish its validity, and will make a persistent effort to see all things, record all things, and remember all things bearing upon his particular problem.

In this last effort he may derive one great practical advantage in respect to field and laboratory methods. Having analyzed the observational process, he will realize that it is almost always doubly defective so long as observations are recorded by memory alone; defective, first, because the mind does not retain all it initially records; but also, and far worse, defective because the exposure of the mind to the facts is normally too rapid for all of them to make an impression upon even that sensitive recording instrument. Once aware of this source of danger, the investigator can, by a simple procedure, remove the double difficulty. By fully recording his observations in writing, in the presence of the facts, he not only insures against the fallibility of memory, but at the same time insures far more complete and effective observation. Both the time required for a written record, and the stimulation and direction of the observing faculties inevitably consequent upon systematic recording, enormously increase the completeness of the record.

II. ANALYSIS IN THE STAGE OF CLASSIFICATION

Is there any need for employing analysis in the second stage of an investigation, the stage of *Classification*? Let us note first that the analysis involved in stage one led to a sort of rudimentary classification, the separation of observed facts into those which were relevant and those of doubtful relevancy. We at once suspect that analysis must likewise be involved in any further effective classification of the facts of observation surviving the analysis of stage one.

The investigator is not content merely to exclude from consideration every doubtful bit of data. For example, our hypothetical student does not rest when he has excluded from his study every topographic form which does not certainly belong in the group of elevated marine benches. He raises the question as to whether all these benches need necessarily have had the same history. He takes cognizance of the fact that if the benches were carved at different times, and some of them were affected by events, such as continental uplift, which did not affect the others, a serious difficulty may be introduced into the attempt to explain their history. The investigation, to be successful, must deal with a group of facts which are comparable. A simple hypothesis competent to explain benches five feet above high tide, may conceivably be quite incompetent to explain benches fifteen feet above high tide.

The investigator therefore seeks to segregate the relevant facts into groups having like characteristics. Nor is he willing to base his classification on superficial resemblances and obvious differences. Just as the biologist classifies the whale with mammals rather than with fishes, so the student in any other field must make a deliberate effort to discover fundamental similarities and differences which escape the ordinary observer, but which may have great significance for the scientific inquirer. To accomplish this task most effectively he brings his analytical faculties to bear upon the data to be classified. He separates his observations into their constituent parts, compares them one with another, and strives to discover their essential elements. Only thus can he hope to see revealed those similarities and differences not immediately obvious.

In the case of our hypothetical study of marine benches, it may happen that one group of benches falls systematically within a limited altitude range, while another group has a different and more variable range. If so, the investigator will discover that fact, and classify the groups separately. He may also find that some of the benches are more weathered than others; that some are mantled with debris while others are not. If so, appropriate groupings will be established.

III. ANALYSIS IN THE STAGE OF GENERALIZATION

Classification paves the way to the third stage of investigation, in which *Generalization* occurs. Thus, in our hypothetical study of marine benches, it is only when Classification has been effected that the investigator can make significant generalizations; such, for example, as that in one group the benches have a narrow range in altitude, are always close to the sea, present no evidence of weathering of the rock surfaces, and are free from debris; that in a second group the benches have higher altitudes but equally narrow range in altitude, present evidence of moderate weathering, and have small quantities of debris near their inner margins; while in a third group the benches are still higher, vary extensively in altitude, and have surfaces which are deeply weathered and prevailingly covered with debris.

As thus stated, the foregoing generalizations involve no explanation of the classified facts, although it is obvious that they represent a step forward toward possible explanations. We may, however, easily imagine an inductive inference, based on the same mass of observed facts, which does involve partial or complete explanation of them. The inference that the benches were carved by wave erosion involves a partial explanation. The inference that the benches represent former shorelines of the sea raised by some upheaval of the land involves a more nearly complete explanation. These are all generalizations based on a number of particular facts, and are essentially inductive in nature; although they include, as do all inductions, an element of deduction in that they assume uniformity in Nature's laws and the repetition of certain shore forms wherever the waves of the sea have operated.

We have now to inquire whether the analytical faculties of the investigator can serve any useful purpose in the stage of generalization. Let us first note that the empirical generalizations stated above, free from any explanatory element, had their roots in the careful analysis that made effective classification possible. But it is when inducing generalizations which involve more or less explanation that the analytical process can render greatest service. Such generalizations, like those free from explanation, must be rooted in facts and represent a normal outgrowth of them. They must be legitimate inferences induced from the facts themselves. As we shall see in a later section, the investigator may formulate conceptions involving explanations which are not an immediate normal outgrowth of the facts, but the products of invention. As these two types of explanation, the induced and the invented, have somewhat different standing before the court of the intellect, it is important that they be not confused. It is in making this discrimination that the investigator must again employ his analytical powers. The nature of the facts, the nature of their distribution, their relationships, and other pertinent elements must critically be examined, to the end that only legitimate inferences may be induced from them.

For example, the first two groups of classified facts described above, in both of which uniformity of altitude above sealevel is a characteristic of the benches, may properly give rise to the generalization that after the benches were carved the sealevel dropped. Uniformity of bench altitude suggests this as a legitimate inference involving explanation. But no such generalization may be based on the third group of facts, in which great variation in elevation is an outstanding characteristic of the benches. Such heterogeneity in elevation does not suggest either uniform drop of sealevel or systematic differential uplift of the land. Perhaps the facts can be explained on the basis of one or the other of these explanations, or on both combined; but if so, the explanations must be deliberately invented and applied to the facts. They are not normal outgrowths of the facts.

If the investigator analyzes the process of generalization a little further he will realize that it is not sufficient that generalizations grow out of the facts. They must grow out of a sufficient number of facts to be truly general. One of the commonest errors in research is to generalize prematurely, on the basis of a wholly inadequate foundation of facts. The investigator who consciously analyzes both processes and their results at each stage of his research, is less likely to fall into this and other errors than is one who thinks without thinking about his thinking.

When one analyzes the grouping of classified facts respecting elevated marine benches, another important consideration becomes clear; generalizations involving explanation, to be effective, must be made separately with respect to each class of facts. Further study may show that a single

XXXII—Bull. Geol. Soc. Am., Vol. 44, 1933

explanation applies equally to all the benches. But it is quite as likely to show the reverse. Hence the necessity of treating the different groups of facts independently, until a common origin and a common history for all have been fully demonstrated.

IV. ANALYSIS IN THE STAGE OF INVENTION

In the fourth stage of our hypothetical investigation the inquirer takes the classified facts and the generalizations concerning their nature, and uses them as a basis for *Invention* of as many explanations of the facts as may be possible. It is here that "The Scientific Use of the Imagination," as Tyndall has happily phrased it, comes most prominently into play. The invention may be deliberate, the result of conscious effort. Often it springs unexpectedly into consciousness, the result of a mind well equipped with pertinent knowledge repeatedly "mulling over" the facts in variable combinations. If others have anticipated our inquirer in the inventive process, as is usually true in greater or less measure, he welcomes every idea of alien origin which offers a possible explanation of the facts, and accords it just as hospitable treatment as those born of his own intellect. If the generalizations of stage three involved explanation of the facts, the investigator seeks additional and independent explanations. Where generalization involved but partial explanation, or merely paved' the way for an explanation, the investigator employs his inventive powers to complete the unfinished task; and then moves forward to the invention of alternative explanations.

As has previously been pointed out, the process of invention is not fully understood. It resembles induction to the extent that the mind starts with concrete facts, and from them passes to conceptions of broad application which take the form of tentative explanations of the facts. But the mind has far greater liberty in the stage of Invention than in the preceding stage of Generalization. The relation to facts is here less close. A proper generalization, being an outgrowth of the facts, must have its roots well grounded in them. An invention may spring from "thin air," the rarefied atmosphere of more abstract reasoning. The stimulus to invention comes from the facts, and the thing invented must not palpably be contradicted by the facts; but it need not be a normal outgrowth of the facts. Thus, the conception of a uniform drop of sealevel is not a normal outgrowth of the third group of facts mentioned above, because scattered remnants of marine benches occurring at a great variety of altitudes do not, of themselves, suggest the idea that the sea surface has been lowered rather than the land surface uplifted. But one may invent the conception of uniform but intermittent drops of sealevel, and offer it as a possible explanation of benches occurring at many different altitudes, since none of the observed facts necessarily contradicts such an explanation.

We have noted the greater liberty enjoyed by the mind in the stage of Invention. But it is not wholly free. Since the explanations induced in stage three, and those invented in stage four, are to serve as working hypotheses in the deductive operations of stage five, they must conform to certain fundamental requirements governing the formulations of hypotheses. In the first place, it is essential that the tentative explanations should be so precisely molded that specific deductions may be derived from them. If an explanation is so vague in its inherent nature, or so unskillfully molded in its formulation, that specific deductions subject to empirical verification or refutation can not be based upon it, then it can never serve as a working hypothesis. A hypothesis with which one can not work is not a working hypothesis.

Again, the explanation as formulated must be possible, for explanations clearly contradicted by well-established natural laws waste time and energy without advancing the investigator toward the true goal of his researches. On the other hand, it is manifestly dangerous to exclude explanations which merely appear incredible or absurd because they run counter to established opinions. Many a door to truth has thus prematurely been closed, and long remained closed. The hypothesis of continental glaciation seemed incredible in 1840, as did the hypothesis of evolution in 1860.

How then shall the inventive mind properly mold its hypothesis, and distinguish between explanations which appear plausible but are really unsound, and those which immediately provoke distrust yet merit hospitable consideration? It seems to me that here the analytical process offers us an instrument of incalculable value. Let us dissect each tentative explanation into its component elements, trace each element back to the assumptions and inferences which lie hidden behind it, and test the reasonableness of the whole by testing the validity of each part. Once more we turn to the shore for concrete illustrations of abstract principles.

Our investigator of marine benches, having first observed and classified his facts, and then derived certain generalizations from these facts, decides that hereafter he will concentrate his study upon the first group of benches earlier described—those lying close to the present shore, ranging from 1 or 2 feet to 5 or 6 feet only above ordinary high tides, showing fresh rock surfaces usually free from debris. To distinguish this group of benches from others, he calls them "shore platforms," thus recognizing their restriction to the immediate vicinity of the shore (SP, figure 1).

He now proceeds to the invention of as many explanations as possible for the facts he has collected concerning these shore platforms. Among some eight or ten explanations which occur to him, or which he appropriates from others, we consider but two. One, already briefly mentioned, is to the effect that the shore platforms, originally carved below sealevel, have been raised by regional uplift of the coast. The second supposes that the platforms were carved in their present position wholly above high tides, by storm waves of the present sealevel. The first explanation is in accordance with current geologic opinions, both regarding the conditions under which waves carve platforms and regarding crustal movements. The mind welcomes it hospitably as a hopeful direction in which to pursue the study. The second runs counter to long established views that the level of platform cutting lies distinctly below sealevel. The mind immediately recoils from this explanation as being inherently improbable. But our investigator, distrusting the mind's intuitive reactions, determines neither to accept the first as a tentative working hypothesis, nor to reject the second, until he has first analyzed them both. Note that his analysis is here directed merely to determining the acceptability of each as a member of his family of working hypotheses. Testing the competence of the hypotheses to explain all facts observed comes later, and involves quite different processes.

As to the first suggested explanation, he asks a number of pertinent questions. What is the fundamental assumption on which the explanation rests? Obviously, that such platforms *must* be cut below sealevel, and that the invention of another factor, such as regional uplift, *must* be invoked to explain their position above the sea. But what is back of these "musts," of these implications of compulsion? First, there is the teaching of standard treatises on geology, which commonly show both wave-cut and wave-built platforms below the surface of the water. Second, there are the reports of various observers who state that they find rock platforms under the sea, and the bases of rocky sea cliffs a number of feet below the level of low tide.

But do the treatises present convincing evidence that text and diagram correctly represent conditions of wave-cutting? They do not. An occasional text represents wave-cutting above sealevel, so that the testimony of the treatises is, in some measure, contradictory. And do the observers submit conclusive evidence that the submerged platforms and cliff bases described by them are in process of normal development at the present time? It must be admitted that they do not. They may believe such to be the case; but the facts they present do not seem to exclude the possibility that platforms and cliffs were carved at a higher level, and recently submerged by subsidence of the land or by rise of sealevel. Thus the investigator pushes his inquiries, and as he does so the tentative explanation which first appealed to him most strongly takes on an increasingly doubtful aspect. He may ultimately be impelled to reject it as lacking the necessary degree of reasonableness. Or caution may dictate that he preserve it as a possible working hypothesis, to be later subjected to the more searching test of deductive reasoning. In any case, critical analysis has assigned the explanation (of wave erosion below sealevel followed by coastal uplift) a position far less dominating than it originally occupied.

The investigator next critically examines that tentative explanation which he first was tempted to reject as being unworthy of serious consideration. What lies back of the suggestion that storm waves may have carved the platforms during the present relations of land and sealevel? For one thing, an assumption that there may be factors involved in wave attack which permit effective erosion above ordinary high tide level. What then are the factors involved in wave attack, and do any of them favor activity at levels relatively high? Observation and memory present certain pertinent facts bearing on this point. The momentum of masses of water hurled up the sloping shore zone by advancing waves insures some erosive activity above still-water level. Furthermore, it is well established that in both waves of oscillation and waves of translation the crests rise higher above still-water level than the troughs sink below it. For this reason the average level of wave attack is above still-water level. Onshore winds are known to pile up the water to levels often a number of feet above normal. Thus, at least three factors in wave attack tend to produce high-level cutting, and it seems not unreasonable to suppose that their combined result might be wave erosion at the level where the low lying platforms are actually observed.

Along these and other lines the investigator pushes the analysis, until he is satisfied that the tentative explanation of shore platforms as a product of storm-wave attack under existing conditions of land and sealevel, while presenting obvious difficulties, is sufficiently reasonable to deserve a place among the tentative hypotheses later to be tested. So in turn he dissects and scrutinizes each tentative explanation which his inventive powers suggest, or which he adopts from the work of others. Throughout this stage of the investigation it is evident that analysis is an invaluable aid to his studies, and the best guarantee of their eventual success.

V. ANALYSIS IN THE STAGE OF VERIFICATION AND ELIMINATION

On entering the fifth stage of the investigation our hypothetical inquirer possesses, let us suppose, some half dozen tentative explanations which have survived the analytical processes applied in the preceding stage. He now treats each of these in turn as a working hypothesis, and, reversing the mental processes employed in the third stage, endeavors by *deductive* reasoning to determine as precisely as possible just what features should characterize the shore platforms in case the particular hypothesis for the moment under scrutiny be the true explanation of these forms. His object is to verify, so far as possible, the competence of some hypotheses, and to eliminate others which critical tests show to be manifestly incompetent.

He secures the required tests by an appeal to facts. After deducing the reasonable consequences of a working hypothesis, he confronts these by such facts as are already in his possession, drawing upon memory, upon his field notes, and upon the published observations of others. If the consequences logically deduced from the hypothesis are in accord with the facts, he accepts this as verification of the competence of the hypothesis to explain the facts already in hand. Where deduced consequences find no counterpart in observed facts, he feels justified in eliminating the hypothesis from further consideration.

Here it should be noted that the success of the deductive operations involved in this fifth stage of an investigation depends primarily upon two things: first, the fullness and accuracy of the logical operations performed by the mind; second, adequate testing. The careful analysis employed in each of the preceding stages has done all that properly can be done to assure that only competent hypotheses shall serve as the starting point for logical deduction. The process of deduction itself, while not directly concerned with the validity of these hypotheses, is vitally concerned with the laws of thought governing a peculiar mental operation to which each hypothesis is now subjected. Obviously, the mind must operate with completeness and precision according to the laws of logic if the results are to be trustworthy.

The second condition of success is adequate testing of the products of deduction. Let us recall that the very essence of the method of multiple hypotheses is hospitable consideration of every possible explanation of a given group of facts, in the full knowledge that not all these explanations can correspond to reality. Let us remember further that the deductive process itself neither adds anything to, nor takes anything from, the original content of any hypothesis, but merely develops and elaborates its inherent possibilities. With these considerations in mind, we can understand how critically important it becomes to test the greatest possible number of deduced consequences for each hypothesis by confronting them with the greatest possible number of facts competent to either verify or contradict them. The whole method stands or falls with the success or failure of this operation.

When all possible consequences of each working hypothesis have been deduced, it will often appear that some of the competing hypotheses have certain consequences in common. These have relatively low critical or discriminative value, since if facts are found corresponding to the deduced expectations, they merely show that any one of several hypotheses may be true. There will usually be found, however, some one or more consequences peculiar to hypothesis A, while certain others are peculiar to hypothesis B, and so on. It is these unlike consequences which have the highest critical value in discriminating between valid and invalid hypotheses, and it is on these that the investigator will most depend when drawing conclusions.

To what extent, if any, can analysis serve the investigator in this stage of *Verification and Elimination*? Since, as we have just seen, effective use of the deductive process depends upon one's ability to make logically correct deductions, it is expedient that the investigator should subject his own mental processes to careful scrutiny. There are fallacies in reasoning which must be avoided. These are treated in works on logic and need not be discussed here. But some study of logic will direct conscious attention to one's mental habits, and help to secure the precision and completeness requisite for careful deduction.

The investigator must realize that if his deductions are false, it matters little whether or not observed facts correspond to them. The conclusions are, in any case, invalidated, and he may either reject a perfectly valid hypothesis or embrace one which is invalid, on the basis of false reasoning. Safety lies only in analysis. Deductions, like inductions, must be separated into their component parts, and each part scrutinized and its validity tested. Let us illustrate what this involves by turning once again to the problem of the shore platforms.

Our investigator has accepted as one working hypothesis the tentative explanation of storm-wave erosion under present conditions of land and sealevel. From this hypothesis he deduces certain expectable consequences, among which let us consider only one. If the platforms are the product of storm waves operating at the present time, he reasons, then there should be found upon their surfaces rock debris constituting the cutting tools with which waves accomplish their erosive work. Turning to his record of field observations, he finds that many of the platforms are remarkably free from debris. Indeed, their outer edges commonly drop abruptly into moderately deep water, so that it is difficult to see how much debris could be cast upon the platforms or maintained there. The facts in hand conspicuously fail to match the expectations deduced from the hypothesis, so our investigator rejects the hypothesis of storm-wave erosion as incompetent. In doing so, he rejects a perfectly valid hypothesis solely because he failed to examine his deductions critically before using them as tests of the competence of the hypothesis.

Let us suppose, however, that our investigator was so fully convinced of the value of analysis as an instrument of research that he would neither accept nor reject a hypothesis until he had subjected each deduction derived from it, and each test of its correspondence with the facts, to such rigid scrutiny as he was able to make. Under these circumstances he would first have examined the deduction relating to the expected presence of wave-cutting tools on the platforms. He would have asked: Upon what does the deduction rest? Evidently it rests upon the assumption that waves of water unarmed with debris cannot effectively erode the coast. And on what, in turn, does this assumption rest? Current geologic opinion may run this way, and geologic textbooks may assert that waves without cutting tools are feeble. But current opinion and the textbooks have often proved fallible guides. So our investigator would have gone back to the sources on which both opinion and textbook should be based; the actual recorded evidence as to the nature and causes of damage accomplished by waves. Here he would have found much information concerning havoc wrought by waves under conditions which seem clearly to preclude the effective intervention of debris. The terrific impact of the water itself, the force of the currents it generates, the direct pressures exerted upon air and water imprisoned in crevices in the rock, the sudden expansion of air in crevices and pore spaces when the rapid retreat of a wave creates a partial vacuum outside, all these are described as effective causes of damage by waves, on the basis of concrete and seemingly reliable evidence. Our investigator must then have concluded that while shore debris is presumably a highly important factor in wave erosion, there is no reason to exclude the possibility that waves relatively free from debris, and striking with a force varying from hundreds to thousands of pounds per square foot, may accomplish much erosive work. He would therefore have rejected, not the hypothesis of storm-wave erosion, but his tentative deduction respecting the necessity of finding debris on the platforms; and would have looked for other deductions from the same hypothesis which he could use as a more reliable basis for testing its validity.

It is not alone in making deductions from hypotheses that one's critical powers should render aid in this fifth stage of an investigation. When confronting deduced consequences with observed facts, the investigator must be equally on the alert. If harmony is apparent, he will be on his guard against too ready acceptance of the pleasing result. He will probe the situation to make sure that harmony is real and not merely apparent. If discordance is revealed, he will dissect and scrutinize the facts to determine whether the discord may not be the result of accidental introduction into the scene of some irrelevant factor not germane to the problem.

The investigator who fully analyzes the process of testing hypotheses by deducing their consequences and confronting these with observed facts, will gain a keen appreciation of one fundamental consideration: the effectiveness of the test will vary with the number and character of the deductions made from any given hypothesis, and the variety and nature of the facts available to match against them. If the deductions are great in number, and some of them at least are peculiar in character; and if they are nicely matched by facts which are equally numerous and peculiar, then the test of the hypothesis is particularly good.

We have now to note certain important limitations of the "verification" involved in this fifth stage of a scientific investigation. "To verify" is to prove to be true. But what is it that we prove to be true in this stage of Verification and Elimination? Not the hypothesis as the true and only explanation of the facts, for seldom, if ever, is a hypothesis directly verified. The thing we prove is the existence of a certain measure of harmony between the deduced consequences of a hypothesis and the observed facts. The "verification" is merely verification of one or more hypotheses as *competent* to explain certain facts, not verification of them as the *true* explanation of those facts. Several hypotheses may be competent to explain a given set of facts, where only one of them can be the true explanation. Obviously, there is need of further study before any hypothesis can be accepted as fully established. This further study is secured in the next following stage, Confirmation and Revision, discussed on subsequent pages. The verification of the present stage is partial and limited.

It is in the present stage that one is especially apt to commit an error in the deductive process which has wrecked many a train of reasoning. I have characterized this as the error of "limiting Nature." It consists in setting narrow and rigid limits to what natural forces can accomplish under hypothetical conditions. "So long as a stream is ungraded through a hard rock barrier, weak rock areas above the barrier cannot be reduced to a peneplane." "A waterfall in homogenous unconsolidated material cannot retreat upstream for any great distance, but must be rapidly flattened out to a gradual slope." So the human intellect rashly sets metes and bounds to Nature. And Nature proceeds calmly to perform the impossible. The deductive process inevitably involves judgment as to what Nature can and can not accomplish. One must see to it, then, that the limits set are sufficiently generous and flexible, and that they are established only after careful analysis of all factors involved. Only thus will deductions be reasonable, and dependence upon them be safe.

One further observation should find place in this part of our discussion. Some who oppose the use of hypothesis and deduction in scientific investigation have asked: "What is the use of setting up men of straw, only to knock them over?" The answer is that men of straw have no place in a scientific investigation. We have already seen that in the earlier stage of Invention only possible explanations are accepted into the family of working hypotheses. Analysis is there deliberately employed to exclude absurd or incredible hypotheses. The men of straw are not set up. They are early cast aside. Deduction has therefore, in the stage of Verification and Elimination, to deal only with such hypotheses as have some valid title to credence. To determine how far they are entitled to credence is the province of this fifth stage. Some hypotheses, when beaten on the anvil of analysis, will break. But this measures the success, not the futility, of the method. Let any who question the efficacy of a method that invents many hypotheses only to destroy some of them, ponder the words penned by Gilbert: "The demolition of hypotheses, instead of testifying to the futility of research, is the method and condition of progress."

We conclude, therefore, that during the fifth stage of the investigation, in the deduction of consequences from the several working hypotheses, and in the confrontation of deduced consequences with observed facts, analysis is both the anvil where falsehood is shattered and the fire where truth is forged.

VI. ANALYSIS IN THE STAGE OF CONFIRMATION AND REVISION

From the fifth stage of the investigation the inquirer should emerge with a much depleted stock of working hypotheses. In some cases only one will remain as apparently competent to explain all the recorded facts. In other cases two or more may survive thus far. If but one, the cautious investigator will seek confirmation of its validity before utilizing it in his ultimate interpretation of the facts. If more than one, he pushes his studies further to discover which one represents the true explanation of the facts, or whether they are jointly responsible for a complex result.

Whether one or several hypotheses survive the tests applied in the preceding stage, the survival is almost always in an "unfinished" state. The tests remain incomplete, and therefore inconclusive, because not all the desired facts are available, and perhaps not all the desirable deductions have been made. The investigator, let us imagine, has deduced five reasonable expectations from a certain hypothesis, of which three only are matched by facts in his possession. On the other hand, he has five categories of facts, of which three only correspond to expectations thus far deduced. He has two deductions having no counterpart in observed facts, and two groups of facts which are disturbing because not related to anything thus far deduced from the hypothesis. Just what these discrepancies mean, he does not know. They may mean that observation was incomplete, and that new facts remain to be discovered. They may mean that deduction was incomplete or imperfect. In either of these two cases the hypothesis may be valid as it stands. But there are other possibilities. The discrepancies may mean that the hypothesis is invalid and must ultimately be rejected; or they may mean that it only requires a certain amount of revision to bring it into harmony with all the facts.

It is the task of the investigator in this sixth stage to find out just what the discrepancies do mean. He searches for the missing facts which, if found, will constitute logical proof of the validity of the generalization involved in the hypothesis. Here, the extraordinary value of the deductive processes of stage five become apparent. The observation of stage one was a wandering, unguided observation. But the renewed observation of stage six is skillfully directed observation. He now knows what to look for, and hence how and where to look. If, like the geologist or geographer, he is dealing with facts to be found in the field, directed observation will materially shorten his field work and greatly increase its fruitfulness. He goes directly to the places where the new facts should be found. If, like the physicist, chemist, or modern student of biology, he must discover new facts through experimentation, skilfully directed experiment will replace the vague gropings of a less advanced stage of research. The deductive process has added new interest to the quest for truth, and has enormously increased the probabilities of finding it through improved induction.

The probative value of facts discovered in this new quest is much higher than that of facts earlier observed. Heretofore facts were observed, and hypotheses adapted to them. Now the investigator has, by a process of deduction, predicted the finding of facts the existence of which is as yet unknown. The prediction is provisional, of course. "If hypothesis A be correct, then I should find, etc." If he finds the facts provisionally predicted for hypothesis A, and does not find those provisionally predicted for hypotheses B, C and D, his confidence in the validity of hypothesis A is doubly strengthened: first, positively by finding the facts appropriate to A; and second, negatively by not finding those which should occur were the other hypotheses valid.

Directed observation has another great value. It is apt to lead to the discovery of facts wholly unanticipated by the deductions which direct the search. As Davis has well said, the investigator's outsight on the facts is sharpened by his insight into the nature of his problem. Things which formerly made no impression on eye or mind now instantly arrest his attention, some because they are expected, but others for the very reason that they are quite unexpected. The discovery of facts that are wholly unanticipated, but which, none the less, find full explanation in a hypothesis under investigation, has exceptionally high value as *Confirmation* of the hypothesis. If many investigators discover many facts, all fully accounted for by the hypothesis, and this is repeated over a long period of time, the confirmation becomes of such high quality that the hypothesis is transformed into a well attested theory.

Where none of the deductions of stage five fully accounts for certain of the facts already in hand; or where newly discovered facts fail to find any counterpart in previous deductions; or, again, where some of the facts which should occur, according to a hypothesis otherwise well substantiated, cannot be found, *Revision* of the hypothesis is strongly indicated. This revision carries the investigator back over all the stages he has previously traversed. Beginning with Observation, he again scrutinizes the facts, especially those which fail to match any deduced consequences. Perhaps they are irrelevant to his problem and should have been excluded; but he dare not exclude them merely because they are inconvenient. They may be critically important for his problem, and show the necessity of revising or rejecting some hypothesis. He must determine whether exclusion of the facts, revision of the hypothesis, or rejection of the hypothesis is called for. Next, the investigator painstakingly reviews his Classification of the facts, to discover if revision is there called for; after which he turns his attention to the

Generalizations based on the classified facts. He then scans his Inventions with care, revising explanations and adding to them where either gives promise of bringing expectations into closer harmony with observed facts. He is particularly careful in his re-examination of the deductions which serve as the basis for Verification and Elimination in stage five, for he realizes the many opportunities for error in deductive reasoning, and hence the great probability that revision may there be required. Lack of completeness in deduction is especially apt to cause difficulty, and proper revision, to include some new consequences not previously foreseen, may bring complete harmony out of partial discord.

The process of securing the requisite confirmation, revision, or rejection of hypotheses makes new and heavy demands upon the critical faculties. Especially is this true of the process of revision and rejection, which carries the investigator back over previous stages of the investigation in a search for some weak link in his chain of reasoning, some factor overlooked, or some item improperly included. If, as we have seen, analysis had a rôle to play in each of these preceding stages, how much more important is it that the analytical faculties should be on the alert in a review of one's reasoning in search for error. If previous work has been carefully performed, the error is not apt to be obvious. More likely it is obscurely involved in some apparently sound procedure, in which case it will continue to escape detection unless each bit of evidence, each argument, and each conclusion are separated into their component parts and subjected to critical scrutiny. It is such analysis as this which must, in the circumstances detailed above, determine whether certain apparently incongruous facts should be rejected as irrelevant to the problem, or included as fully pertinent; and, in the latter case, whether a given hypothesis must be rejected or merely revised, before deduction and observation are brought into harmony.

To illustrate the essential nature of the confirmation and revision envisaged in stage six, let us take one more illustration from the study of shore platforms. Suppose, for example, that after the verification and elimination of stage five, our investigator of shore platforms finds that only one hypothesis offers a reasonable explanation of the facts thus far in his possession. This, let us say, is the hypothesis of storm-wave erosion under present conditions of land and sealevel.

He has previously deduced from this hypothesis certain expectable consequences which are not matched by any facts he has thus far observed. Thus, he has reasoned that if the platforms be the product of storm waves they should be found at higher elevations on exposed headlands where waves are large, and at lower elevations in bays or behind barriers where waves are small. Furthermore, the platforms should be broadest where the most violent storm waves, breaking on a coast, wear it back most rapidly. Here are two perfectly concrete deductions from the hypothesis; but he has never observed any such relation between intensity of wave attack and height and breadth of platform as the deductions call for. Perhaps this is partly because he never gave conscious attention to this relation, and partly because the shores he has studied did not present sufficiently strong contrasts in exposure to produce differences in height and breadth of platforms so great as to force themselves upon his attention.

In any case, the investigator now plans field work on more remote coasts, carefully analyzing the requirements of his problem, and seeking shores where conditions are especially favorable for testing the validity of these two deductions. His powers of observation are now greatly enhanced. He sees many things about shore benches which quite escaped attention before. Among others, he notes the remarkable freshness of the cliff face back of the platforms, the frequent occurrence of a groove or notch eroded in the base of the cliffs, occasional fresh landslides on the platforms, overhanging sod at the cliff top, and marine organisms living in pools of salt water on the platform surface. These are all details he had failed to predict, yet these unanticipated observations find such reasonable explanation in the hypothesis of continued active development of the platforms by present storm waves that he accepts them as particularly valuable confirmation of that hypothesis. When he finds, in addition, that the benches do range higher on the headlands and lower in the bays, he regards the confirmation as doubly strengthened.

There remains the prediction that platforms of maximum breadth should, other things being equal, be found on those coasts subjected to maximum violence of wave attack. Once more our investigator analyzes the variable factors presented by a number of coasts, in the effort to select wisely several localities which may offer him favorable conditions for testing the validity of this last deduction. He chooses his hunting-grounds, visits them in turn, and in each case finds no shore platforms at all. Instead, angry seas beat themselves to foam on a chaos of tumbled rocks.

Much disconcerted by this glaring discordance between deduced expectations and observed facts, our investigator turns to analysis for an answer to the enigma. Again he dissects and scrutinizes in detail the several stages of his work: the reasoning which led to invention of the parent hypothesis, the deduction which caused him to expect the broadest benches where wave attack is most violent, and the facts which fail so conspicuously to harmonize with expectations. Each in turn is probed and tested with greatest care and the most complete detachment and impartiality.

Perhaps the outcome of this analysis is discovery that the hypothesis is valid in its essentials, but requires revision of details. He may conclude that where storm waves attack a coast most violently and undercut it most rapidly, falling debris breaks the waves into confused and surging waters which cannot etch a neatly carved platform at a specific level. If such be the result of his analysis, he will recast the hypothesis in terms of the erosive behavior of storm waves of moderate intensity, realizing that this may well differ widely from the behavior of maximum storm waves. If some other explanation better resolves the difficulty, he will recast the hypothesis accordingly. He will then proceed to test the hypothesis in its new form by inspection of a wide variety of shores where waves of every type and degree of intensity are engaged in consuming the land.

VII. ANALYSIS IN THE STAGE OF INTERPRETATION

Let us suppose that the hypothesis of wave erosion under present conditions of land and sealevel survives, in its modified form, every test applied to it. The investigator has reached the seventh, and final, stage of his labors, in which he proceeds to formulate for publication his *Interpretation* of the origin of shore platforms. Can he at last dispense with analysis, the instrument which has served him so well at every previous step in his labors? Not wholly, I think. In any event, there must remain with him a caution, inspired by his full understanding of the nature of the proofs with which he is dealing, an understanding which comes only from critical analysis of both evidence and arguments. Rarely can geologist or geographer offer mathematical demonstration of the truth of his conclusions. He can, as a rule, go no further than to show that a given proposition has a high degree of probability. Certain facts, carefully observed and properly classified, prompt the invention of various hypotheses. The deduced consequences of one of these hypotheses alone is matched by all the facts, including some which are newly discovered and highly peculiar. The investigator, therefore, has much confidence that an interpretation based on this hypothesis will be valid, but he realizes there is still opportunity for error. He may not have thought of all possible hypotheses. He may not have deduced all the reasonable consequences of those hypotheses considered. He may not have discovered all pertinent facts. Some future contribution to knowledge may throw unexpected light on his problem, and radically affect his conclusions. He therefore wisely regards his present interpretation as highly probably theory, rather than as demonstrated fact.

In previous stages an error led only one investigator astray, and that one could retrace his steps and repair damage by embarking on some new line of reasoning. But in this final stage an error committed to the printed page may lead many astray, and may show wholly unexpected vitality and power for evil. The printed statements must go far enough; but they must not go farther than the evidence and fully tested reasoning warrant. A given interpretation must not be applied to all shore platforms before it has completely been demonstrated that there are no such platforms for which it is not the most satisfactory explanation. It must not even be asserted as the sole cause of any platform, unless and until it has clearly been proven that no contributing cause has played a part in their formation.

In these and other ways the investigator must be on his guard against overstatement, understatement, erroneous statement of his results. Analysis is the weapon with which he defends himself against too broad generalization and other errors which are wont to intrude themselves into the final stage of an investigation. Each sentence he scrutinizes; each conclusion he dissects; checking here and verifying there, to the end that the formulation of his interpretation shall contain no less, and particularly no more, than critically observed facts and carefully tested reasoning may justify. An interpretation, thus cautiously reached and conservatively formulated, will surely command the serious consideration of scientific men.

Conclusion

Have we, in the method of multiple working hypotheses, applied with the aid of rigorous analysis, something which will guide us unfailingly to the discovery of truth? We are compelled to answer this question in the negative. No device, however perfect, can wholly deprive the human intellect of its capacity for making mistakes. De Leon searched in vain for the fountain of youth. Can we hope for a magical fountain of truth?

The most for which we may reasonably hope is by correct methods of research to reduce the chances of error to a minimum, and to raise to its maximum the probability of discovering the real causes and relations of things. This we have done, so far as lies within our power, when we are accurate in observing facts, careful in classifying them, cautious in generalizing from them, fertile in inventing hypotheses, ingenious and impartial in testing their validity, skillful in securing their confirmation or revision, and judicial in formulating ultimate interpretations.

Multiple working hypotheses as a method, employed in connection with critical analysis as an instrument of precision, offer us, in my opinion, the best guarantee of success in scientific research.

10

Copyright © 1963 by the Geological Society of America

Reprinted from *The Fabric of Geology*, C. C. Albritton, ed., Freeman, Cooper. & Company, San Francisco, 1963, pp. 135–163

J. HOOVER MACKIN

University of Texas

Rational and Empirical Methods of Investigation in Geology[1]

Most of us are concerned, and some of us have strong feelings, pro or con, about what has been happening to geology in the past 25 years: greatly increased use of nongeologic techniques in the solution of geologic problems, such as dating by radioisotope methods; the tendency for what were special fields of interest to become nearly or wholly independent disciplines, with separate journals and jargon; and most of all, because it penetrates every field, what may be called the swing to the quantitative.

At meetings of our societies, when the elder brethren gather together in hotel rooms after the technical sessions, the discussion usually comes around to these changes. There are apt to be sad postmortems for certain departments, once powerful, which are now, owing to the retirement or flight of their older stalwarts, largely staffed by dial twisters and number jugglers. It is stated, as a scandalous sign of the times, that in certain departments geologic mapping is considered to be, not research, but a routine operation—something like surveying from the point of view of an engineer—and therefore not suitable as a basis

---

[1] A preliminary draft of this paper was given as an address at the banquet of the Branner Club during the meeting of the Cordilleran Section of the Geological Society in Los Angeles, April 17, 1962. The text has benefitted in substance and form from criticisms by the other authors of papers in this volume. I would like also to express my gratitude to the following, who have read parts or all of the manuscript: Charles Bell, Richard Blank, Howard Coombs, Ronald DeFord, Ken Fahnestock, Peter Flawn, John Hack, Satish Kapoor, William Krumbein, Luna Leopold, Mark Meier, H. W. Naismith, and Dwight Schmidt. Special thanks are due Frank Calkins, who did his best to make the paper readable.

135

for the doctoral thesis.   There is almost always at least one sarcastic remark per evening along the line of what our equation-minded youngsters think is the function of the mirror on a Brunton compass; a comment or two on their ignorance or disregard of the older literature; some skepticism as to whether the author of a new monograph on the mechanism of mountain building had ever been *on* a mountain, *off* a highway; and so on.   This is partly banter, because we are aware that these are merely the usual misgivings of every older generation about the goings-on of every younger generation.   But sometimes there is evidence of real ill-feeling, which in part at least reflects a defensive attitude; and there may be a few who seem to think that the clock ought to be stopped—that nothing new is good.

Though I am one of the elders, I often cross the hall to a concurrent session of another group, our avant-garde, where there is an almost evangelical zeal to quantify, and if this means abandoning the classical geologic methods of inquiry, so much the better; where there are some who think of W. M. Davis as an old duffer with a butterfly-catcher's sort of interest in scenery; where there is likely to be, once in a while, an expression of anger for the oldsters who, through their control of jobs, research funds, honors, and access to the journals, seem to be bent on sabotaging all efforts to raise geology to the stature of a science; where, in the urgency for change, it seems that nothing old is good.

This picture is not overdrawn, but it applies only to a small number: the blacks and the whites, both sure of their ground.   Most geologists are somewhere in the gray between, and are beset with doubts.   As for myself, I have sometimes thought that the swing to the quantitative is too fast and too far, and that, because a rather high percentage of the conclusions arrived at by certain methods of manipulating numerical data are superficial, or wrong, or even ludicrous, these methods must be somehow at fault, and that we do well to stay with the classical geologic methods.   But at other times I have been troubled by questions: why the swing has been so long delayed in geology as compared with physics and chemistry; and whether, with its relative dearth of quantitative laws, geology is in fact a sort of subscience, as implied by Lord Kelvin's pronouncement that what cannot be stated in numbers is not science. (For original wording, and a thoughtful discussion, see Holton, 1952, p. 234.)   Even more disturbing is the view, among some of my friends in physics, that a concern with cause-and-effect relations merely confuses the real issues in science; I will return to this matter later.   If only because of the accomplishments of the scientists who hold these views, we must wonder whether our accustomed ways of thinking are outmoded, and whether we should not drastically change our habits of thought, or else turn in our compasses and hammers and fade away quietly to some haven reserved for elderly naturalists.

Preparation for a talk on quantitative methods in geomorphology, as a visiting lecturer at the University of Texas last year, forced me to examine these conflicting appraisals of where we stand.[2] I suggest that two changes, quite different but closely interlocking, are occurring at the same time and have become confused in our thinking.

One of these changes includes an increase in the rate of infusion of new ideas and techniques from the other sciences and from engineering, an increase in precision and completeness of quantitative description of geologic features and processes of all kinds, and an increased use of statistics and mechanical methods of analyzing data. This change fits readily within the framework of the classical geologic method of investigation, the most characteristic feature of which is dependence on reasoning at every step; "Quantitative Zoology," by Simpson, Roe, and Lewontin (1960) shows the way. In so far as it merely involves doing more completely, or with more refinement, what we have always been doing, it is evolutionary; and it is axiomatic that it is good. Some of us may find it hard to keep abreast of new developments, but few oppose them even privately, and even the most reactionary cannot drag his feet in public without discredit to himself.

The other change is the introduction, or greatly increased use, of an altogether different method of problem-solving that is essentially empirical. In its purest form this method depends very little on reasoning; its most characteristic feature, when it functions as an independent method, is that it replaces the reasoning process by operations that are largely mechanical. Because in this respect and others it is foreign to our accustomed habits of thought, we are inclined to distrust it. By "we" I mean, of course, the conservatives of my generation.

At least a part of the confusion in our thinking comes from a failure to distinguish between the evolutionary quantification, which is good, and the mechanical kind of quantification, which I think is bad when it takes the place of reasoning. It is not easy to draw a line between them because the empirical procedures may stand alone, or they may function effectively and usefully as parts of the classical geologic method; that is, they may replace, or be combined in all proportions with, the reasoning processes that are the earmarks of that method. When this distinction is recognized it becomes evident that the real issue is not qualitative versus quantitative. It is, rather, rationality versus blind empiricism.

---

[2] I was only dimly aware, until some library browsing in connection with methodology in the other sciences, of the extent of the scholarly literature dealing with the history and philosophy of science. And I was surprised, as was Claude Albritton (1961), to find that with a few noteworthy exceptions (for example, Conant, 1951, p. 269–295) geology is scarcely mentioned in that literature. I should like to make it plain at the outset that I am not a scholar—I have only sampled a few anthologies of the history of science. I should emphasize also that I do not presume to speak for geology; what I say expresses the viewpoint of a single field geologist.

Although the timing has been influenced by such leaders as Chayes, Hubbert, Leopold, Krumbein, and Strahler, we are now in the swing to the quantitative because of the explosive increase in the availability of numerical data in the last few decades (Krumbein, 1960, p. 341), and because basic descriptive spadework has now advanced far enough in many fields of geology to permit at least preliminary formulation of significant quantitative generalizations. The quantification of geology will proceed at a rapidly accelerating rate no matter what we do as individuals, but I think the rate might be quickened a little, and to good purpose, if the differences between the two groups on opposite sides of the hall, at least those differences that arise from misunderstanding, could be reduced. An analysis of certain quantitative methods of investigation that are largely empirical will, I hope, serve to bring out both their merits and limitations, and may convince some of our oldsters that although disregard of the limitations may produce questionable results, it does not follow that there is anything wrong with quantification, as such, nor with blind empiricism, as such. But this is not very important—time will take care of the oldsters, soon enough. This essay is for the youngsters—the graduate students—and its purpose is to show that as they quantify, which they are bound to do, it is neither necessary nor wise to cut loose from the classical geologic method. Its message is the not very novel proposition that there is much good both in the old and the new approaches to problem-solving. A brief statement of what I am calling the rational method will point up the contrast between it and the empirical method, with which we are principally concerned.

The Rational Method

I'm sure that most American geologists are acquainted with our three outstanding papers on method: G. K. Gilbert's "Inculcation of the Scientific Method by Example," published in 1886; T. C. Chamberlin's "Method of Multiple Working Hypotheses," published in 1897; and Douglas Johnson's "Role of Analysis in Scientific Investigation," published in 1933. I do not need to describe the so-called scientific method here; for present purposes I need only remind you that it involves an interplay of observation and reasoning, in which the first observations suggest one or more explanations, the working hypotheses, analysis of which leads to further observation or experimentation. This in turn permits a discarding of some of the early hypotheses and a refinement of others, analysis of which permits a discarding of data now seen to be irrelevant to the issue, and a narrowing and sharpening of the focus in the search for additional data that are hidden or otherwise hard to obtain but which are of special diagnostic value; and so on and on. These steps are spelled out in formal terms in the papers just mentioned, and it was useful to do that, but those who use the method all the time never follow the steps in the order stated; the method has become a habit of thought that checks reasoning against other lines of reasoning, evidence against other kinds of evidence, reasoning against evidence, and evidence against reasoning, thus testing both the evidence and the reasoning for relevancy and accuracy at every stage of the inquiry.

It now seems to be the vogue to pooh-pooh this method, as differing in no essential way from the method of problem-solving used by the man in the street. I've been interested in watching the way in which men in the street, including some medical doctors—practitioners, not investigators—arrive at conclusions, and I can only suggest that the scientists who insist that all persons arrive at conclusions in the same way should reexamine their conviction. There are, of course, rare intellects that need no disciplining, but for most of us with ordinary minds, facility in the operations that I have just outlined must be acquired by precept, example, and practice.

The objective of the scientific method is to understand the system investigated—to understand it as completely as possible. To most geologists this means understanding of cause and effect relations within the system (Garrels, 1951, p. 32). Depending on the nature of the problem and its complexity, quantitative data and mathematical manipulations may enter the investigation early or late. In general, the larger the problem, the more many-sided it is, the more complicated by secondary and tertiary feedback couples, and the more difficult it is to obtain the evidence, the more essential it is to the efficient prosecution of the study that the system first be understood in *qualitative* terms; only this can make it possible to design the most significant experiments, or otherwise to direct the search for the critical data, on which to base an eventual understanding in quantitative terms.

A problem—any problem—when first recognized, is likely to be poorly defined. Because it is impossible to seek intelligently for explanations until we know what needs explaining, the first step in the operation of the scientific method is to bring the problem into focus. This is usually accomplished by reasoning, i.e., by thinking it through, although we will see shortly that there is another way. Then, if it is evident that the problem is many-sided, the investigator does not blast away at all sides at once with a shotgun; he shoots at one side at a time with a rifle—with *the* rifle, and *the* bullet, that he considers best suited to that side.

This means that the investigator admits to his graphs, so to speak, only items of evidence that are relevant to the particular matter under investigation, and that are as accurate as practicable, with the probable limits of sampling and experimental error expressed graphically. In reading answers from the graph, he does no averaging beyond that required to take those limits into account. And once an item of information has been admitted to the graph, it cannot be disregarded; as a rule, the items that lie outside the clusters of points are at least as significant, and usually much more interesting, than those that lie within the clusters. It is from inquiry as to why these strays are where they are that most new ideas—most breakthroughs in science—develop.

The scientific method tries to visualize whole answers—complete theoretical structures—at the very outset; these are the working hypotheses that give direction to the seeking-out and testing of evidence. But one never rushes ahead of the data-testing process to a generalization that is regarded as a conclusion. This is not because there is anything ethically wrong with quick generalizing. It is only that, over a period of 500 years, investigators have found that theoretical structures made in part of untested and ill-matched building blocks are apt to topple sooner or later, and that piling them up and building on them is therefore not an efficient way to make progress. The need to test the soundness of each building block *before* it gets into the structure—to determine the quality and the relevance of each item of evidence *before* it gets onto the graph—is emphasized by Douglas Johnson (1933). His approach was the antithesis of that to which we may now turn.

The Empirical Method

What I have long thought of as the engineering method or the technologic method (we shall soon see that it needs another name) deals almost exclusively with quantitative data from the outset, and proceeds directly to a quantitative answer, which terminates the investigation. This method reduces to a minimum, or eliminates altogether, the byplay of inductive and deductive reasoning by which data and ideas are processed in the scientific method; this means that it cannot be critical of the data as they are gathered. The data are analyzed primarily by mathematical methods, which make no distinction between cause and effect; understanding of cause and effect relations may be interesting, but it is not essential, and if explanations are considered at all, there is usually only one, and it is likely to be superficial. All of the reasoning operations that characterize the so-called scientific method depend on a fund of knowledge, and on judgment based on experience; other things being equal, the old hand is far better at these operations than the novice. But the operations of the "engineering method" are much less dependent on judgment; in applying this method the sharp youngster may be quicker and better than the experienced oldster. For this reason and because of its quick, positive, quantitative answers, it makes a strong appeal to the younger generation. I would like now to explain the logic of this method, as it operates in engineering.

Many engineers feel that unless a relation can be stated in numbers, it is not worth thinking about at all. The good and sufficient reason for this attitude is that the engineer is primarily a doer—he designs structures of various types, and supervises their building. In the contract drawings for a bridge he must specify the dimensions and strength of each structural member. Nonengineers may be able to think of a drawing that indicates the need for a rather strong beam at a given place in the bridge. But a young man who has spent five years in an engineering school is incapable of thinking seriously of a "rather strong" beam; all of the beams of his mind's eye have numerical properties. If the strength of a beam cannot be put in numerical terms, thinking about it is mere daydreaming.

The matter of stresses in a steel structure is fairly cut and dried. But the engineer is confronted with many problems for which there are no ready answers; he must deal with them—he must complete his working drawings— against a deadline. If he is charged with the task of designing a canal to carry a certain flow of irrigation water without either silting or erosion of the bed, or with the immensely more complex task of developing and maintaining a 10-foot navigable channel in a large river, he cannot wait until he or others have developed a complete theory of silting and scouring in canals and rivers. It may be 50 or a 100 years before anything approaching a complete theory, in quantitative terms, can be formulated; and his drawings, which must be entirely quantitative, have to be ready within a few days or weeks for the contractors who will bid the job. So he has to make certain simplifying assumptions, even though he realizes that they may be wide of the mark, and he has to make-do with data that are readily available, even though they are not entirely satisfactory, or with data that can be obtained quickly from experiments or models, even though the conditions are significantly different from those existing in his particular canal or river.

He is accustomed to these expedient operations, and he is not much concerned if, in plotting the data, he mixed a few oranges with the apples. In fact, he wouldn't worry much if a few apple *crates* and a few orange *trees* got onto his graph. He cannot scrutinize each item of evidence as to quality and relevancy; if he did, none but the simplest of structures would ever get built. He feels that if there are enough points on a scatter diagram, the bad ones will average out, and that the equation for the curve drawn through the clustered points will be good enough for use in design, always with a goodly factor of safety as a cushion. And it almost always is. This method is *quantitative*, *empirical*, and *expedient*. As used by the engineer, it is logical and successful.

It is of course used by investigators in many fields other than engineering. Friends in physics and chemistry tell me that it accounts for a large percentage of the current research in those sciences. A recent paper by Paul Weiss (1962) with the subtitle "Does Blind Probing Threaten to Displace Experience in Biological Experimentation?" calls attention to its increasing use in biology. The approach and examples are different, but the basic views of Dr. Weiss correspond so closely with those expressed in this essay that I am inclined to quote, not a passage or two, but the whole paper. Because this is impracticable,

I can only urge that geologists interested in this phase of the general problem —whither are we drifting, methodologically?—read it in the original.

In view of its widespread use in science, what I have been calling the engineering or technologic method certainly should not be identified, by name, with engineering or technology as such. And on the other side of the coin, the so-called scientific method is used more consistently and effectively by many engineers and technologists than by most scientists. Besides being inappropriate on this score, both terms have derogatory or laudatory connotations which beg some questions. So, with serious misgivings that will be left unsaid, I will from here on use the term "rational method" for what we are accustomed to think of as the scientific method, and what I have been calling the engineering method will be referred to as the empirical method.[3]

Actually, the method that I am trying to describe is *an* empirical method; it is shotgun or scatter-diagram empiricism, very different from the one-at-a-time, cut-and-try empiricism of Ehrlich who, without any reasoned plan, tried in turn 606 chemical substances as specifics for syphilis. The 606th worked. Both the scatter-diagram and the one-at-a-time types can be, at one extreme, purely empirical, or, if you prefer, low-level empirical. As Conant (1952, pp. 26–30) points out, the level is raised—the empirical approaches the rational —as the gathering and processing of the data are more and more controlled by reasoning.

Use of Examples

The expositions of the rational method by Gilbert, Chamberlin, and Johnson all depend on the use of examples, and having tried several other ways, I am sure that this is the only way to make clear the workings of the empirical method. I have chosen to use actual examples, because these are far more effective than anything I could invent. They could have been selected from any field in geology. My examples are from recent publications dealing with the geologic work of rivers; I know of no other field in which the two approaches to problem-solving stand in such sharp contrast. "Horrible examples" are available, analysis of which would have a certain entertainment value; I shall draw my examples from publications that rank as important contributions. The principal example is from a paper that is unquestionably the outstanding report in this field, "The hydraulic geometry of stream chan-

---

[3] So many friends have objected to these terms that I should say that I am fully aware that they are unsatisfactory, chiefly because they have different connotations in different fields of study. I use them in their plain English meaning. They seem to me to be less objectionable than any other terms, but I will not take issue with those who think otherwise.

nels," by Luna Leopold and Thomas Maddock (1953). I have discussed the methodology of geologic investigation with Leopold on numerous occasions, and we have, in effect, agreed to disagree on some points.[4]

Examples are essential in a discussion of methods, but it is difficult to work with them. The problems of fluvial hydraulics are so complex that if the examples are to be comprehensible they must be simplified, and we must treat them out of context. This may irritate the few who are familiar with these matters at the technical level; I can only ask their indulgence on the ground that I am steering a difficult course between nonessential complexity and over-simplification. I should acknowledge, moreover, that I am an interested party; about 15 years ago I published an article in this field (Mackin, 1948). Finally, and most important, I will be deliberately looking at the way data are handled from the point of view of the conservative geologist, unaccustomed to this manner of handling data and highly critical of it. But I will come around full circle in the end, to indicate that the operations I have been criticizing are those of a valid method of investigation which is here to stay.

Downstream Change in Velocity in Rivers

All of us have seen the white water of a rushing mountain stream and the smooth-surfaced flow of the streams of the plains, and we are prepared by the contrast to suppose that the velocity of the flow decreases downstream. We are aware, moreover, that slope commonly decreases downstream and that velocity tends to vary directly with slope. Finally, we know from observation that the grain size of the load carried by rivers tends to decrease downstream, and that the grain size of the material carried by a river varies directly with some aspect of the velocity. For these reasons, we have always taken it for granted that velocity decreases downstream.

So in 1953, when Leopold and Maddock stated that velocity in rivers *increases* downstream, the statement came as a first-rate shock to most geologists. Three graphs (Fig. 1) from that article are good examples of the sort of evidence, and the manner of handling evidence, on which this generalization is based. They are log-log plots of several parameters; at the top, width of channel against discharge in cubic feet per second; in the middle, depth against discharge; and at the bottom, velocity against discharge. Each point represents data obtained from a U. S. Geological Survey gaging station in the Yellowstone-Big Horn drainage system. The points at the far left, such as 13 and 16, are on small headwater tributaries, and those at the far right, such as 19, are on the main stem of the Yellowstone. The upper and middle graphs show that, as should be expected, both width and depth increase with increase in dis-

---

[4] Leopold states his position elsewhere in this volume.

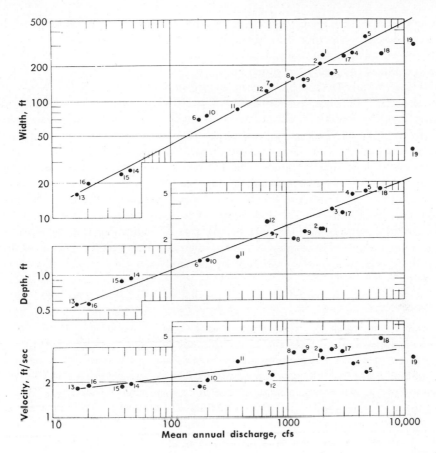

Fig. 1. Width, depth, and velocity in relation to discharge, Bighorn and Yellowstone Rivers, Wyoming and Montana (Leopold and Maddock, 1953, Fig. 6).

charge; the line in the lower graph also slopes up to the right; that is, velocity increases with increase in discharge.

Some may wonder why we have moved over from increase in velocity downstream, which is the exciting issue, to increase in velocity with increase in discharge. While it is true that discharge increases downstream in most rivers, it is at best only an approximate measure of distance downstream—the distance that would be traveled, for example, by the grains composing the load. The answer given in the Leopold-Maddock paper is that there were not enough gaging stations along the rivers to provide a sufficient number of points. Use of discharge, rather than distance, makes it possible to bring onto one graph the main stream and its tributaries of all sizes; or, for that matter, since "main

195

stream" is a relative term, all the neighboring streams in an area large enough to provide enough points to bring out the significant relationships.

This explanation does not quite answer the question, unless expediency is an answer, but it raises another question.

Velocity at any given place—at any gaging station, for example—varies with variations in discharge from time to time during the year; as discharge and depth increase, usually in the spring, velocity at a given place increases very markedly. We may ask, then, *what* discharge is represented by the points on the lower graph? The question is pertinent, because we know that in most rivers much of the year's transportation of bed load—the sand and gravel that move along the bed—is accomplished during a relatively brief period of maximum discharge. But these graphs show mean annual discharges, and the velocities developed at those discharges. The reason for using mean annual discharge is said to be that this parameter is readily available at a large number of gaging stations. This explanation does not answer the question: what is the relevance of mean annual discharge in an analysis of the geologic work of rivers?

This general question, which applies to each of the stations considered individually, takes on another meaning when the relations between mean annual discharge and maximum discharge on streams are considered. Reference to Water Supply Paper 1559 (1960, p. 169) indicates that at point 13 (Fig. 1), which represents a gaging station on the North Fork of Owl Creek, the average annual discharge for the 14-year period of record was 15 cfs (cubic feet per second), whereas the maximum discharge during the same period was 3200 cfs; that is, the maximum was about 213 times the average. The same paper (p. 234) indicates that at point 19, which represents the Yellowstone River at Sidney, Montana, the average annual discharge over a 46-year period was 13,040 cfs, whereas the maximum during the same period was 159,000 cfs; here the maximum was about 12 times the average. The noteworthy thing about this graph—the thing that makes it so exciting—is that it shows that velocity *increases* downstream although we know from observation that grain size *decreases* downstream. The significance of the graph is more readily understood when we remember: (1) that the larger grains move only at times of maximum discharge; (2) that this graph shows mean annual discharge; and (3) that in the small rivers on the left side, the maximum discharge may be more than 200 times as great as the discharge shown on the graph, while in the big rivers on the right, it is less than 20, and usually less than 10 times the discharge shown, that is, that the critical ratios on the two sides are of a different order of magnitude. The slope of the line is an important statistical fact, but it does not bear directly on transportation of bed load by rivers.

One more thought in this connection. The depth at average annual discharge at point 13, on the North Fork of Owl Creek, is shown in the middle graph as being something less than 0.6 foot. I know the general area, and, although I have no measurements at this gaging station, it is my recollection that the larger boulders on the bed of the North Fork are more than 0.6 foot in diameter—the boulders on the bed have diameters that are of the same order of magnitude as the depth at which the very low velocity shown for this point was calculated. Similar relationships obtain for other small headwater streams, the points for which anchor down, so to speak, the left end of the line.

Let us look briefly at one more aspect of the case. The velocity is lower near the bed of a river than near the surface. Rubey (1938) and others have shown that the movement of bed load is determined, not by the average velocity, but by the velocity near the bed. And it has also been established that the relation between average velocity and what Rubey calls "bed velocity" varies markedly with depth of water, roughness of channel, and other factors. We may reasonably ask, then, *what* velocity is represented by the points on the graph? The answer is spelled out clearly by Leopold and Maddock (1953, p. 5).

> Velocity discussed in this report is the quotient of discharge divided by the area of the cross section, and is the mean velocity of the cross section as used in hydraulic practice . . . This mean velocity is not the most meaningful velocity parameter for discussing sediment transport, but it is the only measure of velocity for which a large volume of data is available. Although the writers recognize its limitations, the mean velocity is used here in lieu of adequate data on a more meaningful parameter.

There are various other similar questions about this graph, some of which are discussed by the authors in the clear and candid style of the last quotation. I will not develop these questions, or the secondary and tertiary questions that spring from the answers. Some of you may be thinking: never mind the individual points; what about the trends? It could be argued that if the conclusions are internally consistent; if they match those for other river systems; if, in short, these procedures get results, this alone justifies them.

Let's look at the results. Figure 2 is the velocity-discharge graph of Fig. 1, modified by use of symbols to identify related points and with dashed lines for individual rivers.

Points 1, 2, 3, 4, and 5, are on the main stem of the Big Horn River. Points 1, 2, and 3 are in the Big Horn Basin; point 4 is about 50 miles downstream from 3, and 5 is about 20 miles downstream from 4. The dashed line, which .fits these points quite well, slopes down to the right; it means that on the main stem of the Big Horn, velocity *decreases* downstream.

Points 6, 7, 8, and 9 are on the Wind River, which is actually the upper part of the Big Horn River. I do not know whether points 8 and 9 represent the same types of channel conditions as 1, 2, 3, 4, and 5, as suggested by their positions, or whether they should be grouped with 6 and 7, as called for by the

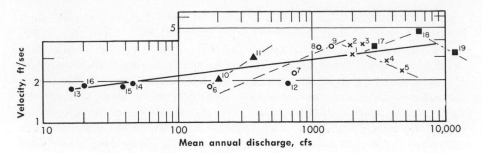

Fig. 2. Same as velocity-discharge graph in Fig. 1, with dashed lines for certain rivers; triangles, Greybull River; open circles, Wind River; X's, Bighorn River; solid squares, Yellowstone River.

geographic usage of the names. Let us say, then, that in what the geographers call the Wind River, velocity at average annual discharge increases downstream.

Points 10 and 11, both on the Greybull River, also suggest by their relative position that velocity increases downstream; the line slopes up to the right. But point 10 is near the mouth of the Greybull, about 40 airline miles downstream from 9; the average annual discharge decreases downstream (Leopold, 1953, p. 612) partly because of withdrawal of water for irrigation. Velocity actually decreases very markedly downstream on the Greybull.

Points 17, 18, and 19 are on the main stem of the Yellowstone. It appears from this graph that velocity increases downstream between 17 and 18, and decreases downstream between 18 and 19.

The generalization that velocity increases downstream, at a rate expressed by the slope of the solid line on this graph, is a particular type of empirical answer. It is what the nonstatistician is likely to think of as an "insurance company" type of statistic—a generalization applying to this group of rivers collectively, but not necessarily to any member of the group. Of the river segments represented on the graph, about half increase in velocity downstream, and about half decrease in velocity downstream. As shown by the different slopes of the dashed lines, in no two of them is the rate of change in velocity the same.

This is really not very surprising. The solid line averages velocity-discharge relations in river segments that are, as we have seen, basically unlike in this respect. Moreover, slope, which certainly enters into velocity, is not on the graph at all. For these reasons the equation of the solid line is not a definitive answer to any geologic question.

But—and here I change my tune—this graph was not intended to provide a firm answer to any question. It is only one step—a preliminary descriptive step—in an inquiry into velocity changes in rivers from head to mouth. This is accomplished by plotting certain conveniently available data on a scatter diagram.

I have indicated earlier how this procedure, which is empirical, expedient, and quantitative, serves the practicing engineer very well in getting answers that are of the right order of magnitude for use in design in deadline situations. Here we see the same procedure operating as a step in a scientific investigation. It is used in this graph to learn something about velocity relations in rivers from a mass of data that were obtained for a different purpose; the purpose of U. S. Geological Survey gaging stations is to measure discharge, not velocity. This gleaning of one kind of information from measurements—particularly long-term records of measurements—that are more or less inadequate because they were not planned to provide that kind of information, is a very common operation in many scientific investigations, and is altogether admirable.

There is another point to be made about this graph. Before the work represented by it was done, there had been no comprehensive investigations of velocity in rivers from head to mouth; this study was on the frontier. In these circumstances, some shots in the dark—some *shotgun* shots in the dark—were quite in order. The brevity with which this point can be stated is not a measure of its importance.

Finally, I wish to emphasize that Leopold and Maddock did not regard the solid line as an answer—its equation was *not* the goal of their investigation. They went on in this same paper, and in others that have followed it, to deal with velocities developed at peak discharges and with many other aspects of the hydraulic geometry of river channels. It is for this reason, and the other two reasons just stated, that I can use the graph as I have without harm to its authors.

But our literature is now being flooded by data and graphs such as these, without any of the justifications, engineering or scientific, that I have outlined. In many instances the graph is simply a painless way of getting a quantitative answer from a hodge-podge of data, obtained in the course of the investigation, perhaps at great expense, but a hodge-podge nevertheless because of the failure of the investigator to think the problem through prior to and throughout the period of data gathering. The equations read from the graphs or arrived at by other mechanical manipulations of the data are presented as terminal scientific conclusions. I suggest that the equations may be terminal engineering conclusions, but, from the point of view of science, they are statements of problems, not conclusions. A statement of a problem may be very valuable, but if it is mistaken for a conclusion, it is worse than useless because it implies that the study is finished when in fact it is only begun.

If this empirical approach—this blind probing—were the only way of quantifying geology, we would have to be content with it. But it is not; the quantitative approach is associated with the empirical approach, but it is not wedded to it. If you will list mentally the best papers in your own field, you will discover that most of them are quantitative and rational. In the study of rivers

I think of Gilbert's field and laboratory studies of Sierra Nevada mining débris (1914, 1917), and Rubey's analysis of the force required to move particles on a stream bed (1938). These geologists, and many others that come to mind, have (or had) the happy faculty of dealing with numbers without being carried away by them—of quantifying without, in the same measure, taking leave of their senses. I am not at all sure that the percentage of geologists capable of doing this has increased very much since Gilbert's day. I suggest that an increase in this percentage, or an increase in the rate of increase, is in the direction of true progress.

We shall be seeing more and more of shotgun empiricism in geologic writings, and perhaps we shall be using it in our own investigations and reports. We must learn to recognize it when we see it, and to be aware of both its usefulness and its limitations. Certainly there is nothing wrong with it as a tool, but, like most tools, how well it works depends on how intelligently it is used.

Causes of Slope of the Longitudinal Profile

We can now turn to a matter which seems to me the crux of the difference between the empirical and the rational methods of investigation, namely, cause-and-effect relations.[5] I would like to bring out, first, an important difference between immediate and superficial causes as opposed to long-term, geologic causes; and second, the usefulness, almost the necessity, of thinking a process through, back to the long-term causes, as a check on quantitative observations and conclusions.

Most engineers would regard an equation stating that the size of the pebbles that can be carried by a river is a certain power of its bed velocity as a complete statement of the relationship. The equation says nothing about cause

---

[5] I am aware that my tendency to think in terms of cause and effect would be regarded as a mark of scientific naiveté by some scientists and most philosophers. My persistence in this habit of thought after having been warned against it does not mean that I challenge their wisdom. Perhaps part of the difficulty lies in a difference between what I call long-term geologic causes and what are sometimes called ultimate causes. For example, a philosopher might say, "Yes, it is clear that such things as discharge and size of pebbles may control or cause the slope of an adjusted river, but what, then, is the cause of the discharge and the pebble size? And if these are effects of the height of the mountains at the headwaters, what, then, is the cause of the height of the mountains"? Every cause is an effect, and every effect is a cause. Where do we stop? I can only answer that I am at the moment concerned with the geologic work of rivers, not with the cause of upheaval of mountains. The question, where do we stop?, is for the philosopher, who deals with all knowledge; the quest for ultimate causes, or the futility of that quest, is in his province. The investigator in science commonly stays within his own rather narrow field of competence and, especially if time is an important element of his systems, he commonly finds it useful to think in terms of cause and effect in that field. The investigator is never concerned with ultimate causes.

and effect, and the engineer might be surprised if asked which of the two, velocity or grain size, is the cause and which is the effect. He would almost certainly reply that velocity controls or determines the size of the grains that can be moved, and that therefore velocity is the cause. To clinch this argument, he might point out that if, by the turn of a valve, the velocity of a laboratory river were sufficiently increased, grains that previously had been at rest on the bed would begin to move; that is, on the basis of direct observation, and by the commonsense test of relative timing, the increase in velocity *is* the cause of the movement of the larger grains. This is as far as the engineer needs to go in most of his operations on rivers.

He might be quite willing to take the next step and agree that the velocity is, in turn, partly determined by the slope. In fact, getting into the swing of the cause-and-effect game, he might even volunteer this idea, which is in territory familiar to him. But the next question—what then, is the cause of the slope?—leads into unfamiliar territory; many engineers, and some geologists, simply take slope for granted.

Our engineer would probably be at first inclined to question the sanity of anyone suggesting that the size of the grains carried by a river determines the velocity of the river. But in any long-term view, the sizes of the grains that are supplied to a river are determined, not by the river, but by the characteristics of the rocks, relief, vegetative cover, and other physical properties of its drainage basin. If the river is, as we say, graded (or as the engineer says, adjusted), this means that in each segment the slope is adjusted to provide just the transporting power required to carry through that segment all the grains, of whatever size, that enter it from above. Rivers that flow from rugged ranges of hard rock tend to develop steep slopes, adjusted to the transportation of large pebbles. Once they are developed, the adjusted slopes are maintained indefinitely, as long as the size of the pebbles and other controlling factors remain the same. Rivers that are supplied only with sand tend to maintain low slopes appropriate to the transportation of this material.

If the sizes of the grains supplied to a given segment of an adjusted river are abruptly increased by uplift, by a climatic change, or by a work of man, the larger grains, which are beyond the former carrying power, are deposited in the upper part of the segment; the bed is raised thereby and the slope is consequently steepened. This steepening by deposition continues until that particular slope is attained which provides just the velocity required to carry those larger grains, that is, until a new equilibrium slope is developed, which the river will maintain thence forward so long as grain size and other slope-controlling conditions remain the same.

Thus in the long view, velocity is adjusted to, or determined by, grain size; the test of relative timing (first the increase in grain size of material supplied to the river, and then, through a long period of readjustment, the increase in velocity) marks the change in grain size as the cause of the change in velocity. Note that because the period of readjustment may occupy thousands of years, this view is based primarily on reasoning rather than on direct observation. Note also that we deal here with three different frames of reference spanning the range from the empirical to the rational.

The statement that grain size tends to vary directly with bed velocity is an equation, whose terms are transposable; neither time nor cause and effect are involved, and this first frame may be entirely empirical. The numerical answer is complete in itself.

The short-term cause-and-effect view, that grain size is controlled by bed velocity, is in part rational, or if you prefer, it represents a higher level of empiricism. As I see it, this second frame has a significant advantage over the first in that it provides more fertile ground for the formulation of working hypotheses as to the mechanical relations between the flow and the particle at rest or in motion on the bed, leading to purposeful observation or to the design of experiments.

The third frame, the long-term view, that velocity is controlled by grain size, has a great advantage over the short-term view in that it provides an understanding of the origin of slope, which the short-term view does not attempt to explain. It is largely rational, or if you prefer, it represents a still higher level of empiricism.

Because I think that the objective of science is an understanding of the world around us, I prefer the second and third frames to the first, but I hope that it is clear that I recognize that all the frames are valid; the best one, in every instance, is simply the one that most efficiently gets the job done that needs doing. The important thing is to recognize that there *are* different frames; and that they overlap so completely and are so devoid of boundaries that it is easy to slip from one to the other.

The difference between the rational and the empirical approach to this matter of river slope, and the need for knowing what frame of reference we are in, can be clarified by a little story. One of the earliest theories of the origin of meanders, published in a British engineering journal in the late eighteen hundreds, was essentially as follows: divested of all geographic detail. Two cities *A* and *B*, both on the valley floor of a meandering river, are 50 airline miles apart. City *B* is 100 feet lower than city *A*; hence the average slope of the valley floor is two feet per mile (Fig. 3). But the slope of the river, measured round its loops, is only one foot per mile. The British engineer's theory was, in effect, though not expressed in these words, that the river said to itself, "How, with a slope of one foot per mile, can I manage to stay on a valley floor with a slope of two feet per mile? If I flow straight down the middle of the valley floor, starting at *A*, I will be 50 feet *above* the valley floor at *B*, and that simply will not do." Then it occurred to the river that it could meet this

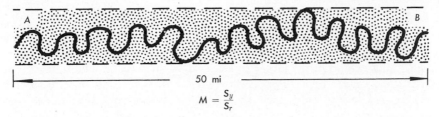

$$M = \frac{S_y}{S_r}$$

Fig. 3. Diagram illustrating an hypothesis for the origin of meanders.

problem by bending its channel into loops of precisely the sinuousity required to keep it *on* the valley floor, just as a man might do with a rope too long for the distance between two posts. And it worked, and that's why we have meanders.

Note that this theory not only explains meandering qualitatively, but puts all degrees of meandering, from the very loopy meanders of the ribbon-candy type to those that are nearly straight, on a firm quantitative basis—the sinuosity or degree of meandering, M, equals the slope of the valley floor, S_y, over the slope of the river, S_r.

There is nothing wrong with this equation, so long as it only describes. But if its author takes it to be an explanation, as the British engineer did, and if he slips over from the empirical frame into the rational frame, as he may do almost without realizing it, he is likely to be not just off by an order of magnitude, but upside-down—to be not only wrong but ludicrous. This explanation of meanders leaves one item out of account—the origin of the valley floor. The valley floor was not opened out and given its slope by a bulldozer, nor is it a result of special creation prior to the creation of the river. The valley floor was formed by the river that flows on it.

Causes of Downvalley Decrease in Pebble Size

It is a matter of observation that there is commonly a downvalley decrease in the slopes of graded rivers, and it is also a matter of observation that there is commonly a downvalley decrease in the size of pebbles in alluvial deposits. A question arises, then, as to whether the decrease in slope is caused in part by the decrease in pebble size, or whether the decrease in pebble size is caused in part by the decrease in slope, or whether both of these changes are independent or interdependent results of some other cause. My third and last example applies the empirical and rational approaches to a part of this problem, namely, what are the causes of the decrease in pebble size? The reasoning is somewhat more involved than in the other examples; in this respect it is more truly representative of the typical geologic problem.

The downvalley decrease in pebble size could be caused by either of two obvious, sharply contrasted mechanisms: (1) abrasional wear of the pebbles as they move along the bed of the stream, and (2) selective transportation, that is, a leaving-behind of the larger pebbles. The question is, which mechanism causes the decrease, or, if both operate, what is their relative importance?

There is no direct and satisfactory way of obtaining an answer to this question by measurement, however detailed, of pebble sizes in alluvial deposits. The most commonly used approach is by means of laboratory experiment. Usually fragments of rock of one or more kinds are placed in a cylinder which can be rotated on a horizontal axis and is so constructed that the fragments slide, roll, or drop as it turns. The fragments are remeasured from time to time to determine the reduction in size, the corresponding travel distance being calculated from the circumference of the cylinder and the number of rotations. This treatment does not approximate very closely the processes of wear in an actual river bed. Kuenen (1959) has recently developed a better apparatus, in which the fragments are moved over a concrete floor in a circular path by a current of water. Whatever the apparatus, it is certain that the decrease in pebble size observed in the laboratory is due wholly to abrasion, because none of the pebbles can be left behind; there is no possibility of selective transportation.

When the laboratory rates of reduction in pebble size per unit of travel distance are compared with the downvalley decrease in pebble size in alluvial deposits along most rivers, it is found that the decrease in size along the rivers is somewhat greater than would be expected on the basis of laboratory data on rates of abrasion. If the rates of abrasion in the laboratory correctly represent the rates of abrasion in the river bed, it should be only necessary to subtract to determine what percentage of the downvalley decrease in grain size in the alluvial deposits is due to selective transportation.

Field and laboratory data bearing on this problem have been reviewed by Scheidegger (1961) in his textbook, "Theoretical Geomorphology," which is about as far out on the quantitative side as it is possible to get. Scheidegger (p. 175) concludes that "... the most likely mechanism of pebble gradation in rivers consists of pebbles becoming contriturated due to the action of frictional forces, but being assigned their position along the stream bed by a sorting process due to differential transportation."

If I understand it correctly, this statement means that pebbles are made smaller by abrasion, but that the downvalley decrease in pebble size in alluvial deposits is due largely (or wholly?) to selective transportation.

On a somewhat different basis—the rate of reduction of pebbles of less resistant rock, relative to quartzite, in a downvalley direction in three rivers east of the Black Hills—Plumley (1948) concludes that about 25 per cent of the reduction in these rivers is due to abrasion, and about 75 per cent is due to selective transportation.

These two conclusions as to the cause of the downstream decrease in pebble size, solidly based on measurements, agree in ascribing it mainly to selective transportation. Let us try a different approach—let us think through the long-term implications of the processes.

Downstream decrease in pebble size by selective transportation requires that the larger pebbles be left behind permanently. The three-inch pebbles, for example, move downstream to a certain zone, and are deposited there because they cannot be transported farther. The two-inch pebbles are carried farther downstream, to be deposited in an appropriate zone as the slope decreases. These zones may have considerable length along the stream, they may be poorly defined, and they may of course overlap, but there is a downstream limit beyond which no pebbles of a given size occur in the alluvial deposits because none could be carried beyond that limit, which is set by transporting power.

Consider a river carrying a bed load of sand and gravel under steady-state conditions such that the slope and altitude in a given segment are maintained indefinitely without change, and let it be assumed for simplicity that the channel is floored and walled by rock (Fig. 4a). The load moves chiefly

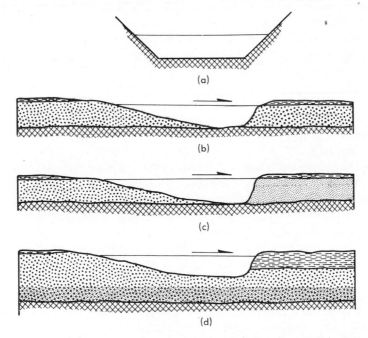

Fig. 4. Diagram illustrating exchange in graded and aggrading rivers.

during high-water stages and lodges on the bed during low-water stages. The smaller pebbles are likely to be set in motion sooner than the larger pebbles during each rising stage, they are likely to move faster while in motion, and they are likely to be kept in motion longer during each falling stage. In this sense, the transportation process is selective—if a slug of gravel consisting of identifiable pebbles were dumped into the segment, the smaller pebbles would outrun the larger, and this would cause a downstream decrease in the sizes of these particular pebbles in the low-water deposits. But in the steady-state condition, that is, with a continuous supply of a particular type of pebble or of pebbles of all types, all the pebbles deposited on the bed during the low stages must be placed in motion during the high stages; if the larger pebbles were permanently left behind during the seasonal cycles of deposition and erosion, the bed would be raised, and this, in turn, would change the condition. A non-aggrading river flowing in a channel which is floored and walled by rock cannot rid itself of coarse material by deposition because there is no place to deposit it where it will be out of reach of the river during subsequent fluctuations of flow; every pebble entering a given segment must eventually pass on through it. The smaller pebbles move more rapidly into the segment than the larger pebbles, but they also move more rapidly out of it. In the steady-state condition, the channel deposits from place to place in the segment contain the same proportions of the smaller and larger pebbles as though all moved at the same rate. Selective transportation cannot be a contributing cause of a downstream decrease in pebble size in our model river because there can be no selective deposition.

In a real river that maintains the same level as it meanders on a broad valley floor, bed load deposited along the inner side of a shifting bend is exchanged for an equal volume of slightly older channel deposits eroded from the outside of the bend. If these channel deposits were formed by the same river, operating under the same conditions and at the same level over a long period of time (Fig. 4b), the exchange process would not cause a reduction in the grain size of the bed load; insofar as selective transportation is concerned, the relation would be the same as in our model river. But if, by reason of capture or climatic change or any other change in controlling conditions, the older alluvial deposits in a given segment are finer grained than the bed load now entering that segment (Fig. 4c), exchange will cause a decrease in pebble size in a downstream direction, at least until the older deposits have been completely replaced by deposits representing the new regime. Exchange also causes a reduction in grain size if the river, maintaining the same level, cuts laterally into weak country rock that yields material finer in grain size than the load that is being concomitantly deposited on the widening valley floor.

The selective transportation associated with the process of exchange in the graded river, while by no means negligible, is much less effective as a cause of downstream decrease in pebble size than the selective transportation that characterizes the aggrading river. The essential difference is shown in Fig. 4(d); some of the deposits formed by one swing of the aggrading river across its valley floor are not subject to reworking in later swings, because the channel is slowly rising. The largest pebbles in transit in a given segment in a high-water stage are likely to be concentrated in the basal part of the deposit formed during the next falling stage. Thus the aggrading river rids itself of these pebbles, selectively and permanently, and there is a corresponding downstream decrease in pebble sizes in the deposits.

If upbuilding of the flood plain by deposition of overbank material keeps pace with aggradational rising of the channel, the shifting meanders may exchange channel deposits for older alluvium consisting wholly or in part of relatively fine-grained overbank material (Fig. 4d). But in rapidly aggrading rivers this rather orderly process may give way to a fill-spill mechanism in which filling of the channel is attended by the splaying of channel deposits over adjoining parts of the valley floor. On some proglacial outwash plains this type of braiding causes boulder detritus near the ice front to grade into pebbly sand within a few miles; there is doubtless some abrasional reduction in grain size in the proglacial rivers, but nearly all the decrease must be due to selective transportation.

Briefly then, thinking the process through indicates that the downstream decrease in grain size in river deposits in some cases may be almost wholly due to abrasion, and in others almost wholly due to selective transportation, depending primarily on whether the river is graded or aggrading and on the rate of aggradation. It follows that no generalization as to the relative importance of abrasion versus selective transportation in rivers—all rivers—has any meaning.

A different way of looking at this problem has been mentioned in another connection. As already noted, selective transportation implies permanent deposition, for example, the three-inch pebbles in a certain zone, the two-inch pebbles in another zone farther downstream, and so on. If this deposition is caused by a downstream decrease in slope, as is often implied and sometimes stated explicitly (Scheidegger, p. 171), then what is the cause of the decrease in slope? We know that the valley floor was not shaped by a bulldozer, and we know that it was not formed by an act of special creation before the river began to flow. As we have seen in considering the origin of meanders, rivers normally shape their own valley floors. If the river is actively aggrading, this is usually because of some geologically recent change such that the gradient in a given segment is not steep enough to enable the river to move through that segment all of the pebbles entering it; in this (aggrading) river, the size of the pebbles that are carried is controlled in part by the slope, and the larger pebbles are left behind. But if the river is graded, the slope in each segment is precisely that required to enable the river, under the prevailing hydraulic conditions, whatever they may be, to carry the load supplied to it. The same three-inch pebbles that are the largest seen on the bed and banks in one zone will, after a while, be the two-inch pebbles in a zone farther downstream.

We cannot wait long enough to verify this conclusion by direct observation of individual pebbles, because the pebbles ordinarily remain at rest in alluvial deposits on the valley floor for very long intervals of time between jogs of movement in the channel. We are led to the conclusion by reasoning, rather than by direct observation. In the long-term view, the graded river is a transportation system in equilibrium, which means that it maintains the same slope so long as conditions remain the same. There is no place in this self-maintaining system for permanent deposits: if the three-inch pebbles entering a given zone accumulated there over a period of geologic time, they would raise the bed and change the slope. As the pebbles, in their halting downvalley movement in the channel, are reduced in size by abrasion, and perhaps also by weathering while they are temporarily at rest in the valley floor alluvium, the slope, which is being adjusted to their transportation, decreases accordingly.

Does this reasoning settle the problem? Of course not! It merely makes us take a more searching look at the observational data. Since it is theoretically certain that the mechanisms which cause pebbles to decrease in size as they travel downstream operate differently, depending on whether the river is graded or aggrading, there is no sense in averaging measurements made along graded rivers with those made along aggrading rivers. However meticulous the measurements, and however refined the statistical treatment of them, the average will have no meaning.[6]

The reasoning tells us that, first of all, the rivers to be studied in connection with change of pebble size downstream must be selected with care. Because a steady-state condition is always easier to deal with quantitatively than a shifting equilibrium, it would be advisable to restrict the study, at the outset, to the deposits of graded rivers; when these are understood, we will be ready to deal with complications introduced by varying rates of aggradation. Similarly, it will be well, at least at the beginning, to eliminate altogether, or at least reduce to a minimum, the complicating effects of contributions from tributaries or other local sources; this can be done by selecting river segments without large tributaries, or by focusing attention on one or more distinctive rock types from known sources. There are unavoidable sampling problems, but some of these

---

[6] I owe to Frank Calkins the thought that, like most hybrids, this one would be sterile. The significance of this way of expressing what I have been saying about the averaging of unlike things is brought out by Conant's (1951, p. 25) definition of science as "an interconnected series of concepts and conceptual schemes that have developed as a result of experimentation and observation and are fruitful of further experimentation and observations. In this definition the emphasis is on the word 'fruitful'."

can readily be avoided; for example, there are many river segments in which the alluvial deposits are not contaminated by lag materials. Any attempt to develop sampling procedures must take into account, first of all, the fact that the channel *deposits* in a given segment of a valley differ significantly in gradation of grain size from the material moving through the channel in that segment in any brief period; the investigation may deal with the bed load (trapped in a box, so to speak), or with the deposits, or with both; but if both bed load and the deposits are measured, the measurements can only be compared, they cannot be averaged. Certainly we must investigate, in each river individually, the effects of weathering of the pebbles during periods of rest.

We must also take another hard look at the abrasion rates obtained by laboratory experiments, and try to determine in what degree these are directly comparable with abrasion rates in rivers. It is clearly desirable to develop other independent checks, such as those given by Plumley's measurement of rates of downstream reduction in sizes of pebbles of rock types differing in resistance to abrasion. Finally, it goes without saying that the reasoning itself must be continuously checked against the evidence, and one line of reasoning must be checked against others, to make sure that the mental wheels have not slipped a cog or two.

When we eventually have sufficient data on rates of downstream decrease of pebble size in alluvial deposits along many different types of rivers (considered individually), it will be possible to evaluate separately, in quantitative terms, the effect of special circumstances influencing the process of exchange in graded rivers, rates of aggradation in aggrading rivers, and the other causes of downstream decrease in pebble size. These generalizations will apply to all river deposits, modern as well as ancient, and it may even be that we can draw sound inferences regarding the hydraulic characteristics of the ancient rivers by comparing their deposits with those of modern rivers, in which the hydraulic characteristics can be measured.

This rational method of problem-solving is difficult and tortuous, but the history of science makes it clear, again and again, that if the system to be investigated is complex, the longest way 'round is the shortest way home; most of the empirical shortcuts turn out to be blind alleys.

Whither Are We Drifting, Methodologically?

I would like now to return to some of the questions asked at the outset. Must we accept, as gospel, Lord Kelvin's pronouncement that what cannot be stated in numbers is not science? To become respectable members of the scientific community, must we drastically change our accustomed habits of thought, abandoning the classic geologic approach to problem-solving? To the extent that this approach is qualitative, is it necessarily loose, and therefore bad?

Must we now move headlong to quantify our operations on the assumption that whatever is quantitative is necessarily rigorous and therefore good?

Why has the swing to the quantitative come so late? Is it because our early leaders, men such as Hutton, Lyell, Agassiz, Heim, Gilbert, and Davis, were intellectually a cut or two below their counterparts in classical physics? There is a more reasonable explanation, which is well known to students of the history of science. In each field of study the timing of the swing to the quantitative and the present degree of quantification are largely determined by the subject matter: the number and complexity of the interdependent components involved in its systems, the relative ease or difficulty of obtaining basic data, the susceptibility of those data to numerical expression, and the extent to which time is an essential dimension. The position of geology relative to the basic sciences has been stated with characteristic vigor by Walter Bucher (1941) in a paper that seems to have escaped the attention of our apologists.

Classical physics was quantitative from its very beginning as a science; it moved directly from observations made in the laboratory under controlled conditions to abstractions that were quantitative at the outset. The quantification of chemistry lagged 100 years behind that of physics. The chemistry of a candle flame is of an altogether different order of complexity from the physics of Galileo's rolling ball; the flame is only one of many types of oxidation; and oxidation is only one of many ways in which substances combine. There had to be an immense accumulation of quantitative data, and many minor discoveries—some of them accidental, but most of them based on planned investigations—before it was possible to formulate such a sweeping generalization as the law of combining weights.

If degree of quantification of its laws were a gage of maturity in a science (which it is not), geology and biology would be 100 to 200 years behind chemistry. Before Bucher (1933) could formulate even a tentative set of "laws" for deformation of the earth's crust, an enormous descriptive job had to be well under way. Clearly, it was necessary to know what the movements of the crust *are* before anybody could frame explanations of them. But adequate description of even a single mountain range demands the best efforts of a couple of generations of geologists, with different special skills, working in the field and the laboratory. Because no two ranges are alike, the search for the laws of mountain growth requires that we learn as much as we can about every range we can climb and also about those no longer here to be climbed; the ranges of the past, which we must reconstruct as best we can by study of their eroded stumps, are as significant as those of the present. Rates of growth and relative ages of past and present ranges are just as important as their geometry; the student of the mechanics of crustal deformation must think like a physicist and also like a historian, and these are very different ways of thinking, difficult to combine. The evidence is hard to come by, it is largely circumstantial, and there is never enough of it. Laboratory models are helpful only within narrow limits. So it is also with the mechanism of emplacement of batholiths, and the origin of ore-forming fluids, and the shaping of landforms of all kinds, and most other truly geologic problems.

It is chiefly for these reasons that most geologists have been preoccupied with manifold problems of description of geologic things and processes—*particular* things and processes—and have been traditionally disinclined to generalize even in qualitative terms. Because most geologic evidence cannot readily be stated in numbers, and because most geologic systems are so complex that some qualitative grasp of the problem must precede effective quantitative study, we are even less inclined to generalize in quantitative terms. Everybody knows the story of Lord Kelvin's calculation of the age of the earth.

These things are familiar, but they are worth saying because they explain why geology is only now fully in the swing to the quantitative. Perhaps it would have been better if the swing had begun earlier, but this is by no means certain. A meteorologist has told me that meteorology might be further ahead today if its plunge to the quantitative had been somewhat less precipitous—if there had been a broader observational base for a qualitative understanding of its exceedingly complex systems before these were quantified. At any rate, it is important that we recognize that the quantification of geology is a normal evolutionary process, which is more or less on schedule. The quantification will proceed at an accelerating pace, however much our ultraconservatives may drag their feet. I have been trying to point out that there is an attendant danger: as measurements increase in complexity and refinement, and as mathematical manipulations of the data become more sophisticated, these measurements and manipulations may become so impressive in form that the investigator tends to lose sight of their meaning and purpose.[7]

This tendency is readily understandable. Some of the appealing features of the empirical method have already been mentioned. Moreover, the very act of making measurements, in a fixed pattern, provides a solid sense of accomplishment. If the measurements are complicated, involving unusual techniques

---

[7] The subtitle of a recent article by Krumbein (1962), "Quantification and the advent of the computer open new vistas in a science traditionally qualitative" makes evident the overlap of our interests. Professor Krumbein's article deals explicitly with a mechanical method of processing data; the fact that there is no mention of the use of reasoning in testing the quality and relevance of the data to the specific issue being investigated certainly does not mean that he thinks one whit less of the "rational method" than I do. Similarly, I hope that nothing that I have said or failed to say is construed as meaning that I have an aversion to mechanical methods of analyzing data; such methods are unquestionably good if they bring out relationships not otherwise evident, or in any other way advance the progress of the rational method of investigation. When mechanical processes *replace* reasoning processes, and when a number *replaces* understanding as the objective, danger enters.

and apparatus and a special jargon, they give the investigator a good feeling of belonging to an elite group, and of pushing back the frontiers. Presentation of the results is simplified by use of mathematical shorthand, and even though nine out of ten interested geologists do not read that shorthand with ease, the author can be sure that seven out of the ten will at least be impressed. It is an advantage or disadvantage of mathematical shorthand, depending on the point of view, that things can be said in equations, impressively, even arrogantly, which are so nonsensical that they would embarrass even the author if spelled out in words.

As stated at the outset, the real issue is not a matter of classical geologic methods versus quantification. Geology *is* largely quantitative, and it is rapidly and properly becoming more so. The real issue is the rational method versus the empirical method of solving problems; the point that I have tried to make is that if the objective is an understanding of the system investigated, and if that system is complex, then the empirical method is apt to be less efficient than the rational method. Most geologic features—ledges of rock, mineral deposits, landscapes, segments of a river channel—present an almost infinite variety of elements, each susceptible to many different sorts of measurement. We cannot measure them all to any conventional standard of precision—blind probing will not work. Some years ago (1941) I wrote that the "eye and brain, unlike camera lens and sensitized plate, record completely only what they intelligently seek out." Jim Gilluly expresses the same thought more succinctly in words to the effect that most exposures provide answers only to questions that are put to them. It is only by thinking, as we measure, that we can avoid listing together in a field book, and after a little while, averaging, random dimensions of apples and oranges and apple crates and orange trees.

Briefly, then, my thesis is that the present swing to the quantitative in geology, which is good, does not necessarily and should not involve a swing from the rational to the empirical method. I'm sure that geology is a science, with different sorts of problems and methods, but not in any sense less mature than any other science; indeed, the day-to-day operations of the field geologist are apt to be far more sophisticated than those of his counterpart—the experimentalist—in physics or chemistry. And I'm sure that anyone who hires out as a geologist, whether in practice, or in research, or in teaching, and then operates like a physicist or a chemist, or, for that matter, like a statistician or an engineer, is not living up to his contract.

The best and highest use of the brains of our youngsters is the working out of cause and effect relations in geologic systems, with all the help they can get from the other sciences and engineering, and mechanical devices of all kinds, but with basic reliance on the complex reasoning processes described by Gilbert, Chamberlin, and Johnson.

REFERENCES CITED

ALBRITTON, C. C., JR., 1961, Notes on the history and philosophy of science (1) A conference on the scope and philosophy of geology: J. Graduate Research Center, Southern Methodist Univ., vol. 29, no. 3, pp. 188–192.

BUCHER, W. H., 1933, The deformation of the earth's crust: Princeton, N. J., Princeton Univ. Press, 518 pp.

———, 1941, The nature of geological inquiry and the training required for it: Am. Inst. Mining Metall. Eng., Tech. Pub. 1377, 6 pp.

CHAMBERLIN, T. C., 1897, The method of multiple working hypotheses: J. Geol., vol. 5, pp. 837–848.

CONANT, J. B., 1951, Science and common sense: New Haven, Yale Univ. Press, 371 pp.

———, 1952, Modern science and modern man: New York, Columbia Univ. Press, 111 pp.

GARRELS, R. M., 1951, A textbook of geology: New York, Harper, 511 pp.

GILBERT, G. K., 1886, The inculcation of the scientific method by example, with an illustration drawn from the Quaternary geology of Utah: Am. J. Sci., vol. 31, (whole no. 131), pp. 284–299.

———, 1914, The transportation of debris by running water: U. S. Geol. Survey, Prof. Paper, 86, 263 pp.

———, 1917, Hydraulic-mining debris in the Sierra Nevada: U. S. Geol. Survey, Prof. Paper 105, 154 pp.

HOLTON, G., 1952, Introduction to concepts and theories in physical science: Reading, Mass., Addison-Wesley, 650 pp.

JOHNSON, DOUGLAS, 1933, Role of analysis in scientific investigation: Geol. Soc. Am., B., vol. 44, pp. 461–494.

KRUMBEIN, W. C., 1960, The "geological population" as a framework for analysing numerical data in geology: Liverpool and Manchester Geol. J., vol. 2, pt. 3, pp. 341–368.

———, 1962, The computer in geology: Science, vol. 136, pp. 1087–1092.

KUENEN, P. H., 1959, Fluviatile action on sand, Part 3 of Experimental Abrasion: Am. J. Sci., vol. 257, pp. 172–190.

LEOPOLD, L. B., 1953, Downstream changes of velocity in rivers: Am. J. Sci., vol. 251, pp. 606–624.

———, and MADDOCK, T., JR., 1953, The hydraulic geometry of stream channels and some physiographic implications: U. S. Geol. Survey, Prof. Paper 252, 57 pp.

MACKIN, J. H., 1941, Drainage changes near Wind Gap, Pennsylvania; a study in map interpretation: J. Geomorphology, vol. 4, pp. 24–53.

———, 1948, Concept of the graded river: Geol. Soc. Am., B., vol. 59, pp. 463–512.

PLUMLEY, W. J., 1948, Black Hills terrace gravels; a study in sediment transport: J. Geol., vol. 56, pp. 526–577.

RUBEY, W. W., 1938, The force required to move particles on a stream bed: U.S. Geol. Survey, Prof. Paper 189-E, pp. 120–140.

SCHEIDEGGER, A. E., 1961, Theoretical geomorphology: Berlin, Springer-Verlag, 333 pp.

Simpson, G. G., Roe, Anne, and Lewontin, R. C., 1960, Quantitative zoology: New York, Harcourt-Brace, 440 pp.

U. S. Geological Survey, 1960, Surface water supply of the United States, 1958; Part 6A, Missouri River Basin above Sioux City, Iowa: U. S. Geol. Survey Water-Supply Paper 1559, 434 pp.

Weiss, Paul, 1962, Experience and experiment in biology: Science, vol. 136, pp. 468–471.

11

Copyright © 1963 by the Geological Society of America

Reprinted from *The Fabric of Geology*, C. C. Albritton, ed., Freeman, Cooper & Company, San Francisco, 1963, pp. 175–183

CHARLES A. ANDERSON

U. S. Geological Survey

Simplicity in Structural Geology[1]

In 1951, I made the statement that "Until more precise correlations of the older Precambrian rocks can be made, based on radioactivity or other methods, the *simplest* explanation is that only one period of orogeny, corresponding to Wilson's Mazatzal Revolution, has occurred in Arizona during early Precambrian time." (Anderson, 1951, p. 1346; italics added) This explicit reference to the use of simplicity in correlating structural events and reconstructing geologic history led the Chairman of the Anniversary Committee to ask for an essay on simplicity in structural geology.

Principle of Simplicity

The principle of simplicity has been called Occam's (or Ockham's) razor, the principle of parsimony, or the principle of economy. Allusions to simplicity in the literature are innumerable and varied in intent and nuance. A revival of interest in this principle among philosophers of science has been partly inspired by the work of Nelson Goodman.

All scientific activity amounts to the invention of and the choice among systems of hypotheses. One of the primary considerations guiding this process is that of simplicity. Nothing could be much more mistaken than the traditional idea that we first seek a true system and then, for the sake of elegance alone, seek a simple one. We are inevitably concerned with simplicity as soon as we are concerned with system at all; for system is achieved just to the extent that the basic vocabulary and set of first principles used in dealing with the given subject matter are simplified. When simplicity of

---

[1] I wish to thank my two colleagues, James Gilluly and Walter S. White, for helpful and critical comments in their review of this essay.

basis vanishes to zero—that is, when no term or principle is derived from any of the others—system also vanishes to zero. Systematization is the same thing as simplification of basis. (Goodman, 1958, p. 1064)

William of Occam, known as *Doctor Invincibilis* and *Venerabilis Inceptor*, was born around 1300 and became a member of the Franciscan order while still a youth. He was an intellectual leader in the period that saw the disintegration of old scholastic realism and the rise of theological skepticism. The famous dictum attributed to him, "*Entia non sunt multiplicanda praeter necessitatem*," has appeared in nearly every book on logic from the middle of the nineteenth century. It is doubtful that the "Invincible Doctor" used these words (Thorburn, 1918), but he did use similar words such as *Frustra fit per plura quod potest fieri per pauciora* (Laird, 1919, p. 321). The "razor" is commonly used now without special reference to the scholastic theory of entities, and Laird believes that the precise form in which Occam stated it is irrelevant.

Russell (1929, p. 113) states that Occam's razor, "entities are not to be multiplied without necessity," is a maxim that inspires all scientific philosophizing, and that in dealing with any subject matter, one should find out what entities are undeniably involved, and state everything in terms of these.

The concept of simplicity is a controversial topic in the philosophy of science according to Kemeny (1953, p. 391). One school believes it involves an assumption about the simplicity of nature, whereas others justify it as a matter of convenience, "a labor-saving device." Jevons (1883, p. 625) objected to the generalization that the laws of nature possess the perfection which we attribute to simple forms and relations, and suggested that "Simplicity is naturally agreeable to a mind of limited powers, but to an infinite mind all things are simple."

Mill (1865, p. 461) questioned Hamilton's belief that "Nature never works by more complex instruments than are necessary," stating that "we know well that Nature, in many of its operations, works by means which are of a complexity so extreme, as to be an almost insuperable obstacle to our investigations." Mill (1865, p. 467) believed that we are not justified in rejecting an hypothesis for being too complicated, but "The 'Law of Parcimony' needs no such support; it rests on no assumptions respecting the ways or proceedings of Nature. It is a purely logical precept; a case of the broad practical principle, not to believe anything for which there is no evidence . . . The assumption of a superfluous cause, is a belief without evidence." Mill (*ibid.*) emphasizes that the principle which forbids the assumption of a superfluous fact, forbids a superfluous law and "The rule of Parcimony, therefore, whether applied to facts or to theories, implies no theory concerning the propensities or proceedings of Nature."

Feuer (1957, p. 121), like Mill, emphasized that the scientific principle of simplicity does not rest on the assumption that the laws of nature are simple,

216

and he pointed out that the simplicity of nature has had a long philosophical history which he would call the *metascientific principle of simplicity*, to distinguish it from the scientific methodological principle of Occam's razor. Verifiability is the important element in Occam's razor; the principle of simplicity is thus a straightforward basis for rejecting theories if they are unverifiable (Feuer, 1957, p. 115). The verified theory is the simplest because every unnecessary component is an unverified item.

Demos (1947) expresses some skepticism about the use of simplicity and suggests that the scientific philosopher tries to evade the charge of fallacious reasoning by introducing the principle of simplicity, thereby enabling him to choose among the several theories consistent with the observed facts. The scientist then selects the theory that explains the greatest number of phenomena with the fewest assumptions. Bridgman (1961, p. 10) regarded Occam's razor:

> . . . as a cardinal intellectual principle, . . . I will try to follow it to the ut-most. It is almost frightening to observe how blatantly it is disregarded in most thinking . . . To me it seems to satisfy a deep-seated instinct for intellectual good workmanship. Perhaps one of the most compelling reasons for adopting it is that thereby one has given as few hostages to the future as possible and retained the maximum flexibility for dealing with unanticipated facts or ideas.

Simplicity and Geology

A recent issue of "Philosophy of Science" contains a symposium on simplicity, but only one paper mentions its application to geology. Barker (1961, p. 164) revives the old problem of the meaning of fossils: are they remains of organisms that actually existed on earth millions of years ago, or were they placed there by the Creator to test our faith? Barker concluded that unless there is independent evidence in favor of a Creator, the simple theory is that plants and animals existed in the past in circumstances similar to those in which we find them today. This leads to uniformitarianism, a topic discussed elsewhere in this volume.

An excellent geologic example of simplicity is given by Woodford (1960) in discussing the magnitude of strike slip on the San Andreas fault. In 1906, the right-lateral slip along this fault was 22 feet in central California, and in southern California offset streams indicate right-lateral movement of thousands of feet during Quaternary time. But for pre-Quaternary movements, it is necessary to distinguish between separation and slip. A structurally complex succession of granodiorite, Paleocene, and Miocene rocks north of the San Gabriel Mountains is offset in a way that seems to require 30 miles of right-lateral slip since the middle Tertiary. Displacements may have been even greater (a range from 160 to 300 miles has been suggested), but Woodford prefers a working hypothesis that includes some dip slip and so he limits strike slip on the San Andreas fault to 30 miles, right lateral. "The tentative choice of short slips, if these will do the business, is an example of the use of the principle *Disjunctiones minimae, disjunctiones optimae.* This rule may be considered a quantitatively parsimonious relative of Ockham's law: *Entia non sunt multiplicanda praeter necessitatem.*" (Woodford, 1960, p. 415)

The preparation of a geologic map is the essential first step in structural geology, and one of the first steps is the building up of the stratigraphic sequence. The law of superposition is vital to the success of this study. Even where we have "layer-cake" stratigraphy, Albritton (1961, pp. 190–191) has pointed out that it is not clear how the principle of simplicity operates in geology:

> . . . [given] two nearby mesas of three formations conformably arranged in similar sequence from bottom to top. Without evidence to the contrary, most stratigraphers would recognize only three formations in all, perhaps on the ground that it is in vain to do with more what can be done with fewer. But if a three-formation column is simpler than a [six-formation] column, would it not be simpler still to lump the three formations into one group, and then have a single entity?

Of course the answer is that to do so is to lose information. The purpose of the geologic study has an important bearing on the choice of stratigraphic units. Fundamentally, selection is made to focus attention on the environment of deposition of the sediments, to indicate the various stages in the geologic history. For structural interpretations, delineation of thin units may help the geologic map to elucidate the structure. The objective is to use the map as a means of showing as many as possible of the elements that bear on the geologic history and structure, and something about the basis on which these elements have been verified.

In regions of complex structure, particularly if the rocks are nonfossiliferous and folded isoclinally, stratigraphy and structure are determined concurrently by mapping distinctive lithologic units whether they are beds, zones, or formations. Structural elements and data on the direction that the tops of beds are now facing are diligently searched for in all exposures. In this manner, the stratigraphy and structure unfold together, and the final interpretation results from the integration of both. The interpretation may be complex, but the principle of simplicity will be followed if no unverifiable facts are essential to the interpretation. In actual practice there are few situations in which a geologic interpretation does not require some unverified assumptions, and in general the use of the principle of simplicity involves the acceptance of the interpretation that has the maximum of verifiable facts and minimum of assumptions.

Prediction is an important facet of structural geology; in mapping, a field geologist commonly predicts what will be found on the next ridge, valley, or mountain range. In a sense, this is a field test of the interpretations developing in the mind of the geologist. No doubt it is a frequent experience with a geologist, mapping in a region of complex geology, to find his predictions erroneous; the geology may be more complex than the interpretation of the moment. But this is a part of the accumulation of field data and in no way conflicts with the use of simplicity in the final interpretation. Prediction is the end product of many studies in structural geology in proposing exploration programs, and in the search for new mineral deposits and petroleum accumulations.

Many examples could be cited where early expositions of the geologic history of a particular region are less complex than later explanations based on additional field studies. Probably most geologists would accept as axiomatic that new information leads to a more complex story. In a sense, it is a mark of progress as we build upon the experience of those who worked before us on similar problems. This is to be expected, particularly in regions where the rocks have been acutely deformed by past tectonic activity. The geologic history becomes more complex as we build up a storehouse of "verifiable elements," even though each succeeding historical account does not introduce entities beyond necessity.

In regions where heavy vegetation and thick soil cover the rocks, it is a time-consuming process to assemble the facts that are needed to reconstruct the story of the stratigraphy and structure. The early interpretations are likely to be simple because of meager data. Trenching, drill holes, and painstaking studies of the saprolites may in time add to the verifiable elements to give a more complete and more complex history.

Older Precambrian in Arizona

I would like to discuss in more detail the older Precambrian geology in Arizona as an example of the workings of the principle of simplicity in structural geology. By 1951, sufficient geologic mapping had been done in the older Precambrian rocks of the Grand Canyon, Globe-Miami, Mazatzal Mountains, Little Dragoon Mountains, Bagdad, and Prescott-Jerome areas to indicate that only one period of orogeny, followed by the intrusion of granitic rocks, could be recognized in each of these areas.

The natural temptation is to assume that the orogenies in these five separate areas occurred at the same time, particularly because of the general parallelism of the folds where the trends were determined, and Wilson (1939) has termed this probable widespread orogenic disturbance, the Mazatzal Revolution. From a purely academic view, one might question this conclusion, for it is well known that the Precambrian covers an immense period of time, and it would be surprising if only one period of orogeny occurred in Arizona during the early Precambrian time. Hinds (1936) has suggested that two periods of orogeny and two periods of granitic invasion occurred in Arizona prior to the deposition of the Younger Precambrian Grand Canyon series and Apache group, the Mazatzal quartzite marking the period of sedimentation between these orogenies. Because no positive angular unconformities have been found between the Mazatzal quartzite and Yavapai schist, some doubt exists regarding the validity of this older period of orogeny and granitic invasion. (Anderson, 1951, p. 1346; thereupon followed the sentence quoted in the introductory paragraph.)

Philip M. Blacet of the U.S. Geological Survey has recently mapped, south of Prescott, a basement of granodiorite gneiss older than the Yavapai Series. This gneiss underlies a basal conglomerate containing angular blocks of the granodiorite and abundant well-rounded boulders of aplite. The basal conglomerate grades upward into feldspathic sandstone, gray slate, pebble conglomerate, and beds of rhyolitic tuffaceous sandstone (now recrystallized to quartz-sericite schist) of the Texas Gulch Formation, described by Anderson and Creasey (1958, p. 28) as possibly the oldest formation in the Alder Group of the Yavapai Series. Similar slate, pebble conglomerate, and tuffaceous sandstone occur in the type section of the Alder Group in the Mazatzal Mountains (Wilson, 1939, p. 1122) and were deformed during Wilson's Mazatzal Revolution. The unconformable relation between the Texas Gulch Formation and the granodiorite gneiss south of Prescott is important as proving the existence of a granitic rock older than the Yavapai Series, and therefore older than the granitic rocks intruded during the Mazatzal Revolution. Therefore two periods of granitic intrusion in the older Precambrian of Arizona are proved by normal stratigraphic relations, superposition, and transgressive intrusive contacts. Hinds was correct in suggesting the two periods, but he placed his second period after the Mazatzal Revolution rather than before. He did not have verifiable data to support his conclusion, and following the principle of simplicity, his contribution had to be ignored.

Progress is being made in the use of radiometric measurements to correlate rocks and structural events in the older Precambrian of Arizona. Some of these data are shown in Table 1; the ages indicated for the gneisses, granites, and pegmatites from which mica samples were collected range from around 1200 to around 1500 million years (m. y.).

Mica from the pre-Yavapai granodiorite gneiss south of Prescott gave K-Ar and Rb-Sr measurements indicating an age of about 1250 m. y. (Carl Hedge, written communication). Measurements of the isotopes of lead in the zircon from the granodiorite gneiss indicate a minimum age of 1700 m. y. (E. J. Catanzaro, oral communication). These data indicate that the mica in the granodi-

TABLE 1

RADIOMETRIC AGES OF ROCKS
FROM THE OLDER PRECAMBRIAN OF ARIZONA

Sample locations*	Ages in million years	
	K-Ar	Rb-Sr
1. Gneiss, Grand Canyon	1390	1370
2a. Lawler Peak granite, Bagdad	1410	1390
2b. Pegmatite in Lawler Peak granite	1410	1500
3. Pegmatite, Wickenburg	1160	1300
4. Pegmatite in Vishnu schist,		1550
Grand Canyon		1530
5. Migmatite zone in Vishnu schist,		
Grand Canyon		1390
6. Granite near Valentine		1300
7. Diana granite, Chloride		1350
8. Chloride granite, Chloride		1210
9. Oracle granite, Oracle		1450

* Samples 1 through 3 are from Aldrich, Wetherill, Davis, 1957, p. 656, and samples 4 through 9 are from Giletti and Damon, 1961, p. 640.

orite gneiss recrystallized during the deformation of the Yavapai Series, corresponding in a general way to the time Aldrich, Wetherill, and Davis (1957) have called the 1350-m.y. period of granitic rocks. The zircon gives an older age, more in keeping with the stratigraphic relations. It should be noted that Silver and Deutsch (1961) have reported an age of 1650 m. y. for zircon from a granodiorite in southeastern Arizona (Cochise County).

It is tempting to assume that the 1350-m.y. period corresponds to the Mazatzal Revolution; unfortunately no reliable radiometric dates have been obtained from granitic rocks clearly related to the Mazatzal orogeny, that is, from the Mazatzal Mountains or adjacent areas. The available age data clearly demonstrate the need for systematic work, for there is much to be learned from radiometric measurements of the Precambrian rocks in Arizona. I predict that such studies will ultimately show that the structural history is more complex than can be documented from present data.

Most structural geologists would infer that there are, in Arizona, metamorphic rocks older than the granodiorite gneiss south of Prescott and that such rocks were deformed prior to or during the intrusion of the granodiorite made gneissic during the deformation of the Yavapai Series. Using the principle of simplicity, we can say with assurance that the simplest explanation in 1962 is that there were at least two periods of orogeny and granitic intrusion in the older Precambrian history of Arizona. As more verifiable elements are discovered, the story may well become even more complex.

Concluding Statement

Much of my discussion has been limited to the use of the principle of simplicity in explanatory or interpretive aspects of structural geology rather than in developing theories or laws. It is appropriate to refer to Mario Bunge, who raises doubts about simplicity in the construction and testing of scientific theories. A theory must at least be consistent with the known facts and should predict new and unsuspected facts (Bunge, 1961, p. 133). Bunge (1961, p. 148) believes that simplicities are undesirable in the stage of problem finding, but desirable in the formulation of problems, and much less so in the solution of problems. His advice is that "Ockham's razor—like all razors—must be handled with care to prevent beheading science in the attempt to shave off some of its pilosities. In science, as in the barber shop, better alive and bearded than dead and cleanly shaven."

It seems to me that when a structural geologist is formulating explanatory hypotheses, the principle of simplicity should not restrict his imagination; complex hypotheses may stimulate and guide the work toward new and different data. For this phase of a study, Bunge has made an excellent point; it is only in the final selection of the hypotheses that the assortment should be pared by Occam's razor.

REFERENCES CITED

ALBRITTON, C. C., JR., 1961, Notes on the history and philosophy of science. (1) A conference on the scope and philosophy of geology: J. Graduate Research Center, Southern Methodist Univ., vol. 29, no. 3, pp. 188–192.

ALDRICH, L. T., WETHERILL, G. W., and DAVIS, G. L., 1957, Occurrence of 1350 million-year-old granitic rocks in western United States: Geol. Soc. Am., B., vol. 68, pp. 655–656.

ANDERSON, C. A., 1951, Older Precambrian structure in Arizona: Geol. Soc. Am., B., vol. 62, pp. 1331–46.

——— and CREASEY, S. C., 1958, Geology and ore deposits of the Jerome area, Yavapai County, Arizona: U. S. Geol. Survey, Prof. Paper 308, 185 pp.

BARKER, S. F., 1961, On simplicity in empirical hypotheses: Phil. Sci., vol. 28, pp. 162–171.

BRIDGMAN, P. W., 1961, The way things are: New York, Viking (Compass Books Edition) 333 pp.

BUNGE, MARIO, 1961, The weight of simplicity in the construction and assaying of scientific theories: Phil. Sci., vol. 28, pp. 120–149.

DEMOS, RAPHAEL, 1947, Doubts about empiricism: Phil. Sci., vol. 14, pp. 203–218.

FEUER, L. S., 1957, The principle of simplicity: Phil. Sci., vol. 24, pp. 109–122.

GILETTI, B. J. and DAMON, P. E., 1961, Rubidium-strontium ages of some basement rocks from Arizona and northwestern Mexico: Geol. Soc. Am., B., vol. 72, pp. 639–644.

GOODMAN, NELSON, 1958, The test of simplicity: Science, vol. 128, pp. 1064–1069.

HINDS, N. E. A., 1936, Uncompahgran and Beltian deposits in western North America: Carnegie Inst. Washington, Pub. 463, pp. 53–136.

JEVONS, W. S., 1883, The principles of science: London, MacMillan, 786 pp.

KEMENY, J. G., 1953, The use of simplicity in induction: Phil. Rev., vol. 62, pp. 391–408.

LAIRD, JOHN, 1919, The law of parsimony: The Monist, vol. 29, p. 321–344.

MILL, J. S., 1865, An examination of Sir William Hamilton's philosophy: London, Longmans, Green, 561 pp.

RUSSELL, BERTRAND, 1929, Our knowledge of the external world: New York, W. W. Norton, 268 pp.

SILVER, L. T. and DEUTSCH, SARAH, 1961, Uranium-lead method on zircons: New York Acad. Sci., Ann., vol. 91, pp. 279–283.

THORBURN, C. C., JR., 1918, The myth of Occam's Razor: Mind, vol. 27, pp. 345–353.

WILSON, E. D., 1939, Pre-Cambrian Mazatzal Revolution in central Arizona: Geol. Soc. Am., B., vol. 50, pp. 1113–1164.

WOODFORD, A. O., 1960, Bedrock patterns and strike-slip faulting in southwestern California: Am. J. Sci., vol. 258A, pp. 400–417.

Addendum by Charles A. Anderson

Continuing geologic studies in the Prescott–Jerome area in Central Arizona have demonstrated that some of the statements in the section "Older Precambrian in Arizona" need revision as a result of isotopic lead–uranium dates from zircons obtained by L. T. Silver and T. W. Stern. The Texas Gulch Formation, which rests unconformably above the granodiorite gneiss (Brady Butte Granodiorite), is not the oldest formation in the Yavapai Series but is younger than this series. The Brady Butte Granodiorite is largely in fault contact with the Yavapai Series, but in part it is intrusive into a volcanic sequence (Spud Mountain Volcanics) of this series. Zircons from the Spud Mountain Volcanics are apparently 1,775 ± 10 m.y. old, whereas the Brady Butte Granodiorite is 1,770 ± 10 m.y. old. Two other plutons intrusive into the Yavapai Series are 1,770 ± 15 m.y. and 1,760 ± 15 m.y. and 1,760 ± 15 m.y. old, indicating an episode of important plutonism younger than the Yavapai Series (Anderson et al., 1971).

Two periods of deformation can be recognized in the Yavapai Series, the first resulting in widespread open folds, followed by the intrusions of several plutons. In a limited area between two major faults, the Yavapai Series, Brady Butte Granodiorite, and Texas Gulch Formation have been deformed into overturned folds and strongly foliated rocks. Younger plutons locally intrude this foliated sequence.

At the present time enough zircon dates are available to confirm that the older Precambrian history in Arizona is complicated. At Bagdad, volcanic and sedimentary rocks 1,760 m.y. old are intruded by granodiorite that is 1,740 ± 15 m.y. old (Silver,

1968). Granite gneiss in the Grand Canyon has an apparent age of 1,725 ± 15 m.y., which gives a minimum age for the Vishnu Schist (Pasteels and Silver, 1966). Th se dates suggest that the older history at Bagdad and Grand Canyon was more or less contemporaneous with the history in the Prescott–Jerome area. In contrast, Precambrian stratified volcanic and sedimentary rocks intruded by granitic rocks and exposed to the southeast of the Prescott–Jerome are younger than the Yavapai Series and associated plutons. In the Mazatzal Mountains, rhyolite has an apparent age of 1,715 ± 15 m.y., and the associated postdeformational plutonic rock is 1,660 ± 15 m.y. (Silver, 1965), similar to the age of the granitic rocks in southeastern Arizona (Silver and Deutsch, 1961, 1963).

The Lawler Peak Granite at Bagdad, which is 1,375 m.y. old, is an example of a thermal event not associated with contemporaneous deformation (Silver, 1968). As far as we know now, orogenic disturbances between 1,370 and 1,450 m.y. have not been documented by careful mapping and dating in Arizona.

References

Anderson, C. A., Blacet, P. M., Silver, L. T., and Stern, T. W., 1971, Revision of the Precambrian stratigraphy in the Prescott–Jerome area, Yavapai County, Arizona: U.S. Geol. Survey Bull. 1324, p. C1–C16.

Pasteels, Paul, and Silver, L. T., 1966, Geochronological investigations in the crystalline rocks of the Grand Canyon, Arizona [abs]: Geol. Soc. Amer. Spec. Paper 87, 124 p.

Silver, L. T., 1965, Mazatzal orogeny and tectonic episodicity [abs]: Geol. Soc. America Spec. Paper 82, p. 185–186.

———, 1968, U–Pb isotope relations and their historical implications in Precambrian zircons from Bagdad, Arizona [abs]: Geol. Soc. Amer. Spec. Paper 101, 420 p.

Silver, L. T., and Deutsch, Sarah, 1963, Uranium–lead isotopic variations in zircons—a case study: New York Acad. Sci. Annals, v. 91, p. 279–283.

12

Copyright © 1967 by the Geological Society of America

Reprinted from *Uniformity and Simplicity*, Geol. Soc. Amer. Spec. Paper 89, C. C. Albritton, ed., 1967, pp. 3–33

Critique of the Principle of Uniformity

M. KING HUBBERT

U. S. Geological Survey, Washington, D. C.
School of Earth Sciences, Stanford University, Stanford, California

In terms of the profundity of their effects in altering the intellectual outlook of the learned world, two scientific developments since the fifteenth century are outstanding. The first of these is the Copernican-Keplerian-Galilean revolution during which the Ptolemaic geocentric universe was displaced by the heliocentric solar system and the foundations were laid for the appropriate system of mechanics perfected during the succeeding half century by Isaac Newton. Inevitably, the demise of the geocentric system carried with it strong repercussions for the philosophical systems and theological dogmas which formed its principal supports. The theological dogma that the earth, being the abode of God's favorite Creation, Man, could not occupy a lesser place than the seat of honor at the center of the universe, was severely shaken by the establishment of the earth, not as the stationary center of the universe, but as only one of the six known planets encircling the sun.

The second major scientific revolution is that which may be referred to as the Huttonian-Lyellian-Darwinian. During this, an earth with a presumed Biblical history of only some 6000 years, whose plant and animal inhabitants were initiated by Divine Creation, was supplanted by an earth the length of whose decipherable history was estimated to be at least hundreds of millions of years, and whose plant and animal inhabitants had evolved during those years from ever more primitive ancestral forms. Man, instead of being God's highest and most favored Creation, was reduced to being a direct biological descendent, in common with all other members of the animal kingdom, of the long animal evolutionary chain.

Our attention will focus upon this second intellectual revolution, and particularly upon the role and significance of one of its major philosophical tenets, the so-called Principle of Uniformity. In examining current or recent English-language textbooks of geology, one will find statements to the effect that the Principle of Uniformity constitutes the foundation of the whole subject of historical geology. In informal discussions among geologists, the question frequently arises whether it is possible for a science such as geology to develop "laws" comparable to the well-known "laws" of physics and chemistry. In reply to such questions, the Principle of Uniformity is most often proposed as a geologic example of a fundamental law or principle of an importance comparable to the major laws of physics.

If the question be asked, however, just what, precisely, *is* the Principle of Uniformity, a variety of nonequivalent answers such as the following is likely to be received[1]:

(1) The present is the key to the past.

(2) Former changes of the earth's surface may be explained by reference to causes now in operation.

(3) The history of the earth may be deciphered in terms of present observations, on the assumption that physical and chemical laws are invariant with time.

(4) Not only are physical laws uniform, that is invariant with time, but the events of the geologic past have proceeded at an approximately uniform rate, and have involved the same processes as those which occur at present.

Despite the variety of its definitions, the fact remains that the use of the Principle of Uniformity has been fundamental in the working out of the history of the earth, and in the evolution of geologic science itself. The object of this inquiry is to re-examine this principle in the light of present scientific knowledge, with the view of determining the extent of its validity.

An understanding of this principle can best be achieved in the historical context in which it originated. In 1785, on March 7 and April 4, before two successive meetings of the Royal Society of Edinburgh, James Hutton presented a paper entitled *Theory of the Earth; or an Investigation of the Laws observable in the Composition, Dissolution and Restora-*

---

[1] For a comprehensive review of the different formulations of the Principle of Uniformity *see* Hooykaas, 1959.

tion of Land upon the Globe. In this paper (1788), in outlining the subject to be treated, Hutton stated (p. 209–210):

"When we trace the parts of which this terrestrial system is composed, and when we view the connection of those several parts, the whole presents a machine of a peculiar construction by which it is adapted to a certain end. We perceive a fabric, erected in wisdom, to obtain a purpose worthy of the power that is apparent in the production of it. . . . We shall thus also be led to acknowledge an order, not unworthy of Divine wisdom, in a subject which, in another view, has appeared as the work of chance, or as absolute disorder and confusion."

The purpose of this machine, Hutton continued, is to produce an habitable world for living forms. He then pointed out that the whole system is composed of three different bodies: (1) a solid body of earth, (2) an aqueous body of sea, and (3) an elastic fluid of air. The interaction of these three bodies forms the theory of the machine which he proposed to examine.

He then stated that soil is nothing but the materials collected from the destruction of the solid rock and traced the cycle of weathering to produce soil, the erosion and transportation of soil by running water, and finally its deposition in the sea in stratified deposits. He noted that the fossils in limestones and marble are evidences of the marine deposition of these rocks and estimated that nine tenths of the visible parts of the earth consist of strata originally deposited in the depths of the ocean. From this he concluded that a mechanism must exist to elevate the bottom of the sea above sea level and produce the observed consolidation of the sediments.

Hutton pointed out that this consolidation had to occur outside the domain of observation. This, he postulated, must have occurred at the bottom of the ocean in response to heat, causing actual fusion of the sediments. The same heat caused expansion of the rock and its uplift above sea level.

Observation of angular unconformities and stratigraphic and structural evidence convinced Hutton that the cyclical sequence of erosion of sedimentary and other rocks now above sea level, deposition in the ocean as new sediments, and renewed uplift had been operative since the earliest geologic time for which evidence exists, a conclusion which he masterfully stated (p. 304):

"But if the succession of worlds is established in the system of nature, it is in vain to look for any thing higher in the origin of the earth.

The result, therefore, of our present enquiry is, that we find no vestige of a begin-ning,—no prospect of an end."

Hutton's philosophical premises are set forth in the following two significant passages:

"In examining things present, we have data from which to reason with regard to what has been; and, from what has actually been, we have data for concluding with regard to that which is to happen here after. Therefore, upon the supposition that the opera-tions of nature are equable and steady, we find, in natural appearances, a means of concluding a certain portion of time to have necessarily elapsed, in the production of those events of which we see the effects." (p. 217)

"But how shall we describe a process which nobody has seen performed, and of which no written history gives any account? This is only to be investigated, *first*, in examining the nature of those solid bodies, the history of which we want to know; and, *2ndly*, In examining the natural operations of the globe, in order to see if there now actually exist such operations, as, from the nature of the solid bodies, appear to have been nec-essary in their formation." (p. 219)

Hutton further stated (p. 285):

"Therefore, there is no occasion for having recourse to any unnatural supposition of evil, to any destructive accident in nature, or to the agency of any preternatural cause, in explaining that which actually appears."

Concerning plants and animals, Hutton speculated that the same species of marine animals as exist now must have existed throughout geologic time (p. 291), "The animals of the former world must have been sustained during indefinite successions of ages"; Man, however, was of recent origin. He also pointed out that fossil wood and coal are evidences of former plant life, and that animal life depends upon plant life.

Concerning the length of geologic time he stated (p. 294), ". . . in nature, we find no deficiency in respect of time, nor any limitation with regard to power." For an idea of the length of time from the deposition of rocks now forming the land, to the present, Hutton pointed out that no perceptible changes in coastlines occur during a lifetime and that comparisons between recent and ancient maps show no significant changes since Greek and Roman times. He then con-cluded (p. 301):

"To sum up the argument, we are certain, that all the coasts of the present continents are wasted by the sea, and constantly wearing away the whole; but this operation is so extremely slow, that we cannot find a measure of the quantity in order to form an estimate. Therefore, the present continents of the earth, which we consider as in a state of perfection, would, in the natural operations of the globe, require a time in-definite for their destruction."

This classical paper of Hutton has been dwelt upon at some length because it represents one of the earlier formulations of what later became known as the Principle of Uniformity. Hutton's method of investigating the geologic past consisted of studying the evidence contained in the rocks themselves, interpreted in terms of processes such as weathering and erosion, transportation, and deposition of sediments in the ocean, and, finally, renewed uplift. Coupled with this was the explicit rejection of any form of supernaturalism.

Among the most important conclusions reached was that concerning the immensity of geologic time, and the lack of any evidence of a beginning or ending of the sequence.

Hutton's inferences which have proved untenable are the fusion of sediments as a means of consolidation and the cataclysmic rate at which he considered the uplifts to have taken place. Also, although he observed the occurrence of marine fossils in sedimentary rocks, he made no special study of these. Consequently his inference that the same species (except man) have always existed was not based on the evidence available.

Despite Hutton's disavowal of "the agency of any preternatural cause," a strong theological flavor permeates his entire treatise. This is to be noted in his apparent conviction that the working of the "machine," which he described, was evidence of a preordained plan "not unworthy of Divine wisdom" to render the earth a suitable abode for its biological inhabitants, particularly man.

Hutton (1726–1797) was a physician by education (but not by practice), a gentleman farmer, and a member of a brilliant scientific group associated with the University of Edinburgh. He was a contemporary of the French chemist, Lavoisier (1743–1794), and was familiar with the chemistry of his day, including the newly established principle of conservation of matter. The properties of energy were yet unknown since the first and second laws of thermodynamics were not formulated until about 60 and 75 years later.

With regard to the knowledge of geology at the time, there had been three centuries of largely sporadic writing by western European authors in which the principal subject of concern had been the origin of stratified rocks and their contained fossils. On this subject, the expressed views (Geikie, 1905, p. 50–73; Lyell, 1875, Chapt. 3, p. 27–66) ranged from the rational inferences of Leonardo da Vinci (1452–1519), Fracastoro (1483–1553), Nicolaus Steno (1613–1683), Robert Hooke (1625–1703), and others, that stratified rocks with their contained shell-like forms found in inland and even in mountainous localities had originated as sediments deposited in the sea, to the opposing views that the fossils in such rocks were nonorganic in origin and represented "sports of nature" created by some mysterious "plastic force," or else were due to the influence of the stars.

By the eighteenth century there was a convergence toward the view that these rocks and their contained fossils were the result of the Biblical flood on an earth which had been divinely created some 6000 years previously and whose subsequent history had been in accordance with Biblical chronology. The dominance of the latter views in seventeenth- and eighteenth-century British geology is readily understandable when it is considered that most of the British writers on geological subjects during that period were orthodox members of the established churches, and that the preponderance of them bore the title of "Reverend" (Gillispie, 1959).

Space here will permit only the briefest account of the geological developments during the 42 years between the publication of Hutton's *Theory of the Earth* in which the Principle of Uniformity was clearly stated and applied, and the publication of Charles Lyell's *Principles of Geology* for which the Principle of Uniformity was taken as the theoretical foundation. The history of this period, however, has been treated in great detail by various authors (Geikie, 1905; Adams, 1938; Gillispie, 1959), on whose writings the following synopsis is based.

In view of the disparity between Hutton's views and those of most of his contemporaries concerning the nature of geological processes, and, particularly, their bearing on the history of the earth and immensity of geologic time, it was almost inevitable that conflicts should arise. The first of these, which raged until about 1820, was that which came to be known as the Neptunist-Vulcanist controversy.

In 1775, Abraham Gottlob Werner (1749–1817) became Inspector and Teacher of Mining and Mineralogy at the Academy of Mines in Saxony (Adams, 1938, p. 209–249; Geikie, 1905, p. 201–240). Werner was perhaps the leading mineralogist of his time, and, as a teacher, he appears to have exercised an almost hypnotic influence over his students. Within a short time, he attracted students from all over Europe and became recognized as the foremost teacher of geological subjects of his generation. In addition to mineralogy, Werner developed what he called "geognosy" (Adams, 1938, p. 207–227; Geikie, 1905, p. 201–236). Although he had never traveled far from Freiberg, he generalized his observations of the rocks in that vicinity into what he supposed to be a worldwide system. In order of descending age, the rocks of the earth, according to Werner (Adams, 1938, p. 217–227), could be classified into five principal systems—Primary, Transitional, Flötz, Alluvial, and Volcanic.

Initially the earth had been covered by a universal ocean rising above the crests of the highest mountains. Out of this universal ocean the Primary rocks, comprising granite and associated crystalline rocks, had been formed by chemical precipitation. As this ocean began to subside, the rocks of the later sequences, a series comprising diminishing proportions of chemical precipitates and increasing mechanical deposits, were formed.

This fanciful and incredible scheme of supposed geology and geological history was taught by Werner to his students, who in turn went out into the world zealous to establish the master's system in whatever part of the world they happened to be. This geological dogma had the virtue of being in approximate agreement stratigraphically with the rocks in the vicinity of Freiberg; it had the virtue of simplicity and easy comprehensibility; it had an implied time scale which was compatible with Biblical chronology; and, finally, it had the sanction of authority—the professor himself had said that it was so!

Werner's system was seized upon by Hutton's critics as the geological basis for his refutation. Because, according to this scheme, all rocks (including granite and basalt) except recent volcanic lavas, were presumed to be chemical or mechanical deposits in the universal ocean, the followers of Werner became known as Neptunists. Because Hutton had postulated subterranean heat as a mechanism for raising the continents, the defenders of the Huttonian system were known as Vulcanists.

The attacks began as early as 1793 when the Irish chemist, Richard Kirwan, with no personal knowledge of geology, coupled the Wernerian system to a literal interpretation of the Bible in support of the charge of atheism against Hutton. In response, Hutton expanded his original paper into a two-volume treatise entitled *Theory of the Earth with Proofs and Illustrations*, which was published in Edinburgh in 1795, 2 years before his death.

In 1797 and 1799 (Gillispie, 1959, p. 49–56), Kirwan's renewal of the attack brought Hutton's friend John Playfair, professor of mathematics at the University of Edinburgh, to his defense with the book

Illustrations of the Huttonian Theory of the Earth, which was published in 1802.

The situation was further complicated when a Scottish student of Werner, Robert Jameson (1774–1854), was appointed professor of natural history at the University of Edinburgh in 1804 (Gillispie, 1959, p. 66–69) and promptly became the master's principal advocate in Great Britain. Jameson organized the Wernerian Natural History Society (with himself as permanent president) whose Memoirs afforded a vehicle for the Neptunist views. He also wrote a *System of Mineralogy*, volume III of which (1808), subtitled *Oryctognosie, Mineralogy, Geognosie, Mineral Geography and Economic Mineralogy*, was an exposition of the Wernerian system.

Throughout its history the Neptunist-Vulcanist controversy was a mixture of geology and theology, and the central issue was the Huttonian versus the Biblical interpretation of geologic history. Gradually, as geologic work progressed, the evidence for great lengths of time became so convincing that even the Neptunists found it necessary to amend the Biblical interpretation whereby Biblical "days" came to mean geologic periods. Then, on the purely geological side, the fundamental issue of whether basalt is an igneous rock or a Wernerian aqueous precipitate, was finally resolved by overwhelming evidence in favor of the Vulcanist view.

No sooner had the Neptunist-Vulcanist controversy begun to wane than a new anti-Uniformitarian storm blew up. In 1812, Georges Cuvier (1769–1832), one of the leading paleontologists of France, published, as a preface to a paleontological treatise, his own comprehensive theory of the earth (Adams, 1938, p. 263–268; Geikie, 1905, p. 363–376; Gillispie, 1959, p. 98–102). This theory, in essence, was that during the long history of the earth, the land had been repeatedly invaded by the sea or by transient floods. These invasions did not come gradually. On the contrary, the majority of the cataclysms that produced them were sudden, as attested by the dislocations, shifting, and overturning of the strata. As a result of each of these cataclysms, the animals and plants of both land and sea were largely destroyed. This, in turn, led to the necessity of a series of successive biological creations in which animals at each successive stage were more advanced than in the one preceding. Furthermore, in stratigraphic successions, the order of fossils is: fish, amphibia, reptilia, mammalia, which also agrees with the order of Creation.

This catastrophic view of geologic history achieved almost universal acceptance among British geologists during the 1820's. By this time it was fairly generally admitted that geological history was much longer than had been assumed by all except the Huttonians. Furthermore, geologic and Biblical evidence were in complete agreement on the one most recent catastrophe, the Mosaic flood. This last became such a firm geological date that geological history was readily divided into the Ante-Diluvial and Post-Diluvial eras; as Gillispie (1959, p. 91) has remarked, "The flood itself was not a speculative matter in 1820."

One of the more solid geological achievements during this period occurred largely without regard to the contemporary geological controversies. This was the work of William Smith (1769–1839) who was a land surveyor engaged in work on canals and drainage projects in various parts of England (Geikie, 1905, p. 381–396). As a by-product of his work, Smith discovered about 1794 that separate strata had distinctive suites of fossils by which they could be recognized. Following up this interest, he traced out and mapped the principal stratigraphic systems of England and, in 1815, published the first geological map of the country.

The period of confusion following the Hutton and Playfair publications was brought to a termination by Charles Lyell's *Principles of Geology* (1830–1833). Lyell (1797–1875), the son of a well-to-do Scottish family, was born in the family mansion, Kinnordy, in Forfarshire, Scotland, in 1797, the year of Hutton's death (Bonney, 1895; Bailey, 1963). From 1816–1819 he attended Oxford University where he prepared for a career in law. During the period 1820–1830 he finished his law studies, was admitted to the bar, and practiced law for 2 years, but apparently without enthusiasm.

As a student at Oxford, Lyell had attended the lectures of William Buckland, the most influential British geologist of that period (Gillispie, 1959, p. 98–120). His interest in geology soon became so overwhelming that he devoted all available time to geological field observations, the study of collections in museums, visits with prominent geologists, and extensive geological excursions in the United Kingdom, and in France, Switzerland, Italy, and Spain.

Until about 1825, Lyell's geological interpretations were largely in accord with the Catastrophic-Diluvial school represented by Buckland and most other contemporary British geologists. As his own field

233

studies were extended into more and more areas, however, he began to be convinced that the postulated catastrophic events could not be supported by field evidence.

By 1827 he had begun to consider writing a book setting forth his new views and countering the prevailing Catastrophic-Diluvial opinion. He began writing about 1828 or 1829 and in 1830 published volume I of the new geological treatise, *Principles of Geology, Being an Attempt to Explain the Former Changes of the Earth's Surface by Reference to Causes Now in Operation*. Volume II appeared in 1832, and volume III in 1833.

The work, an immediate success, went through several printings. In 1832, while volumes II and III were still being published, volume I had already been revised and re-issued.

Except for the brief period 1831–1833, when he was persuaded to accept the newly created Chair of Geology at King's College, London, Lyell devoted his life to his geological studies and to the repeated revisions of the *Principles*. The twelfth and last edition, published in 1875, was being revised when Lyell died.

As its title implies, Lyell's *Principles of Geology* was devoted almost exclusively to the deciphering of the history of the earth on the basis of his own modification of the Huttonian thesis that the former changes of the earth's surface may be explained by reference to causes which are now in operation. Since this involved the assumption of "uniform" operations in geologic processes, the Huttonian-Lyellian philosophy became known as Uniformitarianism, and its basic premise, the Principle of Uniformity.

Because Lyell's entire treatise is an exposition of the uniformitarian philosophy, statements on different aspects of this philosophy are widely distributed throughout the work. The first four chapters of volume I are devoted to historical review of the development of geologic thought. In Chapter V, *Review of the causes which have retarded the progress of Geology*, he stated (p. 75–76):

"We have seen that, during the progress of geology, there have been great fluctuations of opinion respecting the nature of the causes to which all former changes of the earth's surface are referrible. The first observers conceived that the monuments which the geologist endeavours to decipher, relate to a period when the physical constitution of the earth differed entirely from the present, and that, even after the creation of living beings, there have been causes in action distinct in kind or degree from those now forming part of the economy of nature. These views have been gradually modified, and some of them entirely abandoned in proportion as observations have been multiplied, and the signs of former mutations more skilfully interpreted. Many appearances, which for a long time were regarded as indicating mysterious and extraordinary agency, are finally recognized as the necessary result of the laws now governing the material world; and the discovery of this unlooked for conformity has induced some geologists to infer that there has never been any interruption to the same uniform order of physical events. The same assemblage of general causes, they conceive, may have been sufficient to produce, by their various combinations, the endless diversity of effects, of which the shell of the earth has preserved the memorials, and, consistently with these principles, the recurrence of analogous changes is expected by them in time to come.

Whether we coincide or not in this doctrine, we must admit that the gradual progress of opinion concerning the succession of phenomena in remote eras, resembles in a singular manner that which accompanies the growing intelligence of every people, in regard to the economy of nature in modern times. In an early stage of advancement, when a great number of natural appearances are unintelligible, an eclipse, an earthquake, a flood, or the approach of a comet, with many other occurrences afterwards found to belong to the regular course of events, are regarded as prodigies. The same delusion prevails as to moral phenomena, and many of these are ascribed to the intervention of demons, ghosts, witches, and other immaterial and supernatural agents. By degrees, many of the enigmas of the moral and physical world are explained, and, instead of being due to extrinsic and irregular causes, they are found to depend on fixed and invariable laws. The philosopher at last becomes convinced of the undeviating uniformity of secondary causes, and, guided by his faith in this principle, he determines the probability of accounts transmitted to him of former occurrences, and often rejects the fabulous tales of former ages, on the ground of their being irreconcilable with the experience of more enlightened ages."

Thus, after an extensive summary of earlier views which had come to be recognized as untenable, Lyell's own view was embodied in the statement, "The philosopher at last becomes convinced of the undeviating uniformity of secondary causes . . ."

The term "secondary causes" pertains, in the philosophy and theology of that period, to events subsequent to the "First Cause" which ordinarily is synonymous with "Divine Creation." Hence, Lyell's "undeviating uniformity of secondary causes" appears to be equivalent to asserting the permanency of physical laws and a rejection of any form of supernaturalism or interferences by Divine Providence in post-Creational geological phenomena.

Lyell further enlarged on his views by the remarks:

"Our estimate, indeed, of the value of all geological evidence, and the interest derived from the investigation of the earth's history, must depend entirely on the degree of confidence which we feel in regard to the permanency of the laws of nature. Their immutable constancy alone can enable us to reason from analogy, by the strict rules of induction, respecting the events of former ages, or, by a comparison of the state of things at two distinct geological epochs, to arrive at the knowledge of general principles in the economy of our terrestrial system." (p. 165)

"Those geologists who are not averse to presume that the course of Nature has been uniform from the earliest ages, and that causes now in action have produced the former changes of the earth's surface, will consult the ancient strata for instruction in regard to the reproductive effects of tides and currents." (p. 311)

Concerning a beginning and an ending in the history of the earth, Lyell essentially accepted the Huttonian view that it is fruitless to talk of either. He dismissed the earlier cosmological discussions as impediments to understanding the history of the earth.

Regarding the rates of geological processes in the past as compared with those of the present, he stated in the second edition of volume I (1832, p. 73):

"There can be no doubt, that periods of disturbance and repose have followed each other in succession in every region of the globe, but it may be equally true, that the energy of the subterranean movements has been always uniform as regards the *whole earth*. The force of earthquakes may for a cycle of years have been invariably confined, as it is now, to large but determinate spaces, and may then have gradually shifted its position, so that another region, which had for ages been at rest, became in its turn the grand theatre of action."

We infer from these quotations, and from the treatise as a whole, that Lyell essentially accepted Hutton's view that the phenomena displayed by the rocks of the earth may be entirely accounted for by geologic processes which are in operation now, on the assumption that the laws of nature throughout geologic history are invariant with time. Like Hutton's, Lyell's view of the history of the earth involves a panorama of events extending indefinitely into a past of essentially unlimited geologic time.

Lyell's greatest difficulty in the application of the Principle of Uniformity arose in connection with the plant and animal life of the earth, evidently a source of some embarrassment. He rejected the Lamarckian theory of gradual development in favor of permanency of species and cited evidence for the recent origin of man. After reviewing this problem for several chapters in volume II (1832), Lyell's principal conclusion regarding species was as follows (p. 65):

"6thly. From the above considerations, it appears that species have a real existence in nature, and that each was endowed, at the time of its creation, with the attributes and organization by which it is now distinguished."

Concerning the origin of man, Lyell (v. I, 1830, p. 156) commented:

"But another, and a far more difficult question may arise out of the admission that man is comparatively of modern origin. Is not the interference of the human species, it may be asked, such a deviation from the antecedent course of physical events, that the knowledge of such a fact tends to destroy all our confidence in the uniformity of the order of nature, both in regard to time past and future?"

After wrestling with the problem for several pages, Lyell finally reached the somewhat lame conclusion (p. 164):

"To this question we may reply, that had he previously presumed to dogmatize respecting the absolute uniformity of the order of nature, he would undoubtedly be checked by witnessing this new and unexpected event, and would form a more just estimate of the limited range of his own knowledge, and the unbounded extent of the scheme of the universe. But he would soon perceive that no one of the fixed and constant laws of the animate and inanimate world was subverted by human agency, and that the modifications produced were on the occurrence of new and extraordinary circumstances, and those not of a *physical*, but a *moral* nature."

It is thus clear that although Lyell had gone far beyond his contemporaries in emancipating himself from such theological tenets as Providential interference with terrestrial events on the inorganic side and had succeeded completely in eliminating Biblical chronology as a basis for geological history, he was still frustrated when confronted with biological phenomena. To account for the origin of organic species, he found it necessary to resort to Divine Creation, in complete contradiction to his exclusion of supernatural considerations in other respects. His rescue from this difficulty came only when Charles Darwin's *On the Origin of Species* was published in 1859.

Charles Darwin (1809–1882) attended the University of Cambridge from 1827–1831 where he studied geology under Adam Sedgwick and botany under John Stephens Henslow (de Beer, 1964). In 1831, upon Henslow's recommendation, he signed on as the Naturalist on the exploration ship H.M.S. BEAGLE, for a round-the-world voyage which lasted from December 27, 1831, until October 2, 1836. Upon Henslow's recommendation that it should be read, but on no account believed, one of the few books which Darwin took with him was the first volume of Lyell's *Principles of Geology* which had been published only the preceding year. As Darwin subsequently related in his *Voyage of the Beagle* (1839), Lyell's book was fairly digested within the first few days of the voyage and formed the foundation for his own geological work throughout the expedition. In Chapter IX of his master work *On the Origin of Species* (1859, p. 282), Darwin stated, "He who can read Sir Charles Lyell's grand work on the PRINCIPLES OF GEOLOGY, which the future historian will recognize as having produced a revolution in natural science, yet does not admit how incomprehensibly vast have been the past periods of time, may at once close this volume."

It is clear, therefore, that Lyell's work profoundly influenced Charles Darwin in his progression toward his major biological synthesis. In effect, what Darwin did was to develop paleontology from the point where Lyell had left it. By the rejection of Lyell's residual supernaturalism where organisms were concerned, and by an extension of Lyell's Principle of Uniformity to the plant and animal kingdoms, the theory of evolution was an almost inevitable consequence.

In 1859, when *On the Origin of Species* was first published, Lyell was 62 years old. There is no greater measure of the man's intellectual integrity than the fact that he was one of the first to accept the Darwinian theory of evolution and to promptly develop it further than Darwin had by applying it explicitly to the origin of man in *The Geological Evidences of the Antiquity of Man with Remarks on Theories of the Origin of Species by Variation* (1863).

In addition, in 1866, 13 years after the publication of the ninth edition, Lyell published the tenth edition of *Principles of Geology* in which the ninth chapter, on the progressive development of organic life, was entirely re-written, to conform to the Darwinian theory.

Up until the 1860's, the opposition to the Huttonian-Lyellian Uniformitarianism had been largely a rearguard action—often a very powerful one—of the defenders, geological as well as clerical, of the prerogatives of Divine Providence to set aside the laws of nature and to interfere with terrestrial activities in any arbitrary manner. At about this time a criticism arose from a entirely different quarter. In 1862 William Thomson published two related papers: (1) *On the Age of the Sun's Heat* (1862a), and (2) *On the Secular Cooling of the Earth* (1862b), in which entirely new relationships were introduced into the consideration of terrestrial and solar phenomena.

William Thomson (later Lord Kelvin, 1824–1907) was Professor of Natural Philosophy at the University of Glasgow and one of the leading British physicists of the nineteenth century. He was a young man when James P. Joule (1818–1889) was performing his epoch-making experiments on the mechanical equivalent of heat during the 1840's and was one of Joule's strongest supporters in the establishment of the principle of conservation of energy, or the first law of thermodynamics. Through the 1850's, Thomson was one of the leading researchers in further thermodynamic investigation and was one of the formulators of the second law of thermodynamics. He was also responsible for the concept of absolute zero of temperature and for the absolute scale of temperature, now known, in his honor, as the Kelvin scale. Thomson was also fascinated with the mathematical theory of heat conduction introduced in 1822 by Joseph Fourier (1768–1830) in his classical treatise, *Théorie Analytique de la Chaleur.*

This was the background from which Thomson wrote the two papers cited. In the second, *On the Secular Cooling of the Earth* (*in* Thomson and Tait, 1890, pt. II, p. 468–469), he presented his basic thesis as follows:

"(*a*.) For eighteen years it has pressed on my mind, that essential principles of Thermo-dynamics have been overlooked by those geologists who uncompromisingly oppose all paroxysmal hypotheses, and maintain not only that we have examples now before us, on the earth, of all the different actions by which its crust has been modified in geological history, but that these actions have never, or have not on the whole, been more violent in past time than they are at present.

(*b*.) It is quite certain the solar system cannot have gone on, even as at present, for a few hundred thousand or a few million years, without the irrevocable loss (by dissipation, not by *annihilation*) of a very considerable proportion of the entire energy initially in store for sun heat, and for Plutonic action. It is quite certain that the whole store of energy in the solar system has been greater in all past time than at present; but it is conceivable that the rate at which it has been drawn upon and dissipated, whether by solar radiation, or by volcanic action in the earth or other dark bodies of the system, may have been nearly equable, or may even have been less rapid, in certain periods of the past. But it is far more probable that the secular rate of dissipation has been in some direct proportion to the total amount of energy in store, at any time after the commencement of the present order of things, and has been therefore very slowly diminishing from age to age."

Because of the geothermal gradient, Thomson's argument continued, the earth is losing heat by conduction. Since there was no known source of this heat except an earth which must have been originally much hotter—presumably molten—than at present, then it followed by means of the Fourier theory, that it should be possible to calculate within broad limits the time that it had taken for the earth to reach its present geothermal gradient since first becoming solidified. The result obtained by Thomson was that this period could not have been less than 20,000,000 nor more than 400,000,000 years. Later, based on a lower melting temperature of 7000°, this larger figure was reduced to 98,000,000 years.

Thomson particularly challenged the Uniformitarian school, whose views, he pointed out, were equivalent to an earth operating as a perpetual-motion mechanism. Specifically, he cited a passage from the 1853 edition of *Principles of Geology* in which Lyell had postulated that the earth's heat could be generated by internal chemical reactions which could then be undone electrolytically by thermo-electric currents. Concerning this supposed mechanism, Thomson commented (p. 471) ". . . thus the chemical action and its heat continued in an endless cycle, violates the principles of natural philosophy in exactly the same manner, and to the same degree, as to believe that a clock constructed with a self-winding movement may fulfil the expectations of its ingenious inventor by going for ever."

Thomson also challenged the Uniformitarian view that the past diastrophic events of the earth were of the same magnitude and occurred at the same rates as those of the present, in view of the fact that throughout geologic history the earth has been an energy-dissipating system. Concerning this, he stated (p. 472):

"It would be very wonderful, but not an absolutely incredible result, that volcanic action has never been more violent on the whole than during the last two or three centuries; but it is as certain that there is now less volcanic energy in the whole earth than there was a thousand years ago, as it is that there is less gunpowder in a 'Monitor' after she has been seen to discharge shot and shell, whether at a nearly equable rate or not, for five hours without receiving fresh supplies, than there was at the beginning of the action. Yet this truth has been ignored or denied by many of the leading geologists of the present day, because they believe that the facts within their province do not demonstrate greater violence in ancient changes of the earth's surface, or do demonstrate a nearly equable action in all periods."

In the first paper, *On the Age of the Sun's Heat*, Thomson followed a similar line of reasoning. His opening statement was, "The second law of Thermodynamics involves a certain principle of *irreversible action in nature*." He pointed out that the sun is radiating energy at an enormous rate, the most plausible source of which is the infalling of meteorites and gravitational contraction, the theory of which had been developed by Hermann von Helmholtz (1821–1894).

After considering the various magnitudes involved in the sun's activities, Thomson concluded that the sun probably has not illuminated the earth for 100,000,000 years, and almost certainly not for 500,000,000 years. This, however, was qualified by the following concluding statement (*in* Thomson and Tait, 1890, pt. II, p. 494):

"As for the future, we may say, with equal certainty, that inhabitants of the earth cannot continue to enjoy the light and heat essential to their life, for many million years longer, unless sources now unknown to us are prepared in the great storehouse of creation."

This is one of the few cautionary statements by Thomson on record with regard to this subject.

The next 35 years was a period of intermittent strife between Thomson and the British geologists. In geology textbooks of the 1860's, as later cited by Thomson, many of the authors had gone so far as to postulate unlimited time for the history of the earth. Some geologists, impressed by Thomson's estimates of the earth's age since its cooling enough to sustain life, had attempted to compress geological history within 20 to 40 million years; others, including Darwin, considering the slowness of erosion and deposition and the aggregate thickness of strata of different ages, estimated the length of geologic time since the beginning of the Paleozoic era to be of the order of magnitude of hundreds of millions of years.

During this period an almost complete impasse developed because the natural-history estimates of the age of the earth and the physical estimates of Thomson—each within its own premises accurate to about an order of magnitude—differed from one another by a factor of 100–1000.

This situation culminated in 1899, when Lord Kelvin's address on *The Age of the Earth as an Abode Fitted for Life* was published in *Science*. In this address, Kelvin quoted extensively from geological works written from about 1850 to 1870, pertaining to the length of geologic time. These included a statement by Charles Darwin from the first edition (1859) of *On the Origin of Species* that "In all probability a far longer period than 300,000,000 years has elapsed since the latter part of the secondary period."

Statements from other authors were quoted, including the following:

". . . the student should never lose sight of the element TIME, *an element to which we can set no bounds in the past* . . ." (1859, *Advanced Student's Text-Book of Geology*)
"The time required for such a slow process to effect such enormous results must, of course, be inconceivably great." (Jukes, 1862, *The Student's Manual of Geology*, p. 290)

In addition, Kelvin cited a conversation with Andrew Ramsay in 1867 during which he had asked whether Ramsay thought geologic time could be as long as 10^9 years, or 10^{10}. To both questions Ramsay had replied, "Certainly I do."

Kelvin then renewed his attack on the "doctrine of eternity and uniformity" of Hutton, Playfair, and Lyell as representing a system of perpetual motion. Finally, he reviewed his own cumulative studies of the preceding half century. These involved the assumption of an originally white-hot molten earth and its subsequent cooling to its present temperature. He mentioned his own earlier estimates of the time since the earth had solidified, and then cited an 1893 paper by Clarence King, in which King, using measurements of the thermal properties of rocks by Carl Barus and the Kelvin premises, had obtained a figure of 24 million years as the maximum age of the earth. Kelvin stated that his own recent calculations were in substantial agreement with this figure.

In outlining the history of the earth from its supposed consolidation, Kelvin noted that it was doubtful if there could have been any free oxygen in the atmosphere initially. Within a century after consolidation, however, the earth would have been suitable for plant life, and only a few thousand years (or possibly hundreds) of plant life would have been required to produce oxygen for animal life.

Concerning the mode of origin of plant and animal life, Kelvin stated (p. 711):

"Mathematics and dynamics fail us when we contemplate the earth, fitted for life but lifeless, and try to imagine the commencement of life upon it. This certainly did not take place by any action of chemistry, or electricity, or crystalline grouping of molecules under the influence of force, or by any possible kind of fortuitous concourse of atoms. We must pause, face to face with the mystery and miracle of the creation of living creatures."

Obviously the physicist Kelvin had not emancipated himself from the replacement of the laws of nature by Divine Providence.

Later, during the same year, *Science* published a reply to the Kelvin address by the American geologist, T. C. Chamberlin (1843–1928). Chamberlin had spent many of his earlier years studying the geology of Wisconsin and the neighboring states where one of his major interests had been the Pleistocene glaciation, which, in accordance with the views expressed by Kelvin and others, would represent a late stage in the earth's cooling before final refrigeration.

However, the Pleistocene glaciation did not exhibit a monotonic cooling, as might have been expected, but alternate periods of cooling and warming, as evidenced by the successively younger tills that resulted from advances and retreats of the glaciers, with lengthy interglacial intervals. The problem was still further complicated by the discoveries of such earlier continental glaciations during both the Paleozoic and Proterozoic eras.

These observations had led Chamberlin to inquire into the sources of the opinion that the earth had once been a molten white-hot body now cooling off in the manner postulated by Kelvin. At the time Chamberlin wrote his reply to Kelvin he had already been engaged in this critical inquiry for some 10 years.

In the reply, Chamberlin expressed the indebtedness and gratitude of the geological profession to Lord Kelvin for reminding the geologists that unlimited geological time was not a tenable hypothesis. He acknowledged that during the first half of the century, when more sober modes of interpreting geological data were struggling to displace the cataclysmic extravagances of more primitive times, it was not strange that there should have arisen, as a material outgrowth of the contest, an ultra-Uniformitarianism which demanded for the evolution of the earth an immeasurable length of time. He mentioned, however, that "there were other camps in Israel even then." There were the ultra-Catastrophists as well as the ultra-Uniformitarians.

It was Chamberlin's opinion that "The great body of serious geologists have moved forward neither by the right flank nor by the left, but on median lines." These lines, he thought, had lain ". . . rather in the field of a qualified uniformitarianism than in the field of catastrophism," and the body of competent geologists of Chamberlin's time were probably more nearly disciples of Hutton, Playfair, and Lyell than of their opponents.

With regard to Kelvin's reasoning, Chamberlin warned that a physical deduction which postulates excessively short geological history may as easily lead to false views as did the reckless license of earlier times, and added (p. 890):

"The fascinating impressiveness of rigorous mathematical analysis, with its atmosphere of precision and elegance, should not blind us to the defects of the premises that condition the whole process. There is, perhaps, no beguilement more insidious and dangerous than an elaborate and elegant mathematical process built upon unfortified premises."

Chamberlin quoted Kelvin's estimate that the time since consolidation of the earth had been more than 20 and less than 40 million years, and probably nearer 20. He then asked (p. 891),

"Can these definite statements, bearing so much the air of irrefutable truth, be allowed to pass without challenge? What is their real nature and their true degree of certitude when tested respecting their fundamental postulates and their basic assumptions?"

Chamberlin quoted Kelvin's statement that his results were based upon the "very sure assumption" that the solid earth was once a white-hot liquid, and added (p. 891):

"I beg leave to challenge the certitude of this assumption of a white-hot liquid earth, current as it is among geologists alike with astronomers and physicists. Though but an understudent of physics, I venture to challenge it on the basis of physical laws and physical antecedents."

In examining Kelvin's premises concerning the initial white-hot molten earth, Chamberlin pointed out that, according to Kelvin, the heat required for this result was supposed to have been generated by a sudden infall and accumulation of meteorites. Chamberlin admitted that had the earth been formed by such a brief event the heat generated would indeed have been sufficient to produce a white-hot molten earth. He questioned, however, both on physical and astronomical grounds, whether such a sudden infall was likely to have occurred. On the other hand, should the earth have been formed by a slow infall of meteorites over a very long period the heat of fall and impact would largely have been dissipated by outward radiation, so that the earth need never to have acquired a temperature anywhere near the melting point of rocks.

Chamberlin next examined Kelvin's premises regarding the energy of the sun and the length of time during which the sun may have been hot enough to give sufficient radiation to support life on earth. He quoted (p. 12) the Kelvin statement:

"If the consolidation of the earth was finished 20 or 25 million years ago the sun was probably ready, though probably not then quite so warm as at present, yet warm enough to support some kind of vegetable and animal life on the earth."

Concerning this statement Chamberlin added:

"Here is an unqualified assumption of the completeness of the Helmholtzian theory of the sun's heat and of the correctness of deductions drawn from it in relation to the past life of the sun. There is the further assumption, by implication, that no other essential factors entered into the problem. Are these assumptions beyond legitimate question? In the first place, without questioning its *correctness*, is it safe to assume that the Helmholtzian hypothesis of the heat of the sun is a *complete* theory? Is present knowledge relative to the behavior of matter under such extraordinary conditions as obtain in the interior of the sun sufficiently exhaustive to warrant the assertion that no unrecognized sources of heat reside there? What the internal constitution of the atoms may be is yet an open question. It is not improbable that they are complex organizations and the seats of enormous energies. Certainly, no careful chemist would affirm either that the atoms are really elementary or that there may not be locked up in them energies of the first order of magnitude. No cautious chemist would probably venture to assert that the component atomecules, to use a convenient phrase, may not have energies of rotation, revolution, position and be otherwise comparable in kind and proportion to those of a planetary system. Nor would he probably feel prepared to affirm or deny that the extraordinary conditions which reside in the center of the sun may not set free a portion of this energy. The Helmholtzian theory takes no cognizance of latent and occluded energies of an atomic or ultra-atomic nature."

In view of its date, 1899, the foregoing stands as one of the more prophetic statements in the annals of science. The phenomenon of radioactivity had been discovered by Henri Becquerel only 3 years before. The isolation of radium (1902), the discovery that radioactivity involves transmutations of chemical elements (1902), the development of the law of radioactive disintegration (1902) and the measurement of the amount of energy released (1903), the hypothesis of the planetary atom (1911), the working out of the mechanism for the fusion of hydrogen to helium as the source of the energy from the sun (1939), and the achievement of controlled fission (1942) and of uncontrolled fusion in the hydrogen bomb a few years later, were all in the future (Andrade, 1964).

As a result of these subsequent developments, however, the half-century of wide discrepancies between physical estimates of the possible age of the earth and the sun, and the much longer geological estimates, has completely been resolved. The work of Rutherford and Soddy (1902; 1903) on the law of radioactive disintegration, and their partial determination of three radioactive disintegration series, provided the theoretical basis for the subsequent development of radioactive age determinations of geological events. The present estimate by these means that the Paleozoic era began about 600 million years ago is in substantial agreement with the estimate by Charles Darwin (cited by Lord Kelvin in his address) that the time elapsed since the later part of this era must have been about 300 million years.

The anomaly of the source of the earth's heat was largely resolved by Rutherford and Soddy in 1903 by their initial determination of the approximate amount of heat generated during radioactive disintegration. They found that the heat released by the disintegration of 1 gram of radium and its daughter products through five stages of alpha-particle generation amounts to 10^8 gram-calories, whereas the energy released in a typical chemical reaction, such as the combination of oxygen and hydrogen to produce water, is only to about 4000 gram-calories per gram.

The appreciation of these authors for the cosmological implications of their new discoveries is clearly indicated in the concluding paragraph of their paper on *Radioactive Change* (1903, p. 590–591): ·

"All these considerations point to the conclusion that the energy latent in the atom must be enormous compared with that rendered free in ordinary chemical change. Now the radio-elements differ in no way from the other elements in their chemical and physical behaviour. On the one hand they resemble chemically their inactive prototypes in the periodic system very closely, and on the other they possess no common chemical characteristic which could be associated with their radioactivity. Hence there is no reason to assume that this enormous store of energy is possessed by the radio-elements alone. It seems probable that atomic energy in general is of a similar, high order of magnitude, although the absence of change prevents its existence being manifested. . . . It must be taken into account in cosmical physics. The maintenance of solar energy, for example, no longer presents any fundamental difficulty if the internal energy of the component elements is considered to be available, *i. e.* if processes of sub-atomic change are going on."

As has already been noted, these surmises have been confirmed theoretically by Bethe (1939) in his work on the possible hydrogen-to-helium fusion transformation producing the heat of the sun and other stars; they have been confirmed experimentally by the first nuclear pile at Chicago in 1942 (Smyth, 1945) in which controlled fission was achieved, and by the subsequent development of the hydrogen bomb involving the explosive release of energy from the fusion of hydrogen or other light elements.

In looking back at the dissensions between Kelvin and the geologists concerning the age of the earth, what we now principally remember is how enormously wrong Lord Kelvin was. At the same time, with regard to another fundamental aspect of earth history, that is, that the various geological processes, past and present, are all energy-dissipative processes and involve continuous losses of energy by the earth, we tend to forget how right he was.

The energy inputs into the earth's surface environment consist of radiation from outer space—overwhelmingly from sunshine—tidal energy from the kinetic and potential energy of the earth-moon–sun system, and thermal, chemical, and mechanical energy from the earth's interior. In view of the approximate constancy of the earth's surface temperature when averaged over a year or more, and the nearly fixed quantity of stored surface energy, it follows that the outward flux of energy from this system must be very nearly equal to the inward flux. Because of the temperature increase with depth, thermal energy from the outside can penetrate only to shallow depths beneath the earths' surface. Hence, the outward flux must be by means of radiation from the earth into outer space. In this process, solar energy is used principally to produce a continuous circulation of the atmosphere and the oceans which not only degrades the solar energy but also, by erosion, continuously dissipates into heat the potential energy of the earth's topographic configuration.

Thermodynamically, erosion and transportation of sediments are irreversible processes in which the initial mechanical energy is converted by friction into low-temperature heat. This heat is added to the thermal input of the surface environment of the earth which it leaves by long-wave-length radiation. Thus the topography of the earth represents a large reservoir of mechanical energy which the erosional process continuously dissipates. The energy of one orogeny, therefore, can never be used to produce another, since that energy is completely dissipated and discharged from the earth by one cycle of erosion. Hence, after peneplanation of any area of the earth, new mountains can be formed only from a new source of energy from inside the earth. Because there have been repeated orogenies and episodes of vulcanism throughout the earth's history, these can only have occurred at the expense of a continuous diminution of the earth's initial supply of energy.

In addition to the energy lost from the earth by orogenies, there is also the energy transported to the earth's surface from its interior by heat conduction in virtue of the geothermal gradient, and that convected by mass transport resulting from the activities of volcanoes and hot springs.

It is for these reasons, that the Huttonian-Lyellian view of an earth on which the same processes have been and always will be operative, and at about the same rates, is, as Lord Kelvin pointed out, equivalent to a perpetual-motion mechanism and a physical impossibility. Because of its involvement in thermodynamically irreversible processes, the earth history, despite the long timescale, can only be in the long run a unidirectional progression from some initial state characterized by a large store of available energy to a later state in which this energy has been discharged from the earth. In this latter state, if the earth continues in its planetary orbit about the sun, and if the solar energy has not also been exhausted by that time, we may anticipate the continuance of atmospheric and oceanic circulations, but an ultimate cessation of diastrophic and vulcanic activities, with a corresponding permanent peneplanation of the land areas.

The opposite result is obtained when we extrapolate backward in time. Considering that the radioactive isotopes of uranium and thorium have been declining exponentially and producing radiogenic isotopes of lead throughout geologic history, backward extrapolation requires an exponential increase in the amounts of these radioactive materials and of their heat-generation rates. A limit is set to the time to which such extrapolations can be carried by the fact that the disintegration of one atom of uranium-238 produces eventually one atom of the isotope lead-206, and one atom of thorium-232 degrades to one atom of lead-208. Hence the sum of the atoms of lead produced plus the atoms of uranium (or thorium) remaining must represent the number of atoms of uranium (or thorium) present initially.

For any given rock, backward extrapolation cannot be extended beyond the time required for all the radiogenic lead to be restored to its original state as uranium or thorium. This limiting time, in fact, is said to represent the "age" of the rock considered. Such measurements are commonly made on igneous rocks in which the radiogenic lead produced since crystallization from an igneous melt has been trapped inside the crystal structure of the minerals containing the radioactive element. The limiting time so obtained represents, therefore, the time since the consolidation of the igneous rock.

The oldest age found so far for any terrestrial rock identified by this procedure is 3.6 to 3.7 $\times$ 10^9 years. However, similar measurements made on meteorites give a consistent limiting time of 4.5 $\times$ 10^9 years. Numerous rocks have dates of approximately 3.1–3.2 $\times$ 10^9 years, and these rocks are associated with ancient marine sediments. (Personal communication from Thomas W. Stern, Isotope Geology Branch, U. S. Geol. Survey.)

We are thus led to the conclusion that backward extrapolation of the decay of the long-life radioactive isotopes cannot be extended indefinitely. On the assumption that the meteorites of the solar system are products of the same cosmic event which produced the earth and other planets, the time of 4.5 $\times$ 10^9 years before the present appears to be the approximate time at which this event occurred. Because the half-life period for U-238, the common isotope of uranium, is 4.5×10^9 years, and that for Th-232 is 13.9×10^9 years, it would appear that the maximum amounts of these two elements which the earth, as a planet, could ever have contained would have been about 2 and 1.3, respectively, times the earth's present content of these radioactive materials. The rate of heat generation from these sources, with allowance for their relative geochemical abundances, would have been about 1.5 times that of the present.

From such considerations based on present knowledge, we may conclude that the earth originated in an as yet indefinite manner from some cosmic event which occurred about 4.5×10^9 years ago and has been undergoing a unidirectional evolution ever since. In the early stages this must also have involved the origin and development of the oceans (Rubey, 1951) and atmosphere. It also appears unlikely that the earth, during its formation and earliest history, could have afforded an environment in which life in any form would have been possible.

When we trace back the paleontological evidences of life on earth we see the effects of Darwinian evolution in reverse. The forms become simpler and the branches less numerous as the record recedes in time, until finally we find only traces of living organisms in some of the older Precambrian sedimentary rocks. We are thus led to the logical necessity of a spontaneous origin from initially inorganic materials of the primitive oceans and atmosphere of amino acids and other organic-type molecules of progressively increasing complexity. Present evidence indicates (Cloud and Abelson, 1961; Holland, 1962; Rutten, 1962) that the atmosphere of a lifeless, primitive earth was markedly different from that of the present. It probably was highly reducing, consisting principally of water vapor, methane, and ammonia, but devoid of free oxygen. Apparently, such an anoxygenic atmosphere is essential for the chemical steps necessary for the initial stages of evolution from inorganic materials to progressively more complex organic systems. One of the major events in the history of the earth must have occurred, therefore, when organic evolution in an anoxygenic environment had advanced to the stage of photosynthesis and the generation of free oxygen in the atmosphere. This not only made further evolution of the more complex forms of plant and animal life possible, but it also terminated the period during which life could originate from inorganic constituents. Hence, the origin of life appears to have been a self-terminating geologic event which could have oc-

curred only during a limited period in the earth's history. Further biological evolution appears to have occurred by successive elaborations
of the chemical and biological mechanisms developed during the
initial anoxygenic phase.

Succeeding stages in biological evolution of the first order of importance are represented by the transition from an aqueous to a land
habitat, first about Silurian time by plants and shortly thereafter by
animals. It would also appear that once the continents were blanketed
by a protective cover of plants, the rate of subaerial erosion, under
otherwise similar conditions, must have been markedly decreased.
Still another major nonuniformitarian (in the strict Lyellian sense)
biological change was the large series of extinctions which occurred
at the end of the Mesozoic era. This has recently been reviewed by
Bramlette (1965) who has presented evidence indicating that a possible cause of this major biological event may have been a worldwide
impoverishment of the oceans of essential plant nutrients at the time
of the widespread deposition of Cretaceous chalk.

Returning now to the Principle of Uniformity as understood by
Hutton and Lyell, it is appropriate to ask to what degree we may still
regard this principle as valid. In our historical review, we have traced
a somewhat tortuous, but essentially unidirectional, progression
toward emancipation from the idea that so-called natural laws could
be set aside arbitrarily and terrestrial affairs manipulated at will by
the dictates of a Divine Providence. Hutton stated that the deciphering
of the earth's history could be accomplished without "having recourse
to any unnatural supposition of evil, to any destructive accident of
nature, or to the agency of any preternatural cause," yet Hutton could
not refrain from regarding the whole activity of the earth as being the
result of "Divine wisdom" in adapting the earth as an abode for man.

Lyell wrote repeatedly about the dependence of the events of the
earth upon "fixed and invariable natural laws," yet with regard to
the origin of biologic species he was forced to fall back on previously
existing views of Divine Creation of living forms. Only in Darwin
(and his friend Thomas Huxley) do we find an almost complete escape
from the postulates of Divine Providence. In fact, in 1869, when Darwin's friend and earlier collaborator, Wallace, concluded that
". . . some higher intelligence may have directed the process by which
the human race was developed," Darwin, according to his biographer,

de Beer (1964, p. 215), remarked sadly that his friend had deserted science and taken refuge in mysticism.

This first postulate of the Principle of Uniformity, namely, that the laws of nature are invariant with time, is not peculiar to that principle or to geology, but is a common denominator of all science. In fact, instead of being an assumption or an *ad hoc* hypothesis, it is simply a succinct summation of the totality of all experimental and observational evidence.

The second part of the Principle of Uniformity, namely that the events of the geologic past involved essentially the same activities as those occurring on the earth at present, and proceeded at essentially the same rates, with no evidence of a beginning or an ending, rests upon much less secure grounds. As Kelvin pointed out, and we agree, this is equivalent to a perpetual-motion mechanism, and hence is physically impossible. The earth throughout its history must have been an evolving system undergoing a continuous loss of energy.

Despite this, however, for the period of geological history in which Hutton and Lyell were most interested—that since the beginning of the Paleozoic era—the rate of heat generation from the radioactivity of uranium and thorium has declined so slightly that the resulting average rate of thermally induced diastrophism and vulcanism need not have declined perceptibly. Therefore, the assumption of Hutton and Lyell that the events of the geologic past, at least during this period, consisted of the same kinds of activities as those of today, and occurred at about the same rate, is probably a valid approximation. For much longer periods of time, the assumption cannot be regarded as generally valid; neither can it be taken too literally even for recent geological eras. Otherwise, we should have to face the necessity of postulating continental glaciation during all geologic history. Or, conversely, were there no glaciation of any kind at present, we should probably still be embarrassed to find a rational explanation for the phenomena now explained by the hypothesis of continental glaciation.

Is it possible, therefore, that the Principle of Uniformity, having played a strategic role in the development of a valid history of the earth, has by now largely lost its usefulness? Perhaps in answer to this we should consider what are the logical essentials in the deciphering of history, not just geological history, but any kind of history. Because it is impossible for us to observe anything except the present, our interpretations of prior events must necessarily consist of inferences based upon present observations. *History, human or geological, represents our hypothesis, couched in terms of past events, devised to explain our present-day observations.*

What are our assumptions in such a procedure? Fundamentally, they are two:

(1) We assume that natural laws are invariant with time.

(2) We exclude hypotheses of the violation of natural laws by Divine Providence, or other forms of supernaturalism.

These are not arbitrary assumptions, nor are they peculiar to geologic science. Rather, they represent the distilled essence of all human experience, and are common to all sciences. Were the first assumption not valid, it could not be assumed that the freezing temperature of water would be the same tomorrow as it is today, nor that oxygen and hydrogen, which today combine under given conditions to form water, might not in the future, under the same conditions, combine to form alcohol or even sulfuric acid.

The second assumption is actually a corollary of the first, but it requires to be explicitly stated in view of the fact that for centuries the failure to accept this assumption has been one of the principal hindrances to the advancement of scientific understanding.

Thus, we conclude that books and other documents, fragments of pottery, cuneiform tablets, flint tools, and temples, pyramids and similar structures, which were in existence prior to our arrival, have all been the work of man, despite the fact that these postulated past activities have been outside the domain of any possible present-day observations. Having excluded supernaturalism, we draw these conclusions because man is the only known agent capable of producing the effects observed. Similarly, in geology, we conclude the ripple-mark-like forms in folded quartzites were in fact formed by wave action on a sandy beach, or, that sea-urchin-like forms found in inland limestones are indeed the skeletons of sea urchins which lived in a now nonexistent sea, because in neither instance are there other known means of producing these results.

At the same time we must be mindful of the fact that deficiencies in present knowledge will also be reflected in our interpretations of past events. This, in fact, was the principal weakness of the interpretations of geological history by Hutton and Lyell, neither of whom was aware of the limitations imposed by the laws of thermodynamics.

For a contemporary example, it has been only within the last quarter of a century that we have learned of the fact that when more than a few tons of sufficiently rich uranium minerals become concentrated into a small space, spontaneous fissioning and a nuclear explosion will result. It appears highly probable that the same kinds of geologic processes which have produced concentrations of other metallic minerals, may also during geologic time have produced critical concentrations of uranium ores, resulting in nuclear explosions. Were the effects of such an explosion to have been discovered prior to the 1940's, hypotheses to account for them would of necessity have had to be couched in terms of such known processes as volcanic explosions, meteoritic impacts, and the like. Only since the 1940's could the real cause of the observed features have been given.

Historical chronology, human or geological, depends also upon comparable impersonal principles. If one scribes with a stylus on a plate of wet clay two marks, the second crossing the first, another person on examining these marks can tell unambiguously which was made first and which second, because the later event irreversibly disturbs its predecessor. In virtue of the fact that most of the rocks of the earth contain imprints of a succession of such irreversible events, an unambiguous working out of the chronological sequence of these events becomes possible.

During the last five hundred years, the tortuous evolution of geological science has been characterized by a progressive emancipation from the constraints and impediments imposed by assumptions of special Creations and interferences by Divine Providence in geological (and human) affairs. A major part of this emancipation has been accomplished by the employment of the Principle of Uniformity, but this rests upon insecure grounds due in large part to its having been formulated in ignorance of the later-developed laws of thermodynamics.

It has been recounted (Bell, 1937, p. 181) that when the French astronomer Pierre Simon Laplace had completed one of the volumes of his great work, *Mécanique Céleste*, a copy was presented to the Emperor Napoleon. Upon leafing through the volume Napoleon took the author to task for an apparent oversight, "You have written this large book on the system of the world without once mentioning the author of the universe." "Sire," Laplace replied, "I had no need for that hypothesis." It may be that the time has now arrived when geologists too may explicitly declare their lack of necessity for that particular hypothesis, as well as for a vaguely formulated Principle of Uniformity.

References Cited

ADAMS, FRANK DAWSON, 1938, The birth and development of the geological sciences: Baltimore, Md., Williams and Wilkins Co., 506 p. (Reprinted in 1954 by Dover Pub., Inc., New York)

ANDRADE, E. N. DA C., 1964, Rutherford and the nature of the atom: Garden City, N. Y., Anchor Books, Doubleday & Co., Inc., 218 p.

BAILEY, EDWARD, 1963, Charles Lyell: Garden City, N. Y., Doubleday & Co., Inc., 214 p.

BELL, ERIC TEMPLE, 1937, Men of mathematics: New York, Simon & Schuster, 592 p.

BETHE, H. A., 1939, Energy production in stars: Phys. Rev., v. 55, p. 434–456

BONNEY, T. G., 1895, Charles Lyell and modern geology: New York, Macmillan & Co., 224 p.

BRAMLETTE, M. N., 1965, Massive extinction in biota at the end of Mesozoic time: Science, v. 148, p. 1696–1699

CHAMBERLIN, T. C., 1899, Lord Kelvin's address on the age of the earth as an abode fitted for life, Part I: Science, new ser., v. 9, p. 889–901; Part II: Science, new ser., v. 10, p. 11–18

CLOUD, P. E., JR., and ABELSON, P. H., 1961, Woodring conference on major biologic innovations and the geologic record: Natl. Acad. Sci. Proc., v. 47, p. 1705–1712

DARWIN, CHARLES, 1839, Journal of researches into the natural history and geology of the countries visited during the voyage round the world of H.M.S. 'Beagle' under the command of Captain Fitz Roy, R. N.: London, Henry Colburn, 615 p.

—— 1859, On the origin of species by means of natural selection; or, the preservation of favoured races in the struggle for life, 1st Edition: London, John Murray, 511 p. (Facsimile reprint, 1964, Cambridge, Mass., Harvard Univ. Press)

DE BEER, GAVIN, 1964, Charles Darwin: Garden City, N. Y., Doubleday & Co., Inc., 290 p.

FOURIER, JOSEPH, 1822, Théorie analytique de la chaleur, Translated by Alexander Freeman, 1878, The analytical theory of heat: Reprinted in 1945 by G. E. Stechert & Co., New York, 466 p.

GEIKIE, ARCHIBALD, 1905, The founders of geology: London, Macmillan and Co. (Reprinted in 1962 by Dover Pub., New York, 486 p.)

GILLISPIE, CHARLES COULSTON, 1959, Genesis and geology: New York, Harper Torchbooks, Harper & Brothers, 306 p.

HOLLAND, HEINRICH D., 1962, Model for the evolution of the earth's atmosphere, p. 447–477 in Engel, A. E. J., James, Harold L., and Leanord, B. F., Editors, Petrologic studies: A volume in honor of A. F. Buddington: Geol. Soc. America, 660 p.

HOOYKAAS, R., 1959, Natural law and divine miracle: Leiden, E. J. Brill, 237 p.

HUTTON, JAMES, 1788, Theory of the earth, or an investigation of the laws observable in the composition, dissolution, and restoration of land upon the globe: Royal Soc. Edinburgh Trans., v. 1, p. 209–304

—— 1795, Theory of the earth with proofs and illustrations: Edinburgh, William Creech; London, Cadell, Jr., and Davies, v. I, 620 p.; v. II, 567 p. (Facsimile reprint, 1959, Weinheim/Bergstr., Germany, H. R. Engelmann [J. Cramer]; Codicote, Herts, Wheldon & Wesley, Ltd.)

JAMESON, ROBERT, 1808, System of mineralogy comprehending aryctognosie, mineralogy, geognosie, mineral geography and economic mineralogy, 1st Edition: Edinburgh, v. III

LORD KELVIN (WILLIAM THOMSON), 1899, The age of the earth as an abode fitted for life: Science, new ser., v. 9, p. 665–674, 704–711

KING, CLARENCE, 1893, The age of the earth: Am. Jour. Sci., 3rd ser., v. 45, p. 1–20

LYELL, CHARLES, 1830–1833, Principles of geology, 1st Edition: London, John Murray, v. I, 1830, 511 p.; v. II, 1832, 330 p.; v. III, 1833, Text 398 p., Appendices and Index 109 p.

—— 1863, The geological evidences of the antiquity of man: Philadelphia, Pa., Geo. W. Childs, 518 p.

—— 1875, Principles of geology, 12th Edition: London, John Murray, v. I, 655 p.; v. II, 652 p.

PLAYFAIR, JOHN, 1802, Illustrations of the Huttonian theory of the earth: Edinburgh, William Creech; London, Cadell, Jr., and Davies, 528 p. (Facsimile reprint, 1956, Univ. Illinois Press, with an introduction by George W. White)

RUBEY, WILLIAM W., 1951, Geologic history of sea water: Geol. Soc. America Bull., v. 62, p. 1111–1148

RUTHERFORD, E., and SODDY, F., 1902, On the cause of nature of radioactivity.—Part I: Philos. Mag., ser. 6, v. 4, p. 370–396

—— 1903, Radioactive change: Philos. Mag., ser. 6, v. 5. p. 576–591

RUTTEN, M. G., 1962, The geological aspects of the origin of life on earth: Amsterdam and New York, Elsevier Pub. Co., 146 p.

SMYTH, HENRY DEWOLF, 1954, Atomic energy for military purposes: Princeton, N. J., Princeton Univ. Press, 264 p.

THOMSON, WILLIAM, 1862a, On the age of the sun's heat: Macmillan's Mag., March, p. 388–393 (Reprinted in Thomson and Tait, 1890, Treatise on natural philosophy, Part II: Cambridge Univ. Press, p. 485–494)

—— 1862b, On the secular cooling of the earth: Royal Soc. Edinburgh Trans., v. 23, p. 157–169 (Reprinted in Thomson and Tait, 1890, Treatise on natural philosophy, Part II: Cambridge Univ. Press, p. 468–485)

13

Copyright © 1970 by Prentice-Hall, Inc.

Reprinted from *Essays in Evolution and Genetics in Honor of Theodosius Dobzhansky*, M. K. Hecht and W. C. Steere, eds., Appleton-Century-Crofts, New York, 1970, pp. 43–96

Uniformitarianism. An Inquiry into Principle, Theory, and Method in Geohistory and Biohistory

GEORGE GAYLORD SIMPSON

Harvard University and the University of Arizona

Introduction

Many geologists have been selfconsciously perturbed by the fact that their science is extensively descriptive, and by the possibility that it may have little theoretical structure of its own. It has been supposed that its

43

nondescriptive content may be largely, or even wholly, the application of the principles of other sciences to the special objects and events that geologists only describe. (For example, see Bradley, 1963; Kitts, 1963a.)

If asked to specify a principle that is more than a simple generalization of observations and that is strictly geological, geologists may come up with the *principle of superposition.* That is stated in a standard textbook (Dunbar and Rodgers, 1957, p. 110) as follows: "In a sequence of layered rocks, any layer is older than the layer next above." As a matter of fact, that statement is frequently untrue, as the quoted authors know and go on to explain, in part. It is true if restated in some such form as this: "If sediments are deposited in sequence each above the last and not thereafter changed in attitude, lower beds are older than higher beds." Although invariably true, the statement in that form seems both trivial and tautological. Nevertheless it was not obvious to early observers; its notice was an essential element in the rise of a science of geology; it is implicated in a truly basic and strictly geological concept, that of temporal succession of rocks and consequently of geological history; and it specifies one of a rather large number of methods for investigating rock sequence and earth history. That interpretation may seem so obvious as to have no alternatives and to need no statement, but there are alternatives, now almost but not quite universally discarded. For example, two once popular alternatives are: that the superposed rocks were simply created in that form; or that they were deposited all at once in a deluge. Regardless of whether one wants to call the temporal interpretation of superposition a principle or not, it is not really trivial either in the history or in the present practice of geology.

Something still more general lies back of the criterion (let us call it that) of superposition. It is not, in fact, observational and it is one of many things that show that geology is not basically more observational than other sciences, all of which do also depend on observation, and more. The events of deposition of a sequence of rocks, the forces involved, and the segment of time represented are not observed or observable. They are inferred on the basis of similar but quite distinct events, forces, and lapses of time that can be observed at present. The connection involves at least two still more general principles: that the properties of matter and energy have been the same in the past as in the present, and that no additional properties should be postulated unnecessarily. Those are the closely associated principles of uniformity and of simplicity. The latter principle, sometimes called *Ockham's* (or *Occam's*) *razor,* is not particularly associated with geology, although as pertinent there as anywhere. The former principle, which might be but is not called *Hutton's razor,* has been historically connected with geology, but the validity of that association has been challenged.

The principle of uniformity or the doctrine or school of uniformitarianism can be conventionally, although not precisely, dated from 1795. Ever since then it has been under constant attack. A principle that has so often been belabored and left for dead on the field and that still arises from the ground to fight again has, at least, extraordinary vitality. Discussion is as lively now as ever, including discussion by those who consider the principle outmoded, unnecessary, or simply wrong. Opinions are almost as numerous as those who express them. For some, uniformitarianism is the most important geological principle, and they may add that it is geology's major contribution to science. For others, regardless of source or value, it is not to be considered geological but a requisite of science in general, and they may add that even as such it has become merely banal. In the latter case, it appears that what interest it has is entirely historical and that it may be, and should be ignored at present. (Those who continue to write about uniformitarianism in that vein evidently would apply the conclusion rather to others than to themselves.) A currently popular opinion is that uniformitarianism comprises two quite distinct principles. Those who hold that opinion add that one of the principles is false, and the other is superfluous.

I hope to show that from time to time, and even to the present, the term "uniformitarianism" has covered not only two principles but many. Some are indeed of only historical interest, but some are highly pertinent to modern geology or science in general. Some are definitely geological, and some are not. Some are false, some are still debatable, and some have been firmly established. Some are superfluous or banal; some are indispensable. Interest in why uniformitarianism will not die, and hopes that even now its protean nature can be further clarified, are the excuse for adding to its studies.

Origins; Hutton

The history of uniformitarianism has been discussed in detail and at length, for example by (in alphabetical order): Adams (1938), Bailey (1967), Cannon (1960), Gillispie (1951), Hooykaas (1956, 1959, 1963), Hubbert (1967), Rudwick (1967), and Wilson (1967). (Full citations are given in the appended bibliography, and the works there cited in turn cite many others.) In spite of those excellent studies, the definitive history has yet to be written. For present purposes many details are unnecessary, but enough historical consideration is needed to determine what has been understood as uniformitarianism, to show how the subject has become so confused, and to suggest a conceivably better analysis.

The conventional starting date, 1795, is that of publication of the two volumes of *Theory of the Earth* by James Hutton although the first part of this had already appeared in a journal (Hutton, 1788) and a still earlier abstract had been printed (1785). A third volume was written, but part of it was lost and the remainder was not published until 1899, more than a century after Hutton's death. (A summary of all three volumes, much more readable than they are themselves, is given by Bailey, 1967.) Inevitably Hutton built on the basis of still earlier work (e.g., Toulmin, 1783) and some of his points had been anticipated by others, not only British but also Italian and Russian. However, it was Hutton's publication of 1795 which led toward modern geology through the lineage of Playfair (1802) and Lyell (1830-1833).

Although Hutton is now generally considered the founder of uniformitarianism, that was only one aspect of his work and not, at the time, the most important one. His theory also involved a number of other issues, all controversial at that time and some still so. Thus from the start uniformitarianism was intricately interwoven with varied principles and opinions with which its relationships even now have not been sufficiently analyzed. The core of the Huttonian theory was a geological cycle in which parts of the earth were supposed to be uplifted by crustal forces, essentially heat, and then worn down by erosion, especially by rivers. The crustal movements were postulated as catastrophic; they were literally called "catastrophes" by Hutton. Erosion, especially as seen in the formation of river valleys, was inferred by an admirable combination of observation and reasoning to be gradual. There later arose a debate between catastrophists and gradualists, and gradualism came to be associated with uniformitarianism. In fact Hutton was *both* a catastrophist and a gradualist, and so are most geologists to this day. Some geological processes *are* more rapid and more radical and others *are* slower and less disruptive. Hutton's belief that regional uplift belongs entirely in the first class was an oversimplification, but not a bad one in the state of knowledge of that time.

A subsidiary point that nevertheless loomed large for Hutton was his belief that the consolidation, or lithification, of sediments took place by heat and partial fusion following deep immersion in the crust. Although that can occur by what were later distinguished as metamorphic processes, as a general thesis Hutton's view was simply wrong. That is no longer an issue today, and it has had little involvement in uniformitarianism.

Another major point was Hutton's conclusion that rocks now called igneous are in fact igneous, having solidified from a hot, molten state. A related conclusion was that such rocks have commonly been injected into or below sediments, and are then younger than those sediments. Here essentially began the Vulcanist-Neptunist or Huttonian-Wernerian battle, which raged for decades after Hutton's death. As is well known, a basic difference between the schools (although by no means the only one) was the Wernerian, Neptunist view that rocks in general and granite and basalt in particular were precipitated from the waters of a worldwide primeval sea. The Vulcanist side of the argument became associated with uniformi-

259

tarianism, but this would seem to be largely because late 18th and early 19th century uniformitarians (by any usual criterion) were generally also Vulcanists. There is some connection with what later came to be called actualism and is a basic element in uniformitarianism. Vulcanism does jibe better with the actual (in the sense of "present") state of the world, where lava is seen in molten state and seas are not universal, but presently observed processes of nature do not inevitably exclude the possibility of Neptunism. In fact, of course, the Neptunists were simply wrong on other grounds, and the bearing of their views on uniformity is not crucial.

Still another Huttonian theoretical element has been called providentialism. From his first brief extract onward, Hutton repeatedly insisted on a final cause, on the cycles of uplift, erosion and the rest having for their *purpose* the maintenance of a habitable and living world, occupied by plants, by animals, and preeminently by man. A final cause implies a first cause and a first cause implies a creation and a Creator. Those implications were accepted by Hutton, but he saw no reason to dwell on them. About as far as he went was to say that the purposeful orderliness of nature is "not unworthy of Divine wisdom." That must be viewed in the light of Hutton's conclusion, still often quoted today and considered scandalous in his day, that "we find no vestige of a beginning." Hutton, carefully pious, did not mean to deny that God created the universe. He simply considered it largely irrelevant to his inquiry. What he did imply by seeing neither beginning nor end is a theory of cyclic change without a secular trend. He could not identify a beginning in the record of the rocks, but only an unending repetition of change ever returning on itself, and he predicted the continuation of that process into an indefinite future. Here, then, is indeed a doctrine of uniformity, of a historical steady state overall, although it describes a dynamic, cyclic equlibrium and not a static one.

From the point of view of later geological theory and of then current theological objections, an important corollary of Hutton's "no vestige of a beginning" is its allowance for an indefinitely long span of geological time. In Hutton's day the creation of the world was generally dated at 4004 B.C., as fixed by Bishop Usher in his *Annales Veteris et Novi Testamenti* (1650-1654). Obviously that would not allow time for the erosional formation of valleys by the rivers flowing in them. Therefore that aspect, at least, of uniformitarianism, actualism, and gradualism demanded unspecified but great spans of time. Although by 1859 Usher's chronology had been abandoned by competent scientists, the same issue arose in regard to Darwin's gradualist theory of evolution. The debate was still being carried on by William Thomson (who by then had become Lord Kelvin) after Darwin and more than a century after Hutton (Kelvin, 1899; see also Hubbert, 1967). Thomson was willing to grant millions of years rather than Usher's thousands, but Darwin and other geologists were quite correctly unwilling to settle for mere millions. (See e.g., Thomson, 1862, and Darwin, 1959, Peckham reference XIV, 55.1:*f*; Darwin referred to Thomson = Kelvin as Thompson, and did not cite him until the 6th edition of the *Origin* although Thomson has published pertinent objections well before the 4th edition.)

Given the system of the earth, which, however or whenever it came to be, had been cycling for uncountable aeons, Hutton was very definite that its operation excludes the preternatural. That aspect of what later came to be called uniformitarianism in a broad sense was particularly objectionable to theologians and also to pious scientists in Hutton's day. Its formal exclusion of preternatural or miraculous intervention in history is still not universally accepted (see Hooykaas, 1959). As noted above, Hutton's theory included catastrophic events, but he considered them to be naturalistic and actualistic, that is excluding the miraculous or preternatural and involving only second causes, defined as forces now extant in nature. There was, however, still a general belief in the biblical deluge preternaturally caused. Especially at the hands of Cuvier (principal publication on this subject, 1812) and mostly after Hutton's time, that was expanded into a theory of successive revolutions, usually considered nonnaturalistic and nonactualistic. That particular version of catastrophism came to be called simply catastrophism, and even today the whole complex of issues is generally oversimplified and confused as a simple contrast of uniformitarianism and catastrophism. The confusion has been compounded by application of the term "neocatastrophism" to modern *naturalistic* theories analogous to revolutionism at least to the extent that they involve supposedly worldwide episodes of heightened tectonic or evolutionary activity. (See Schindewolf, 1963, and for a more moderate naturalistic revolutionist view of evolution see Newell, 1967; this is further discussed below.)

There is, finally but perhaps most fundamentally of all, throughout Hutton's work the conception that the rocks of the earth's crust constitute a historical record and that they can be interpreted as such.

In summary, Hutton's theory was far from simple but constituted a complex of stands taken on a variety of then, and to some extent still, controversial questions. It involved interlocking principles, not always clearly distinguished by Hutton himself and yet multiple. It is not clear in Hutton's works, and it is not always clear today, which of these can be, should be, or have been included in uniformitarianism, a term not proposed until long after Hutton's death.

Lyell; Conybeare

Although historians have lately agreed in ascribing the origin of uniformitarianism to Hutton, the body of theory later loosely understood under that name is both inchoate and somewhat incoherent in Hutton's own work. It was carried further by Playfair (1802) but became fully developed and consistently expressed only in Lyell (1830-1833), with whom modern geology can be said to begin even though he of course had numerous forerunners, Hutton among them. It was to Lyell's work that the term "uniformitarianism" was first applied, probably initially by Whewell (1832, 1837) in a derogatory sense (see Gould, 1965; Wilson, 1967).

The stark essential of Lyellian uniformitarianism appeared in his earliest subtitle (1830): "An attempt to explain the former changes of the earth's surface by reference to causes now in operation." When Lyell spoke of uniformity between past and present geological forces, he meant more than constancy of applicable physical laws or invariance of immanent characteristics of the universe. He did mean that, but he also meant that present causes are sufficient to explain past physical changes. He hedged a bit on biotic changes.) Thus he espoused Hutton's naturalism against the preternaturalism of catastrophists contemporaneous with Lyell. Furthermore he considered it probable that the average intensity of geological forces, "the energy of a cause," has tended to remain constant.

The last point was sharply attacked in Lyell's day and now, long after most of the arguments about immanent constancy and naturalism have died down, uniformity of intensity is still being argued in various guises. It is true that Lyell's original statements (1830–1833) on that principle would probably not be endorsed in detail by anyone today; in fact, they were later modified by Lyell himself. Nevertheless much of the criticism of Lyell and of uniformitarianism through the years (e.g., by Geikie, 1897) and even more recently (e.g., Gillispie, 1951) involves some misunderstanding of Lyell's views and intentions. Lyell of course knew from the start that the action of geologic forces varies greatly from time to time and place to place and in fact that catastrophes occur. That is true in present times, and therefore the conclusion that it was true at earlier geological times may be an application, not a contradiction, of actualism. Lyell's basic issue with the catastrophists was not on that point but on his firm rejection of preternaturalism. In that he was clearly right at least heuristically (and also, I believe, in principle) because the future progress of geology demanded that then still unpopular postulate.

Lyell did hold that the average intensity of geological forces, taking the earth as a whole, has tended to be approximately constant. His contemporaneous critics and later detractors have correctly insisted that this is not precisely true: intensities of such forces as mountain building, vulcanism, and glaciation, averaged for the whole earth, have been more intense at various times in the past than at others or than at present. But here Lyell's basic intention and more important point was to oppose the view, generally held by contemporaneous catastrophists and some others, that there has been a radical, unidirectional, secular reduction in the intensity of geological forces. That view was wrong, and Lyell's opposition to it was necessary and, again, highly heuristic, at least. It is related to the argument about geological time, for unless geological forces were formerly much more intense and have undergone secular reduction a short time scale cannot be accepted. In fact this was relevant to the argument for constancy of kinds of forces (actualism of Hooykaas and others, methodological uniformitarianism of Gould; see later pages) for it could be, and it was then argued that radical changes in intensities and their secular reduction implied inconstancy of forces and hence also preternaturalism.

In a broader, less literal, but essential sense, moreover, Lyell was right about this element of uniformitarianism. There have been revolution-like and catastrophe-like tectonic, volcanic, glacial, and other episodes, but they occur both early and late in geological time. They do not have a definite secular trend, and in the recorded parts of earth history there is no evidence of overall decrease (or increase) in energy flux (see Hubbert, 1967). These considerations led Lyell to conclude, much as Hutton had, that geological history has been in a dynamic equilibrium or steady state for which a beginning cannot be determined or an end predicted. Here there really is a serious weakness in Lyellian uniformitarianism, not only because later knowledge does enable a beginning to be inferred but also because of anomalies involved in this nonhistorical attitude toward history. That, too, will be more fully discussed here.

Major issues raised by Lyell can be specified in the light of criticism at that time. For present purposes that can be done in reference to a recently published (by Rudwick, 1967) letter to Lyell from Conybeare, one of Lyell's most friendly but decisive and well-informed critics. The date of Conybeare's letter is 1841, and it relates to the sixth edition (1840) of Lyell's *Principles.* By that time Lyell had modified his own views, as is indicated by the interesting weakening of the subtitle, quoted above for the first edition. In the sixth it had become: "The modern changes of the earth and its inhabitants considered as illustrative of geology." As will be further noted below, Lyell later retreated still further from the strictest or primitive uniformitarianism partly under the influence of Darwin.

It is not surprising or at this point even particularly interesting that Conybeare, a clergyman but also a geologist of first caliber, was a providentialist to the point of taking for granted that view and its argument from design. It is surprising that Conybeare was willing to accept a time scale of almost any length "short of the infinite." In fact he speaks of "Quadrillions of years" between the "oldest Cambrian and the Newest Pliocene," which is vastly too much. He did not grasp how damaging a long time scale is to the whole catastrophist argument. Gradualistic implications largely escaped him, and much of his letter, as of earlier published criticisms,

263

consequently seems merely mistaken or irrelevant and need not now greatly concern us. Most noteworthy is his continued rejection of the Huttonian theory that rivers have formed the valleys they drain. For Conybeare, "In every single instance the evidence [has] proved this hypothesis to be totally inadequate and contradictory to the whole phaenomena." He even supposed, incorrectly, that Lyell had somewhat come around to his point of view in this respect.

On that subject Conybeare was a catastrophist and diluvialist of then typical stripe. He ascribed major erosional and depositional phenomena to periodic great floods. At that time (1840–1841) recognition of Pleistocene continental glaciation was just coming before geologists through the work of Agassiz, Buckland, and others. This shook the diluvialists because it otherwise explained phenomena they had ascribed to the Noachian deluge and upset then current historicism (see below) because the coming and going of an ice age contradicted the idea of a strictly unidirectional trend in earth history. On Lyell's ideas of climatic change, Conybeare wrote, "Almost thou persuadest me"—but he made it clear that "almost" was far from "quite." In general, the glacial theory exemplifies the uncertain status of both uniformitarianism and catastrophism at the time. Uniformitarians, in their role as actualists, could point to present glaciers as demonstrating the natural forces involved in Pleistocene continental glaciation. On the other hand, catastrophists in the role of (not necessarily unidirectional) historicists could point out that the Pleistocene phenomena exceeded the present ones in intensity or, at least, extent by several orders of magnitude.

The most forceful parts of Conybeare's arguments were historicist. He expounded the thesis of secular, unidirectional trends in earth history in opposition to the Hutton-Lyell "no vestige of a beginning,—no prospect of an end" (Hutton) or, in later terminology, dynamic steady state theory. Conybeare agreed "in believing [in] the absolute uniformity of the laws of nature and general physical causes"; in order words this clergyman was not a preternaturalist as regards the postcreational course of geological history. He agreed that earth history is to be read from the rocks in terms of second causes. But he believed that the causes have acted at different intensities at different times and—this is a main point—that the intensity has in general decreased. He argued for such a trend in physical geology, involving vulcanism (both volcanic and plutonic), tectonics, and sedimentation. He further and most interestingly argued not exactly for decreasing force but for a definite trend in biotic change, the sequence of fossils in the rocks, as a particular example (in reverse order): Man-Mammalia-Saurians-Cartilaginous fish. That is, in Conybeare's term, a converging series, or (in the opposite direction) a progressive one. Therefore it is "perfectly fatal" to Lyell's "system of a continually recurring series of identical terms." "The terminal point cannot be conceived indefinitely or even very remotely distant." Since Conybeare was ready to concede quadrillions (and British quadrillions at that —$n10^{24}$) of years for the ages of rocks then already known to be fossiliferous, his idea of "remotely distant" must have been rather extreme, but the point as opposed to "no vestige of a beginning" is clear enough.

Although Lyell had by this time (6th edition, 1840) conceded that nature might produce new species at successive intervals, he still denied the existence of directionalism or progression in organic as in inorganic history. He explained away the apparent progression on the grounds that the later, "higher," or "more perfect" organisms existed in earlier times but just had not been found. As evidence he pointed to the fact that mammals had been found in the Stonesfield "slate" (a limestone), Jurassic in age, hence long antecedent to the recognized Age of Mammals and contemporaneous with the saurians. That discovery had been made in 1764, but it was not published until 1824, only a few years before the first edition of Lyell's *Principles.* (For that segment of history see Simpson, 1928.)

Conybeare's riposte merits quotation: ". . . You surely cannot consider the exception of the wretched little marsupials of Stonesfield to counterbalance the general bearing of the whole evidence—for all that it would lead to is only this, that in the secondary [Mesozoic] strata a clan of Vertebrata intermediate in their plan between true Mammalia and the lower classes first shewed themselves." Although Stonesfield mammals are not marsupials, the essence of Conybeare's remark is perfectly correct.

Lyell did make one exception to the system of unbounded existence or continual recurrence: he acknowledged that man appeared late in geological history and is neither an original (from the first creation) nor (like groups periodically recreated) a recurrent phenomenon. Conybeare and others reproved him for this inconsistency, although it is not true that it vitiates the Lyellian system as a whole.

As Conybeare's commentator Rudwick (1967) points out, the truly basic disagreement between the archetypal uniformitarian Lyell and the nominal catastrophist Conybeare was not in fact on the subject of catastrophism as usually understood, but on that of a steady state versus a historical model of earth and life history. It complicates matters still more that the system of Cuvier, the truly archetypal catastrophist, in fact was closer to a dynamic steady state than to a really historical model. His catastrophes (like Hutton's for that matter) were cyclic, not secular in trend; he believed that changes in the fossil record were due to migration, not to appearance of really new species; and he opposed the *scala naturae* concept of progression in the animal kingdom and its history. Conybeare's concept was equally nonevolutionary, but it was based on the (supposed) *scala naturae,* which was later taken over, not altogether happily, into evolutionary thought. (On the *scala naturae* see Lovejoy, 1936.)

It has been noticed before, especially by Cannon (1960), that in several respects it was a version of catastrophism and not uniformitarianism that was more nearly appropriate and favorable for development of a theory of organic evolution. Even now this seems not to have been generally understood or sufficiently studied. Yet it is clear that the Hutton-Lyell steady state model excludes the *possibility* of evolutionary interpretation, while the historical model of Conybeare and others is just as consistent with an evolutionary explanation as with their nonevolutionary views. In fact it is more consistent with the former.

Lyell; Darwin

Darwin often referred to his admiration of Lyell and indebtedness to him, and this has been noted in every biography of Darwin and most commentaries on his theories. Nevertheless the nature of the intellectual relationship and the role of uniformitarianism in it have often been misunderstood. As a recent example, Hubbert (1967), in an otherwise impeccable essay on some aspects of uniformitarianism, has written: "In effect, what Darwin did was to develop paleontology from the point where Lyell had left it. By the rejection of Lyell's residual supernaturalism where organisms were concerned, and by an extension of Lyell's Principle of Uniformity to the plant and animal kingdoms, the theory of evolution was an almost inevitable consequence."

Even the preceding brief exposition here suffices to show that this interpretation is untenable. These points cannot be fully discussed here, and further study of them would be profitable, but a few citations from Darwin's own work may clarify them sufficiently for present purposes.

In the *Autobiography* (Lady Barlow's edition, cited here as Darwin, 1958; see p. 101) Darwin wrote, "The science of Geology is enormously indebted to Lyell—more so, as I believe, than to any other man who ever lived." Darwin learned and applied *geological* concepts from Lyell's *Principles* during the voyage of the Beagle. Darwin's first extensive, important research and reputation as a scientist were in geology, and his direct indebtedness here to Lyell is clear and was acknowledged. That was useful background, but it has no direct bearing on the origin of his evolutionary views. Later in the *Autobiography* (p. 119) Darwin wrote, "After my return to England, it appeared to me that by following the example of Lyell in Geology, and by collecting all facts that bore in any way on the variation of animals and plants under domestication and nature, some light might perhaps be thrown on the whole subject [of the explanation of adaptations]." Lyell's contribution here acknowledged is to *method*. The applications of this method to similar subject matter as regards the history of life by the two men led not only to different but also to opposite conclusions.

There, again, the contribution of Lyell, or of uniformitarianism, to Darwin's evolutionary views, as distinct from his way of reaching them, was nil. Incidentally, here and elsewhere Darwin indicates greater application of a purely inductive method, at least in principle, than is generally accepted by recent historians, philosophers, and methodologists of science. (That will be mentioned again in a different connection.)

In the *Origin* (Peckham's variorum edition, cited here as Darwin, 1959) there are 17 references to Lyell in the (variorum) index, but most of these have no direct bearing on the present subject and one (VII.382.65.0.50.51-63:f, by Peckham's reference system) relates not to support but to an objection by Lyell. One, however, is relevant (Morse's IV.124-125). There Darwin points to the then (1859 and subsequently) almost general acceptance of Lyell's views on the gradual erosional excavation of valleys and concludes: ". . . . So will natural selection, if it be a true principle, banish the belief of the continued creation of new organic beings, or of any great and sudden modification in their structure." That is an appeal to gradualism, not any other aspect of uniformitarianism. It is not clear and would be most interesting to know to what extent Darwin derived the idea of gradualism from Lyell or how much he only seized upon this as a support or parallel for his original views in a different field. Darwin also insisted on a long time scale and cited Lyell in support (Peckham reference IX.31-33). That is a requisite for gradualism, and on this point there is reason to believe that Darwin's thought was indeed strongly influenced by Lyell if not entirely derived from him. Incidentally, it is curious that although that passage was indexed in the first edition of the *Origin* and was retained in all later editions with only insignificant modification, the index reference disappeared and has even been missed by the variorum editor.

As far as I have learned, those are the most direct contributions of Lyellian uniformitarianism to Darwinian evolutionism. As is well known, Lyell was a determined opponent of organic evolution until long after the first publication of the *Principles* (1830). However, in the first (1859) edition of the *Origin* Darwin could already say that he had "reason to believe that one great authority, Sir Charles Lyell, from further reflexion entertains grave doubts" as to the immutability of species. In his fourth edition (1866) Darwin could say that Lyell "now almost gives up this view," and in the fifth (1869) triumphantly that "Sir Charles Lyell now gives the support of his high authority" to the mutability of species, as we have seen a *non-* and even *anti-*uniformitarian concept according to the original Lyellian system. Thus instead of Lyellian uniformitarianism leading toward Darwinian evolutionism, Darwinian evolutionism led Lyell away from what had been a major part of his uniformitarian concept. (It is interesting that Lyell, who had agreed that scientists should be destroyed at age 60 because thereafter they could not entertain new ideas, was 62 in 1859, when he had grave doubts, was 69 when he had almost given up his view of 39 years before, and was 72 when he had definitely subscribed to the radically different view that he had fought as false and subversive for most of his life. It is true that a younger man might have taken less than ten years to change his mind, but many younger men did not change their minds at all. Richard Owen and Louis Agassiz, the most eminent lifelong scientific opponents of Darwin, were only 55 and 52, respectively, in 1859, little older than Darwin himself.)

Since Darwin

As has now been indicated, by the later part of the 19th century numerous different concepts and theories had become related in some manner and degree with uniformitarianism in the broadest sense. Still others have arisen or, at least, have become more prominent since then.

One aspect of increasing importance is involved in the subdisciplines named "paleobiology" and "paleoecology" in the present century, although they were already being considered in a primitive way by Hutton and even before him. Both are basically applications of the aphorism that the present is the key to the past. That uniformitarian principle has been criticized as embracing so much that it does not denote anything definite. Yet in these connections it does indicate definite procedures and principles, used by historical geologists and paleontologists every day and taken for granted by them. Thus a major treatise on paleoecology (Ladd, 1957) applies uniformitarian methods and reasoning throughout without any of the contributors specifying them as such. Yet it is not true that what can be taken for granted requires no explicit statement. Statement is necessary for comprehension of the nature and philosophy of a science, and also on occasion for the practical reason that too much or (more rarely) too little may be taken for granted.

The very first requirement for a science of paleontology was an application of the principle that the present is a key to the past: the long-delayed general realization that a fossil resembling part of an organism now alive was, in fact, part of a living organism at some time in the past. The interpretation has come to involve not only the taxonomic affinities of past and present organisms, but also the physiology, behavior, and environmental relationships of past organisms. As a simple example, an extinct animal with skeleton and teeth generally like a recent dog is reasonably assumed to have been warm-blooded (more correctly, endothermic), with a coat of hair, living on land, and eating mainly vertebrate prey that it actively hunted. Although not really a pioneering production, Abel's *Palaeobiologie* (1911) was a milestone in the development of that aspect of paleontology. It is now a normal part of almost any paleontological study and of what is coming to be called general paleontology (exemplified by Simpson, 1953a; Brouwer, 1959).

The broader subject of paleoecology, equally based on application of the principle that the present is the key to the past, has developed largely out of paleobiological studies. Like neoecology, it involves communities and environments, hence the paleobiology of numerous associated fossil species and also clues to physical environments both from those species and from uniformitarian studies of the rocks in which they occur.

It is interesting that one of the earliest paleobiological-paleoecological problems is still discussed and that the name of uniformitarianism is, one might say, taken in vain in this connection. It was known at least as early as 1692 that extinct elephants (which, like most extinct species of elephants, are usually called mammoths) are found in Siberia and that a few of them were frozen before they had completely decayed. As living elephants occur only in warm temperate to tropical regions, it was an obvious but crude paleobiological inference that these extinct elephants also lived in warm climates. Their preservation by freezing thus seemed inexplicable in any but catastrophic terms: it was supposed either that they had really lived to the south and that their remains had been washed to Siberia by a (or *the*) flood, or else that the climate of Siberia, warm when elephants lived there, had catastrophically turned glacially cold. The latter idea was already espoused by the archcatastrophist Cuvier and has ever since been involved in arguments against various concepts of uniformitarianism. In our own day it has been enthusiastically supported and highly unusual mechanisms for it have been hypothesized by some writers who, it is fair to say, are not considered as high scientific authorities by the generality of professional geologists, notably Velikovsky (1955) and Hapgood and Campbell (1958).

Farrand (1961) has reviewed some of the now very extensive literature on the Siberian elephants. Its own characteristics show that this particular species (*Mammuthus primigenius*) was adapted to a cold climate, and the whole ecology as shown in fauna, flora, and matrix is also that of a cold climate, consistent with the presence of permafrost while the elephants were alive in that region. For those and other reasons too extensive for restatement here, Farrand concluded that the freezing of Siberian elephants, to the slight extent of its established occurrence, was an event expectable in their situation according to "uniformitarian concepts." Lippman (1962) sharply attacked Farrand's view, which he characterized as "gradualism," and supported Cuvierian catastrophism as the only reasonable explanation, "possibly by means of the mechanism suggested by Hapgood" cited above).

269

Farrand (1962) replied that Lippman "has apparently confused [gradualism] with uniformitarianism. Uniformitarianism ('the present is a key to the past') is the geologist's concept that processes that acted on the earth in the past are the same processes that are operating today, on the same scale and at approximately the same rates." Farrand added that catastrophes do happen today and that the sudden death and early freezing of a Siberian elephant was indeed catastrophic, but "such catastrophes are in accord with the doctrine of uniformitarianism," while catastrophes of the sort and on the scale envisioned by Lippman ([after Cuvier, Hapgood, and others]) are not.

That example is instructive especially because it is a debate involving uniformitarianism that has gone on literally for centuries and still continues in the 1960's. Few will disagree that Farrand's interpretation of the facts was right in the dialogue cited, and yet his concept of uniformitarianism is still equivocal and calls for further attention.

Some of the concepts that we have seen arising within the general sphere of geological uniformitarianism have also become involved in some of the broadest theoretical and philosophical questions of modern science. That is particularly true of the debates between the steady-state and big-bang models of cosmology. They carry aspects of the Lyellian debate about the earth out into the universe as a whole, although I do not happen to have seen this conceptual relationship noted before in so many words. The steady-state cosmological model is congruent with, or at least analogous to, the Hutton-Lyell steady-state geological model, both with "no vestige of a beginning." There is also a distinct but perhaps less close analogy between Lyell's last efforts to save his steady-state model of biohistory by continual spontaneous creation of species and the maintenance of a steady-state in an expanding universe by continual spontaneous creation of matter. Both are open to the objection that if preternatural causation of continuous "creation" is rejected, it must be considered uncaused, a concept that many of us find almost as uncongenial as that of preternatural causation.

On the other hand big-bang cosmology is a historical model analogous to that of the Conybearean geological catastrophists, and indeed has a beginning that can fairly be considered the greatgrandfather of all catastrophes! However, big-bang cosmology does not involve a Cuverian succession of revolutions. After initial "creation" it follows a course more Huttonian in aspect and yet more directional than strictly Huttonian geohistory. Both in geology and in cosmology the historical model now dominates theoretical opinion, but in both cases with certain modifications and reservations. There is also a parallel between Lyell's view that, although marked local and temporary fluctuations occur, the worldwide average of geological forces is relatively uniform and the cosmological view that although local and temporal concentrations of matter and energy occur, the universe in the large is fairly homogeneous. (A convenient review of the present state of cosmology is given by McCrea, 1968.)

Immanence and Configuration

Thus various aspects of uniformitarianism, most of them originating in the field of geology, continue to ramify through that and other sciences. It has of course long and often been noticed that concepts embraced by, or related to, uniformitarianism are multiple and as a consequence are often ambiguously stated or wrongly understood. Important efforts to clarify the situation and provide more restricted definitions have lately been made successively and in part independently by Hooykaas (1956, 1959, 1963), Visotskii (1961), and Gould (1965). All three would subsume the principles or issues involved in uniformitarianism under two headings. One, called "actualism" by Hooykaas and Visotskii and "methodological uniformitarianism" by Gould, is the proposition that natural laws are invariable (e.g., Gould) and that those now ("actually," in European usage) observable are sufficient (e.g., Hooykaas). The second, called simply, "uniformitarianism" by Hooykaas and Visotskii and "substantive uniformitarnianism" by Gould, is the proposition of geohistorical uniformity in the intensities and rates of natural processes and in material conditions.

Those dichotomies of uniformitarianism, in the classical sense, can considerably clarify various issues. For instance, the dispute between Farrand and Lippman about the Siberian elephants would have been more nearly clear and decisive if Farrand (1961) could have been explicit that he was relying on actualism or methodological uniformitarianism and had not gone on (1962) to confuse the issue by defining his "uniformitarianism" as what Gould later labeled "substantive uniformitarianism," which for Farrand is an extreme form of Lyellian uniformitarianism, perhaps more extreme than Lyell's. And Lippman's objection would have been more pertinent and cogent if he had been clear that what he was opposing was *that* aspect of uniformitarianism and not actualism or methodological uniformitarianism. In fact he considered the issue to be one of gradualism versus catastrophism, an issue not raised by Farrand and not directly involved in the Visotskii-Hooykaas-Gould dichotomy.

The example just given illustrates how the dichotomy can be useful, but also one reason why it is insufficient. There are a number of issues, such as that of gradualism-catastrophism, historically important and still debated in one form or another that are not taken into account. An example even more important is the complex of issues involving historicism, retrodiction, and explanation, here to be referred to later. Another source of possible continuing confusion is that the two sides of the dichotomy as sometimes presented are not clear-cut. Moreover, each of the alternatives offered is still somewhat ambiguous. For instance the invariability of natural laws is indecisive about such basic problems as their sufficiency or as to whether the actions of all historically relevant laws are currently ("actually") observable. The definitions of uniformitarianism by Hooykaas and of substantive uniformitarianism by Gould, further, raise what is to some extent, at least, a *Scheinproblem,* because they are more rigid or extreme than the views of Hutton, Lyell, or most of their followers.

Because Gould's own extreme statement of it "is false and stifling to hypothesis formation," Gould proposes that this concept of uniformitarianism should simply be abandoned. Nevertheless there are aspects of uniformitarianism that as between "methodological" and "substantive" sides must be placed in the latter but that are real and current problems: especially matters of historicity, directionalism, revolutionism, and the like. Gould then proceeds also to demolish "methodological uniformitariansm" on grounds that it "amounts to an affirmation of induction and simplicity." He concludes that it is "subsumed in the simple statement: 'geology is a science.' " Therefore uniformitarianism in this sense is said to have no other than historical interest, and even that is so only in connection with banishing the supernatural from geology, and Gould would have the term dropped from current use. But here again there are methods and principles that must fall under, or at least are closely related to, "actualism" or "methodological uniformitarianism," that are current as to method and involve current problems, that are especially characteristic of geology even if not wholly confined to it, and that are not applicable to or do not involve all "the empirical sciences together." That is most obviously true of the methods and problems of retrodiction and historical explanation, which are peculiarly geological (including biohistorical) and which certainly are not now anachronistic, as Gould also recognizes.

"The present is the key to the past," although so broad a statement as to require analysis and specification, is still highly meaningful and is a basic principle especially arising from and important within studies of geohistory and biohistory.

The conclusion here is not that the dichotomy of actualism versus uniformitarianism or methodological versus substantive uniformitarianism is useless, unimportant, or passé, but quite the contrary, that with some restatement it is even more useful, fundamental, and correct than its proponents have claimed, or admitted. Most of the issues that have arisen in the long years of discussion of uniformitarianism do indeed fall into two classes that differ in an essential way. One has to do with the inherent properties of the universe (it is too restrictive and debatable to call those properties "laws"), that is, with what is *immanent* in it. The other has to do with the *configurations* that have arisen, and continue to arise, in historical sequence and in accordance with those immanent properties. Actualism and methodological uniformitarianism (as a general principle) belong to the former class, and uniformitarianism *sensu* Hooykaas and substantive uniformitarianism to the latter. Methodology in the more usual sense of the word does not strictly enter into either class, because its application to problems relevant to uniformitarianism involves the relationship between the two classes, that is, between immanence and configuration in geohistory and biohistory. (For further discussion of this broader and, I believe, more meaningful dichotomy see Simpson, 1963 and 1964.)

Classification of Current Issues

Burning issues to the founders of uniformitarianism, and not then always clearly distinguished from it, are now of no relevance in this connection. Thus we have seen that two of Hutton's major concerns were Vulcanism, in opposition to Neptunism, and a theory of lithification. We now consider that on the former subject he was mostly but not entirely right and on the latter mostly but not entirely wrong. We no longer argue those points, at least not in anything like Hutton's terms, and we see no special relationship between them and uniformitarianism. It is not surprising that some issues alive nearly two centuries ago are now dead. It is surprising that so many are not.

What remains to be done in the present essay is to specify and classify the principles and issues that do still have some interest within the general topic of uniformitarianism and to consider the present status of each. The following broad classification is proposed:

A. Concerned mostly with immanent properties
1. Naturalism and its alternatives
2. Actualism in its various different aspects or senses and its alternatives
B. Concerned mostly with configurations
3. Historicism with its concomitants, aspects, and alternatives
4. Evolutionism or biohistoricism, as a special, exceptionally important aspect of the foregoing
5. Gradualism, catastrophism, neocatastrophism, revolutionism
C. Concerned mostly with special methodology
6. Historical inference, retroduction, extrapolation, and related principles

The first words of those numbered topics will be used below as labels although the contents of each are more diverse.

Naturalism

Naturalism is a basic postulate of science as now almost always construed, a necessity of method and procedure in science regardless of what theological or philosophical stand may be taken on it. If only on heuristic grounds, scientific explanation must not invoke the supernatural, non-natural, noumenal, or any other preternatural factor. Although Hutton was not the first to banish preternaturalism from consideration of the earth's history, we have seen that he did so firmly, and that was perhaps his greatest contribution. He sharply distinguished First Cause from second causes. He was a providentialist in that he considered the First Cause as ordaining a terrestrial system the final cause of which is the benefit of its inhabitants, especially man, but he believed that the operation of that system, once it had been caused, was by entirely rational second causes, with no preternatural intervention.

That is still a widely accepted view, and it contradicts no scientific canons as long as investigation and explanation in science are confined to naturalistic second causes. Many scientists as a matter of individual faith or opinion reject providentialism or final causes and believe that the First Cause (whatever they may call it) is literally ineffable and incomprehensible, not susceptible either to discussion or to investigation. Others of course follow a simple axiomatic naturalism, without concern for first, second, or final causes. For almost all present scientists naturalism, in reference to their work *as* scientists, is no longer a really active issue. It is a basis for uniformitarianism but not special to the latter.

Actualism

The term "actualism" is widely used in the present connection and is used here, but it is ambiguous, particularly in English, unless given special definition. It has been applied to the philosophical doctrine that the essence of the existent is activity, and it is so defined in some English dictionaries. Not in customary English but in the Romance languages, derivatives of Late Latin *actualis*, itself from classical *actus*, "motion," have come to mean not "actual" in the English sense (that is, "real" or "existent") but "present" *"now"* existing": French *actuel*, Spanish *actual*, Portuguese *atual*, and Italian *attuale* all have that sense, and cognates appear in Germanic and Slavic languages as technical or adopted foreign terms. In those languages their variants of the term "actualism" refer to a doctrine or ism having to do with the present, the currently existent, and the term is now also used in English in that originally non-English sense.

274

The significance of the term here is that what exists now is postulated as having also existed in the past. It thus refers to the principle that the present is the key to the past and in loose usage is virtually a synonym of uniformitarianism in a broad sense. However, as has been noted, Hooykaas, Visotskii, and others have contrasted it with uniformitarianism by confining the latter term to *configurational* aspects of what is present ("actual") and using actualism to refer only to what is *immanent*. In that usage, actualism is the postulate or principle that the so-called laws of nature have been and are unchanging. That is made clearer and some serious difficulties are avoided if, as heretofore in the present study, for "laws" one substitutes "properties" or "inherent characteristics," that is, all immanent aspects of matter and energy as distinct from their position, arrangement, and activity at any one time.

Even in that restricted sense actualism involves more than one aspect or issue. In the usual connection with naturalism, an assumption is made that the immanent characteristics of the universe are now observable, a proposition not self-evident or provable as a complete generalization but obviously true in part, at least. It is almost always implied, although rarely stated, that actualism involves not only that present immanent characteristics have all existed throughout the past (always excepting First Cause or, if one likes, big-bang) but also that past immanent characteristics all exist (with the same exception) and probably are all observable at present. The latter distinct principle might be but, as far as I know has not been, called preteritism (*praeteritus*, "past"). The two principles are complementary but not necessarily equivalent. There is a good reason for preferring actualism to preteritism: science is necessarily based on the observable; the present is observable; the past is not.

Actualism in the full sense of the preceding paragraph is not an obvious *a priori* necessity, for conflicting principles readily can be and in fact have been proposed; but neither is it an arbitrary axiom. There is a large amount of observational evidence bearing on it and agreeing with it, even though in the nature of things its absolute, complete validity cannot be proved. Geologists and paleontologists have now accumulated a truly vast number of observations of recent configurations that have been visibly affected by immanent characteristics over periods up to more than three billion years. These are all consistent with actualism. That is the source and principal support of the canon of actualism, and it is generally taken to justify the acceptance of actualism where relevant in other sciences as well.

Further support, extensive in both space and time, is found in astronomy. Here, too, only what is now present (the actual in that sense) on or near the earth is observable, but the radiation observed here and now contains information about the far distant places and long past times whence and when it emanated. Much of that information is configurational; for example electromagnetic radiation tells us where various other bodies were relative to the present position of the earth when the emanations occurred, and if the red shift is correctly interpreted as a Doppler effect it tells us how rapidly that body was then receding relative to our present position. However, that radiation also contains information about immanent chemical and physical characteristics, and cosmic-ray or particle emissions bring us matter from points extremely distant in place and time. None of that information gives any indication of nonuniformity. It is all consistent with the view that electrons, protons, atoms, and so on, are identical everywhere and at all times and that "the universe follows the same physical laws throughout" (McCrea, 1968). That information is far less extensive than has been obtained by geology here on earth, but it is confirmatory.

McCrea (1968), just quoted as to the apparent bearing of the astronomical evidence on actualism (a term not used by him), nevertheless considers it naive to imply "that the universe suddenly came into existence and found a complete system of physical laws waiting to be obeyed." He adds, however, "Less crudely, according to this view, the notion that a changing universe should change in accordance with unchanging laws is regarded as acceptable." That is an excellent statement of the view of the geohistorian or cosmologist who is a uniformitarian as regards the immanent characteristics of the universe; in other words is an actualist as here defined, but who is not a uniformitarian as regards configurations, that is, not a uniformitarian *sensu* Hooykaas or a substantive uniformitarian *sensu* Gould. McCrea himself raises doubts as to whether the laws of physics as known here and now are applicable to the whole universe and whether they would permit prediction rather than only description of its behavior, hence (I would assume) whether they would permit retrodiction of history. (McCrea's discussion is here oversimplified and his paper is already an oversimplification of extended discussions there cited.) However, as there is no counterindication in any observations within our scope, most of us will be willing to accept actualism as a working principle, at least.

Actualism, or uniformitarianism in this sense, as considered up to this point is a *historical* principle. It refers to uniformity in all four dimensions of space and time. In fact it involves another premise which is rarely stated and is usually confused with historical actualism, from which it is nevertheless distinct: that the natural universe, as regards its immanent characteristics, is a single, consistent system at any one time, in other words that it makes what we humans consider as sense. That is probably the most fundamental or, at least, most necessary of all scientific principles, and, by the way, I think it is the only one we may really owe to the Greeks.

It is that aspect of uniformity and not actualism as a four-dimensional concept which is relevant to the nonhistorical aspects of science. Replications of physical and chemical observations over the few years that these have been made in rigorous fashion have been identical within the limits of experimental error. This comfortably confirms the assumption that here and now we are in a system that does make sense for practical purposes, at least. It also provides the basis for the application of actualistic principles: it establishes a set of probable immanent characteristics that can then be taken as also having four-dimensional uniformity and hence used for the interpretation of history. It does not, in itself, exclude contrary, nonactualistic postulates. For example: changes of immanent characteristics might be too slow to be detected in only a few centuries; quite different, additional immanent characteristics that existed in the past might not now exist and hence would not be detectable by nonhistorical methods; a switch from one consistent system to another might be undetectable because observations would continue to be consistent within each currently existing system; immanent characteristics of the present system might have arisen in the course of history and hence not have occurred at earlier times.

We thus see that actualism as here defined is in fact a principle special to geohistory, biohistory, and astrohistory and of limited application to nonhistorical aspects of science. Within its sphere, the question arises whether changes in immanent characteristics of the universe (or our segment of it) would be detectable by historical methods, whether actualism can be taken as a testable hypothesis. There is no possibility of testing every possible contradiction of the hypothesis, but that is a general disability of scientific theory-making. It might, for example, be difficult or impossible to distinguish a past small change in the immanent law of gravitation from a purely configurational change in the masses and distances involved. However, a major change in the immanent, such as abrogation of the inverse-squares law, would almost certainly be detectable. Many observations do indeed raise the chance of contradicting the hypothesis, and as none of them do so, considerable confidence is justified.

Further attention must now be given to the view that uniformity of the immanent, actualism, or methodological uniformitarianism is equivalent to the principle of simplicity, the principle of induction, or both. It would then have no significance as a distinct principle but would be a tautological description of science in general. "The Principle of Uniformity dissolves into a principle of simplicity that is not peculiar to geology but pervades all science and even daily life" (Goodman, 1967a). "The assumption of spatial and temporal invariance of natural laws is by no means unique to geology since it amounts to a warrant for inductive inference which, as Bacon showed nearly four hundred years ago, is the basic mode of reasoning in empirical science. . . . Methodological uniformitarianism amounts to an affirmation of induction and simplicity. But since those principles belong to the modern definition of empirical science in general, uniformitarianism is subsumed in the simple statement: 'geology is a science' " (Gould, 1965).

"Simplicity is by no means a simple notion, and what is simplest in one sense may not be simplest in another. There is no single and unique ideal of simplicity" (Beck, 1953). The whole subject is in danger of slipping off into a semasiological morass (see, further, Goodman, 1967a, 1967b). By the complex criterion of such recent discussions, it is not clear that actualism is, in fact, in accordance with the principle of simplicity. It would be possible to agree with McCrea (1968) that it is more natural (simpler?) "to expect that, if the universe changes in the large, then its laws might also change in a way that could not be predicted," or with Russell (1953 [1929]) that "there must, at every moment, be laws hitherto unbroken which are now broken for the first time." However, it is possible and, in my opinion, desirable to adhere to a more naive concept of simplicity, nearer to the archaic canon of William of Ockham, and to postulate that immanent features of the universe now existing are valid in all dimensions until or unless contrary observations are made—which has not occurred. That is support for accepting actualism as a working principle, at least. I fail to see how it can be considered that such a principle "dissolves into" a quite different principle that is advanced as a possible criterion for accepting it. Indeed, in this context we would have no reason for traffic with the principle of simplicity if it were not for its possible but not ncessary relevance to actualism. Which principle dissolves into the other may become a mere semantic quibble; I submit that neither does.

In the relationship of actualism, or some other aspect of uniformity, to induction, the situation is reversed. The validity of induction depends on uniformity. Among the many attempts to justify induction is the postulate that the future will resemble the past, which is left-handed actualism or what I have here called preteritism. Black (1967) has recently provided a summary of this and other proposed justifications for induction and has rejected them all in spite of recognition that induction is to some extent, at least, justified by its works. It is used as a method in science and does work, but it works precisely to the extent that there really is uniformity in each particular case, a uniformity, however, that is as often contingent as immanent and as often statistical as absolute. Faith in induction arises from the experience of uniformity, which is its basis (e.g., Pap, 1953 [1949]).

Thus if we admit induction as a valid principle, it depends on a prior admission of a distinct principle of uniformity of one sort or another, and the latter cannot in any real sense be considered as the derivative or secondary principle of the two. It should also be noted that it is not now universally or even usually believed that induction is the basic mode of reasoning in science. Medawar (1967), for example, has reviewed the development of deductive reasoning and explanation in 19th century England and, as a devoted follower of Popper (e.g., 1935), has proclaimed that, "Induction is a myth." Perhaps even more generally accepted is the deductive system of Hempel and Oppenheim (1953 [1948]). In its original form that system is, indeed, inappropriate for historical explanation and hence for actualism, but Beckner (1967) has demonstrated that it can be modified, rather radically to be sure, in such a way as to apply to historical explanation without calling in induction. Thus induction cannot now be taken as definitive of empirical science, and in the opinion of some scientists and philosophers induction is not necessary or not even appropriate in science. In any case, it is quite distinct from actualism.

Confidence in a consistent system of the universe, as it now exists, is indeed a basic norm for almost all science. Actualism, as here more precisely defined, arose mainly from the study of geology and is specially involved in the historical aspects of geology. To that extent, it may properly be considered a geological principle. It does also apply to historical aspects of other sciences, especially biology and astronomy. Its relevance to the largely nonhistorical aspects of geology and to chemistry and physics as a whole is restricted.

Historicism

The term historicism is here used, with some stretching, as a tag for various principles and problems that arise from consideration of the configurations of the earth and the observable universe in relationship to time. Directly under this tag only some of the oldest and most general issues will be discussed. Evolutionism and gradualism, also aspects of historicism, will be considered subsequently, as they are exceptionally important and have their own special problems and principles.

The idea of a literally static, that is, unchanging earth has long been a *Scheinproblem.* From Hutton or before to today it has never been supported by anyone, but it has been repeatedly, brilliantly attacked; in still another figure, it has been a well-battered windmill for generations of Quixotes. The great virtue of the Hooykaas-Visotskii-Gould dichotomy of uniformitarianism is that it removes actualism from the arena of those foolish attacks. It has also somewhat, at least, clarified the usual but false alternatives of uniformitarianism *versus* catastrophism.

Real issues of historicism indicated in the preceding part of the present essay have often been neglected. Alternative views about the sequence of configurations include:

Steady-state models:
    A cyclic steady state, with important, even catastrophic, changes in time but nevertheless with more or less regular return to essentially the same configurations (Hutton and followers).

279

A statistical steady state, also with important changes but these so localized and so distributed as to maintain a more-or-less constant average in space and time (Lyell and followers).
Historical model:

An irreversible sequence changing in a constant direction (Conybeare and many others).

Those three classic views have been stated in their extreme forms. It is not surprising that none can now be accepted in so extreme a form and that the present consensus includes features of all three.

As noted above, the classical form of the historical model, and the one that Lyell opposed, was extreme and involved directionalism or progressionism in the sense that the intensity of geological processes was supposed to have decreased continuously throughout earth history. As geologists and paleontologists are among those who seek to "prove all things" but do not invariably manage to "hold fast that which is good" (I Thessalonians, 21), it is not surprising that exactly the opposite form of directionalism has also been proposed, and that in fairly recent times. Schuchert (especially 1931) and others have concluded that geologic processes have accelerated in the course of geologic time. The same view as regards biological evolution has been even more widely advanced, for example in considerable but largely fallacious detail by Meyer (1954). The question of directionalism in biohistory will be discussed as an aspect of evolutionism. Directionalism of geohistory in either of its extreme forms is not now tenable. It is now clear that such processes as orogeny, vulcanism, and glaciation have varied greatly from time to time and place to place. At particular places and times in the past they have been both more and less active than at present. There is no evident regular progression either of decrease or of increase in their force. To that extent, Lyell's contention of configurational uniformity is confirmed.

On the other hand, it is no longer possible to accept the Hutton-Lyell model that involved no essential difference between early and late states of the earth: "No vestige of a beginning,—no prospect of an end" (Hutton, 1795). Hubbert (1967), in one of the clearest discussions of this aspect of directionalism (not his term), has pointed out that this is primarily a question of thermodynamics. When it was realized that geological processes involve the degradation of energy (increase of entropy, we say now), the principal source of that energy was supposed to be the heat of an originally molten earth (Thomson, 1862; Kelvin [= Thomson], 1899). Other known sources included solar radiation (both directly and as converted into winds and river flow), tides, and perhaps gravitation in other forms (e.g., compaction of the earth, impact of meteorites).

It was correctly maintained that secular cooling of the earth, with no internal generation of further heat, would not suffice to support observed geological expenditures of energy for more than a few hundred million years, at most. That was the basis for the geologically short time scale. In the present connection, it also seemed to indicate that geohistory must be directional and that its energy must have decreased rather steadily.

It now seems improbable that the earth was originally molten, and in any case that source would be entirely inadequate to balance the energy budget of geohistory on the much longer time scale now known to be more nearly correct. The source of energy missing in Thomson's calculations is now known to be radioactivity. But this, too, is a wasting resource. All energy transactions for the earth increase entropy and there is no possible way of reversing the thermodynamic flow, for the same reasons that a perpetual motion machine is impossible. Therefore Thomson was quite right in indicating an overall directionalism in earth history, although, in ignorance of radioactivity, he was wrong about the time scale. "Because of its involvement in thermodynamically irreversible processes, the earth history, despite the long time scale, can only be in the long run a unidirectional progression from some initial state characterized by a large store of available energy to a later state in which this energy has been discharged from the earth" (Hubbert, 1967). The sources of solar and tidal energy also have unidirectional progression. Thus there is (rather more than) a vestige of a beginning and there is a (clear) prospect of an end. The beginning can now be dated, as far as concerns this planet, as more or less 4½ billion (4.5×10^9) years ago, an approximation almost certainly of the right order of magnitude, at least. The date of the end cannot now be approximated, but it will surely come.

Although geological energy flow has a beginning and an end, it has been roughly constant or, at least, as previously indicated, has shown no secular trend over long periods of time. "Since the beginning of the Paleozoic era [about 6×10^8 years before present] . . . the average rate of thermally induced diastrophism and vulcanism need not have declined perceptibly," as Hubbert's (1967) conservative phrase puts it. However, this degree of configurational uniformitarianism does not preclude marked, even catastrophic, local and temporal geological changes, and furthermore this geohistorical conclusion does not necessarily apply to biohistory.

Evolutionism

An idea lately popular in some circles is that there is a grand process of evolution that is unitary through the whole sweep of things from the origins and transmutations of the elements through the histories of galaxies, stars, and planets to the. origins of life and its progress to man and the other humanoids of the universe. Insofar as those things did occur, which is probable for all but not known for some of them, there is such a process. Its name is history. That it is a single process, following a predictable path, under the same principles and forces throughout is not science but inordinate popularization (e.g., Jastrow, 1967), forced analogism (e.g., Shklovskii and Sagan, 1966), or mushy mysticism (e.g., Teilhard, 1959). (The three books cited as examples—not horrible ones—are especially interesting together, as all deal in complete independence and in highly diverse ways with many of the same aspects of astrohistory, geohistory, and biohistory; they do not treat uniformitarianism in any more explicit way.) Transition was gradual and division is arbitrary, but evolution of organisms really has proceeded by quite different ways from changes in the inorganic—ways not contradictory, not even new in their basis in the immanent; just different. Use of the word "evolution" for inorganic changes tends to be misleading even if clearly defined, as it seldom is. In any modern usage, the term should imply historicity—change from one element to another in a physics laboratory is not evolution in any acceptable sense of the word.

For clarity in the present study evolution is confined to biohistory. Evolutionism refers to a historical model or theory of life as changing directionally and irreversibly in the course of descent. An alternative historical model is that of successive extinctions and creations, with the created species progressively different. It will be recalled that this was Conybeare's view, and it was widely shared in that period.

Three main steady-state models of biohistory have been proposed, two creationist and one, oddly enough, evolutionist. The only one to have wide acceptance—unfortunately it is still widely accepted—is that the present species of organisms or, in one variant, genera, were created once and for all and have not since changed significantly. The other steady-state creationist view, adopted for a time by Lyell before he became an evolutionist but never widely accepted, is that there have been repeated creations but within such repetitive taxonomic scope as essentially to maintain the status quo.

The evolutionary steady-state theory, as far as I know never adopted by anyone else, certainly not by the Neo-Lamarckians, was due to Lamarck (1809). According to him, organisms form a continuum, apart from a few nonessential perturbations (seized on by Neo-Lamarckians as the whole idea). Evolution occurs in the sense that the mass of beings constantly moves upward in the continuum from, so to speak, amoeba to man, yet essentially nothing becomes extinct and nothing new arises; amoebas and men continue as the flux passes from one to the other. Evolution simply flows like the water in a river along an established bed, with no new headwaters and no new mouth. (Simpson, 1961, 1964.)

Choice among those models has been so decisively settled as to require no discussion; not even designation is needed here. There are, however, two points special to the present topic that call for some notice: irreversibility and directionalism in evolutionism.

In biohistorical studies irreversibility usually refers to the generalization that in any given line of descent later populations do not return exactly to the condition of an ancestral population. This is commonly called "Dollo's law," after Dollo's statement of it, first in 1893. In fact, he was anticipated by Scott (1891), among others. Restated more rigorously as a generalization, more or less as in the first sentence of this paragraph, exceptions still cannot be completely ruled out *a priori* or on acceptable theoretical grounds, but they are extremely improbable and none are definitely known. In broader view, that is one side of a historical principle of the (usual) irrevocability of evolution, discussed, for example, in Simpson (1953b, especially pp. 310-312; see also Simpson, 1964, Chapter 9). The point to be made here is that when one designates evolutionism as a directional and irreversible model of biohistory, its irreversibility, although rooted in part in the phylogenetic generalization defined above, is a broader phenomenon of history in general.

In any historical model, as opposed to a steady-state model in which maintenance of or return to a given state is postulated, there is a difference between any earlier and any later state in the system as a whole. In the present application the system is that of all organic beings on the earth throughout time. In this model that system as a whole changes irreversibly through time, in addition to or regardless of what happens to any single element in it, simply because time itself is irreversible. The observational data show that this model is unquestionably the one to adopt. The known fossil record shows beyond any doubt that throughout its span, now over 3×10^9 years, change has been continuous, no fauna or flora of any one time being identical with any of a previous time. It may be questioned whether this would hold if the time intervals considered were made infinitesimally short. The answer is that by extrapolation it could be assumed to hold, but that the question has no real scientific meaning because the pertinent observations cannot be made.

A second point about evolutionism in the context of uniformitarianism follows from that just made. As regards directionalism there is a difference between biohistory and geohistory. It was previously noted that although geohistory is directional overall, there are long spans of time for which definitely directional change, as distinct from fluctuation or cycling, is not conspicuous. That is particularly true of the time from Cambrian to Recent, which is just the time for which both geohistory and biohistory are best known. During that time biohistory has been conspicuously directional in some respects. Thus we have a directional biohistory going on within a geohistorical scene that is, at most, distinctly less directional. It was this difference, hardly expectable *a priori*, which confused Lyell and long made him an antievolutionist. Again we see that evolutionism is not an outcome or an application but in this respect a contradiction of Lyellian uniformitarianism. (Many geologists now maintain that some configurational geological changes were occurring in an irreversible directional way from Cambrian to Recent, but this modification does not essentially invalidate the preceding remarks.)

For that general conclusion it does not matter whether biohistory is unidirectional or multidirectional, nor does it matter just what the direction or directions have been. Adequate discussion of those extremely complicated topics is impossible here, but brief notice is of interest as bearing on questions about directionalism as an aspect of general configurational uniformitarianism.

It must be assumed that earliest organisms were of few kinds and comparatively simple structure. In spite of the great inadequacy of the Precambrian fossil record, it does support that inference. The earliest known fossils, believed to be more than 3×10^9 years old, are indeed little varied and simple (Schopf and Barghoorn, 1967). Although biotas were not static, organisms thereafter as seen in the scant record did not become either highly varied or notably complex for a period of well over 2×10^9 years (e.g., Cloud, 1968). Marked increase in variety and organization first appears in fossils of probable but somewhat dubious late Precambrian age (e.g., Glaessner, 1962) and becomes strong and indubitable with the beginning of the Cambrian, about 6×10^8 years before present.

The number of now living species of organisms is not known, but it is probably in the millions, and they are stunning in their diversity. Some, such as ourselves, are vastly more complicated that any known or at all likely in the Precambrian. Thus we can make several definite directional statements about biohistory:

> The number of kinds of organisms has increased enormously.
> Their distinctions in structure, function, and ecology have become much greater.
> The structural complexity of some of them, or the average complexity for the whole, has strongly increased.

Those overall directional generalizations do not apply to all phases or all subdivisions of biohistory. In later phases of the history it is not clear or probable that the number of organic taxa in any one environment has increased significantly once adaptations to that environment had been well established. Increases since some time in the early Paleozoic seem to have resulted mainly from incursions into new environments (notably those of the land in the later Paleozoic) and from increased, usually local, variety in the kinds of existing environments. It follows from the second generalization above that the kinds of evolutionary changes have been far from unidirectional. They have been extremely multidirectional, and the concept of a mainstream or central line of evolutionary development is artificial. It is also noteworthy that within simple superspecific taxa reduction in number of species has been common. Indeed it is the usual outcome for taxa of high rank and ancient origin.

As regards complexity, evolution from unicellular organisms to multi-cellular organisms with organ differentiation certainly represents an increase in complexity. But once the latter stage was fully established, supposed further increase in complexity becomes largely a matter of definition, semantics, subjective opinion, or *ad hoc* criteria. It is questionable whether *Homo sapiens* is more complex than an Ordovician vertebrate by an objective, structural measure. In any case, increase in complexity has not been a universal characteristic of evolution. In some respects and in some taxa a definable decrease in complexity has occurred. There are also organisms now living that closely resemble the oldest known fossils in structure and that are unlikely ever to have had more complex phases in their ancestry.

It is probable that rates of evolution in the late Precambrian and early Paleozoic were more rapid, on an average, than in the early Precambrian. Since the early Paleozoic rates have varied greatly from one group to another and for particular groups at different times. Many examples both of acceleration and of deceleration are known. There is no clear indication of overall increase or decrease in average rates since the early Paleozoic. Meyer (1954), for example, has produced various graphs showing great acceleration in evolutionary rates since the early Paleozoic, but his placing of points along the time scale is arbitrary. Placing of the same data by equally or more defensible criteria eliminates regular acceleration. Meyer and others have evidently been influenced by the real acceleration of human technology (e.g., graphs in Meyer, 1954, Figs. 5, 6), but the analogy with rates of organic evolution is—obviously, one would suppose—false. Incidentally, all Meyer's graphs show rates becoming infinite at a finite time not now far in the future.

Supposed identifications of overall trends in later phases of biohistory are thus dubious, at best. Nevertheless all phases of biohistory, and perhaps especially these late phases, are directional in a more general sense. There is constant, irreversible change of biotas, which neither remain static nor fluctuate about a constant mean. (For more extended discussion and other references see Simpson, 1967.)

Gradualism

As usually conceived, the issue here is between gradualism, the doctrine that processes, originally geological processes, occur at slow rates and in small increments; and catastrophism, the doctrine that they occur at fast rates or instantaneously and in large increments. As noted in the preceding historical review, that issue was long confused by association of naturalism with the former, and preternaturalism with the latter view. Preternaturalism is now excluded from scientific consideration, but that does not settle the matter because there is no reason *a priori* why catastrophism should not be naturalistic.

Indeed the issue, if expressed in such simple and categorical terms, is ambiguous to the point of being either false or meaningless. "Slow," "fast," "small," and "large" are relative terms. How slow must a process be and how small its increments in order to be gradual rather than catastrophic? How fast and how large to be catastrophic rather than gradualistic? Moreover, it is obvious to any reasonably extensive observation that both gradual and catastrophic events do occur in nature, as was of course well known to the earliest proponents of uniformitarianism. To a present-day geohistorian or biohistorian these questions become meaningful only if they refer to specific processes or events in terms of probable ranges of rates in increments. For example, Hutton and all the uniformitarians were certainly right that most valleys have been formed by erosive action of their rivers at rates and with increments that all catastrophists would consider non-catastrophic. Nevertheless, rates and increments within any one valley differed at different times, and those of different valleys have differed still more greatly both at the same and at different times. Moreover there are valleys, such as the east African rifts, that were not primarily caused by erosion and that did sometimes have relatively large and exceedingly rapid increments, even though the rate of formation averaged over their whole history may have been moderate.

The possibility for any major developments of geohistory or biohistory to occur gradually depended on expansion of the time scale from the prescientific approximation of about 6×10^3 by biblical exegesis and even beyond the probable 1×10^8 and utmost limit of about 5×10^8 allowed by Thomson on thermodynamic estimates while physicists were still ignorant of radioactivity. That is no longer a question now that rocks have been dated with fair accuracy at about 3.5×10^9 years and reasonable evidence suggests about 4.5×10^9 years for the age of our planetary system. The possible lower limit for the rate of any historical process is very low indeed. There is no general theoretical upper limit, and investigations of rates for particular processes and events may follow the evidence without *a priori* restrictions.

(Thomson's arguments for a short time scale were previously mentioned [Thomson, 1862; Kelvin, 1899]. Recognition of a longer scale made possible by radioactivity may be dated from Rutherford [1904], although he had less definite predecessors and more definite followers. It was later found that radioactivity not only made the long time scale possible but also made its measurement possible. The substitution of the long for the short scale and the origins of radiometric dating are interestingly treated by

Badash [1968]. Results have been summarized by Kulp [1961], and various dating methods are conveniently discussed by Faul [1966]. See also Hubbert [1967].)

The present inquiry is not concerned with particular investigations on geological time except for the bearing of their accumulated results on more general questions related to uniformitarianism. In the field of geohistory, it is now clear that great, sudden, and worldwide catastrophes on the diluvialist-Neptunist-Cuvierian model have not occurred. Nevertheless it is clear that the intensities of all geological processes as seen in any one region have differed greatly at different times. Apparently or hypothetically contemporaneous maxima of such processes, especially tectonics or orogeny, within western Europe were early taken as division points in geological time. Division into Primary, Transition, Secondary, Tertiary, and Quaternary, names dating mostly from the 18th century, had that basis. A different major division into "Palæozoic, Mesozoic, and Kainozoic" (current American spellings: Paleozoic, Mesozoic, Cenozoic) was proposed in 1840 by John Phillips, who was a nephew of William Smith, "the father of geology." That marked a shift from a geohistorical to a biohistorical basis, to be discussed later. In consequence, the names Primary, Transition, and Secondary are no longer used, and Tertiary and Quaternary, although still accepted by the U. S. Geological Survey and other conservatives, are obsolescent. (For discussion of geological time classification with citations and quotations of original definitions of era, period, and epoch names see Wilmarth, 1925.)

During the 19th and the first half of the 20th centuries, a general belief developed that the major tectonic episodes first identified in Europe, along with other supposedly associated major geological and biological events, were worldwide. What were believed to be geologically brief and world-wide events of that nature came to be called revolutions. That view has been anathematized as neocatastrophism, but in fact geological revolutions have never been ascribed either the scope or the intensity of the events postulated in classical catastrophism, and of course no preternatural element was seen in them. This school or doctrine is more clearly designated as revolutionist in, of course, a geological sense. Although there were a few dissenters, it had become the consensus among geologists by 1948, when it was vigorously assailed by Gilluly (1949) in one of the extraordinarily rare presidential addresses (this one to the Geological Society of America) that was both original and influential. He demonstrated that tectonic and other events have occurred locally or regionally in a virtually random manner and that the supposed worldwide and synchronous association of the most intense of them in revolutions is largely spurious.

Revolutionism in geohistory has not been completely abandoned as a working principle or eliminated from all textbooks, but it is no longer the consensus. The term "revolution" is still in wide use, but revolutions are now usually named and understood as regional, not worldwide, episodes. Thus it is still common to refer to a Laramide revolution as occurring at the Mesozoic-Cenozoic boundary in the Rocky Mountain region of North America. However, studies made without that *parti pris* have shown that this was not a single, short tectonic episode but several over a long period of time, that their acme was not at what is accepted on other grounds as the Mesozoic-Cenozoic boundary, and that there is no reliable correlation with regional "revolutions" elsewhere (e.g., Eardley, 1951; L. S. Russell, 1951). More recently Gilluly (1967) has reviewed data for western United States from the Cambrian onward. (Precambrian data are still too scanty for significance, and really adequate data start with the Devonian.) His tabulations show that there is little evident synchronism at any time within this limited region (Montana-Wyoming-Colorado-New Mexico and hence west to the Pacific) and that there was no time when tectonic activity was not going on somewhere in that region.

For our present subject, the great interest of all this is that we seem finally to be confirming and returning to essentially the Lyellian concept of configurational (or substantive) uniformitarianism for the Paleozoic and Cenozoic, at least. (Lyell's data for western Europe covered approximately the same time range as Gilluly's for western United States.) This aspect of uniformitarianism is not a "law" or principle but is a generalization from observations. The indication is that revolutionism in its usual application to geohistory is dubious, at best.

That applies *a fortiori* to theories popular until recently and still not wholly abandoned that geologic activities have been rhythmic as a result of a cycle or of two or more superimposed cycles of constant period. A small classic by Holmes (1927) reviewed that and other theories of geologic time current rather more than a generation ago. The cyclic theory reached a culmination in a book by Umbgrove (1942; still pre-Gilluly!) appropriately titled *The Pulse of the Earth*. In essence, these are also uniformitarian dynamic theories which are steady-state rather than historical as they envision oscillation about a mean rather than directional change. This is somewhat Huttonian, but Hutton was less insistant on rhythmicity. They are even more susceptible to criticism that is revolutionism, which need not involve fixed periodism, and their postulate of such periodism rests on an even more slender basis of acceptable observation. There has been little serious discussion of a general rhythmic theory of the earth in the last quarter-century.

So much for those aspects of geohistory. Because evolution occurs in the setting of geohistory, it seems probable *a priori* that geological events have affected biohistory. That is unquestionably true as regards biogeography and many adaptive sequences. For example, the making and breaking of connections between continents and between oceans has greatly affected the distributions of organisms, with profound repercussions in the extinctions of some and progressive changes of others of the organisms involved. Adaptations to the special conditions of epicontinental seas or of cordilleras obviously have occurred when and where such seas spread and such cordilleras arose in the course of geohistory. Nothing that follows contradicts the importance of such interactions between geohistory and biohistory, but they are not in question here except as some of them bear on questions of gradualism and evolutionism. (On modern concepts and some conflicts about historical biogeography see, for example, Darlington, 1957, 1965; Simpson, 1965.)

Early in the 19th century it was already becoming evident that major changes in biotas had occurred around the times that we now designate as the Precambrian-Paleozoic, Paleozoic-Mesozoic, and Mesozoic-Cenozoic transitions. As noted above, those terms in–zoic (i.e., "pertaining to animals") followed faunal criteria, thus shifting the basis for major time divisions from geohistorical to biohistorical. There was, indeed, an assumption of synchronism between major geohistorical and biohistorical episodes as in the original historical models of the catastrophists. Diluvial catastrophes were believed to coincide with great biotic changes and to have some causal relationship to them. Strictly biogeographic and creationist interpretations of faunal change were replaced by evolutionism, and diluvial-catastrophist interpretations of geohistory were replaced largely by revolutionism. The theory of synchronism and causal relationship between major biohistorical and geohistorical events, however, continued to be supported by most geologists until relatively recently and still has advocates.

This subject required reconsideration as doubts arose about the revolutionist interpretation of geohistory and as more was learned about biohistory. Compilations of increasingly numerous and reliable data have shown that originations of new organic taxa and extinctions of old have occurred continuously throughout the time of most nearly adequate fossil record (Cambrian-Recent), at least, but that there have been peaks of maximal first and last appearances. For raw data on animals see Moore (1953 seq.), Piveteau (1952 seq.), and Romer (1966); for compilation and discussion see especially Henbest (1952a), Newell (1952, 1962, 1966), Schindewolf (1950, 1954, 1963), and Simpson (1952, 1953b, 1967). Unfortunately there are no fully comparable compilations for plants, and the detailed data are too scattered for citation here. Their general trend is indicated in Arnold (1947), Delevoryas (1962), Emberger (1944) and Mägdefrau (1967).

The sequence of terrestrial faunas across the conventionally agreed Mesozoic-Cenozoic boundary can now be more nearly followed in the

North American high plains and Rocky Mountain region than anywhere else. There is a test of the supposed synchronism of a tectonic (orogenic or diastrophic) revolution and a major faunal change. As previously noted, the supposed tectonic event, the "Laramide revolution," was not brief or clearly definable, and as far as a climax can be specified it evidently occurred millions of years after the major faunal change. The example is typical. Later study has confirmed and made more positive the moderate conclusion voiced by Henbest in 1949, in published version (1952b): "Clear, simple connections of evolutionary history with so-called diastrophic rhythms are not supported by unambiguous evidence." The symposium then led by Henbest was held at the annual meeting of the Geological Society of America following that at which Gilluly gave his epochal presidential address (published in 1949), previously mentioned. The weight of the evidence was quite clear; for example two other participants, Cooper and Williams (in Henbest, 1952a), showed "that the history of the brachiopods is characterized by evolutionary bursts and that these are distributed in time serially and without very clear relation to the geologic periods." Newell showed that among invertebrates "there is no evidence that there is increased evolutionary activity during diastrophic disturbances," and from a study of all the vertebrates I concluded that "little support is found . . . for the theory of simultaneous, worldwide physical and biological climaxes at the period and era boundaries."

Nevertheless, there have been times of particularly marked biotic change over much, or, apparently, all of the earth. Besides the three previously mentioned, one can be seen sometime in late Cambrian to early Ordovician, one in or shortly after the Devonian, one at or about the conventional Triassic-Jurassic boundary and one around the Pleistocene-Recent transition. The times are necessarily stated broadly because these are not sudden episodes pinpointed in geological time. Most of them involve high extinction rates followed by high origination rates. However, the first that can be definitely recognized (late Precambrian-early Cambrian) had high origination rates not, as far as known, preceded by high extinction rates, and the last (Pleistocene-Recent) had high extinction rates not yet, at least, followed by a clear rise in origination rates.

Schindewolf (1963) dislikes the label "neocatastrophist" and prefers to speak of these events as "anastrophes," mainly because he wants to emphasize origination (Greek *ana-*, "upward" or "anew") over extinction. Nevertheless his view (of course not preternatural) is catastrophic. He considers both extinctions and originations in these episodes as occurring suddenly, essentially synchronously over short periods of time, as a result of some factor intermittent at long intervals in geologic time and distinct from any causes of less pronounced continual extinction and origination between the major events. A suggestion as to cause, already made by Schindewolf in 1950 and recently independently advanced by others (see discussion in Simpson, 1968), is bursts of radiation from explosions of supernovae.

The observed data are quite conclusively opposed to that (or any other) form of neocatastrophism. The extinctions and, to still greater degree, the originations shown by Schindewolf (and, to be sure, many others) as if they were instantaneous and absolutely simultaneous show in actual observation as slow changes, ebbs and flows over geologically long periods of time. Just one example may be given to substantiate that criticism: Schindewolf (1963, Fig. 3) shows ten groups (all but one are orders) of mammals as appearing simultaneously exactly at the beginning of the Cenozoic. In fact the first occurrences of these mammals in the known fossil record are spread over a span of about 15 million years and only a single one (Taeniodonta) of the ten groups does appear in the record just at the conventional beginning of the Cenozoic.

The data (in works previously cited) show that extinctions and originations have been going on at all times for which the record is reasonably good and strongly suggest that that has been true ever since life originated. Peaks or accelerations and decelerations for different groups are not notably synchronized through most of the record. The multiple peaks associated in the major episodes are composed of individual peaks spread over considerable spans of time and each probably long drawn out. In short, there is a high probability that these events are gradualistic and not catastrophic in any usual sense of those words. The larger events are not rhythmic, but seem to be distributed more or less at random in geological time. It is possible, although not altogether probable, that the heaping up of rise and fall, or usually in reverse order, fall and rise, of various groups in the major episodes is largely coincidental. It is unlikely that it involves factors quite absent in intervening periods when extinction and origination were also going on. Nevertheless coincidences of accelerations and decelerations do occur, and it must be concluded that some causal factors have been more intense at certain times than at others.

Since the groups involved in the major, more or less revolutionary episodes are highly varied in structure, physiology, and ecology, it seems unlikely that the intensified factors are the same for all of them. Newell (1967) has argued for a "general explanation" of biotic revolutions, as he calls them, citing the principle of simplicity in support. It would not necessarily be contradictory but would indeed be in line with Newell's own discussion of the main proximate factors intensified in an episode of widespread extinction were different for different groups but themselves were related to epeirogenic movements, oscillations in sea level perhaps of only a few feet. Such movements could have quite different effects in different environments, for instance in broadest scale on land and in the sea, but everywhere there would be more or less radical and potentially deadly environmental changes of one sort or another. In any case the emptying of environmental niches and the appearance of new ones is in itself adequate to lead to an eventual increase in evolutionary rates, especially rates of taxonomic origination. No other factor, such as increased mutation rates from radiation or other causes, is required, and here it seems well to follow simplicity by not postulating what is not required.

There is an enormous and, on the whole, diverting literature on causes of extinction, some of it seriously argued from evidence, much more of it speculation, and no inconsiderable amount well beyond the fringe of lunacy. It need not be cited or reviewed here, where the real point is simply that a modified, relatively mild and gradualistic form of revolutionism is in accord with our present knowledge of biohistory, but that neo-catastrophism is not. Probably this provisionally accepted model should be labeled neorevolutionism or biorevolutionism, because besides being gradualistic it negates or, at most, regards as nonproven the association of biohistorical with geohistorical revolutions.

Another aspect of gradualism has become involved in theories of evolutionary processes, as distinct from their results in the course of biohistory. In this field Darwin was a gradualist. He was aware of mutations with clearly discontinuous phenotypic effects, "sports" in the terms of his day, but he believed that evolution proceeded in the main by natural selection acting on "slight modifications," especially as more complex structures and characters are involved. "If it could be demonstrated that any complex organ existed, which could not possibly have been formed by numerous, successive, slight modifications, my theory would absolutely break down" (Darwin, 1959 [1859], Peckham reference VI 137). Terms such as "transitional gradations," "insensible steps," "finely graduated steps," and the like appear repeatedly in *The Origin of Species*. In the early 20th century pioneer geneticists challenged that view, believing that genetic variation is inherently discontinuous, that mutations usually have marked somatic effects, and that selection did not act by successive accumulation of slight variations but only by eliminating certain mutants.

Schindewolf, a paleontologist, developed on that basis a theory that organic taxa, at all levels of the taxonomic hierarchy, appear abruptly as such by single mutations, the higher or lower taxonomic levels being determined by the greater or lesser somatic effects of each mutation (Schindewolf, 1936, 1950a). This is related to Schindewolf's version of catastrophism, previously discussed, as it would make plausible the sudden rise of new high level taxa as part of the postulated catastrophes or, as Schindewolf would put it in this connection, anastrophes. Schindewolf's original evolutionary theory was a brilliant but, as it turns out, premature effort at synthesis of genetics and paleontology. In earlier studies I devoted some attention to its refutation (Simpson, 1944, 1953b). As far as I know, the Schindewolf theory is not now supported by any other paleontologist or biologist, and it need not be further discussed here. (It has more recently been supported by a philosopher, Grene, 1958, not on evidence or logic but because it better fitted a philosophical preconception; her views were sufficiently refuted by two biologists, Bock and von Wahlert, 1963, and also need no discussion here.)

It is now known that mutations, broadly speaking, an ultimate basis of variation, are indeed discontinuous. That requires some modification of Darwinian gradualism, but not abandonment of its essentials. The somatic effects of mutations vary from great to barely preceptible or, quite likely, to imperceptible by usual methods of observation. The probabilities that a mutation will survive or eventually spread in the course of evolution tend to vary inversely with the extent of its somatic effects. Most mutations with large effects are lethal at an early stage for the individual in which they occur and hence have zero probability of spreading. Mutations with small effects do have some probability of spreading and as a rule the chances are better the smaller the effect. Thus the usual, although not quite the only, materials for evolution are indeed "slight modifications" at the somatic level, as Darwin's acumen perceived. Moreover, despite the fact that a mutation is a discrete, discontinous event at the cellular, chromosome, or gene level, its effects are modified by interactions in the whole genetic system of an individual (oddly enough, there is no generally accepted term for that important concept). They are also modified by varying environmental factors. The results are that for many mutations, the somatic effects in different individuals vary in an essentially continuous manner. Even an expression that is a marked modification in some individuals may be only the extreme of what is a gradual sequence in the population.

There is, finally, now another sense in which evolution is known to be usually gradualistic. The instantaneous origin of a new species by a single genetic event can occur but is unusual. It is practically confined to cases of increase in individual chromosome numbers happening to produce a system both viable and capable of reproduction but not capable of backbreeding into the parental population. In usual, or one might even say "normal," cases distinct evolutionary change involves the increase or decrease of proportions of genetic factors in whole populations, and that is a gradual process occurring in successions of generations. The prevailing modern theories of evolution are essentially, although not dogmatically, gradualistic. Stepanov (1959) as plaintively cited by Schindewolf (1963) has claimed that the evolutionary theory supported, in Stepanov's exemplification, by Newell, or by me is anti-Darwinian because it is not gradualist. That is an absurd misinterpretation of our position and that of the majority of modern evolutionists.

(For entrance into the vast literature on topics quickly skimmed here, see J. D. Watson, 1965, and Beadle and Beadle, 1966, at the molecular and cellular level; Dobzhansky, 1951, 1962, at the individual and populational level; and Dobzhansky, 1962, Grant, 1963, Mayr, 1963, and Simpson, 1964, 1967, at the level of general evolutionary theory.)

Historical Inferences

The present is the key to the past in more senses than one. What we know (or theorize) about the immanent characteristics of the universe is derived from observation of the present. If actualism is accepted as a working principle, it is then assumed that throughout time those characteristics have been the same as now, no more and no less. We also observe present configurations and from them infer configurations that preceded them. The principle of actualism is essential for such inferences, and that is its main interest. This, like other aspects of uniformitarianism, arose from the desire to read the history of the earth, and it has become an essential part of the methodology of all historical science.

Actualism is necessary for successful research into history, but it is rarely sufficient. Also involved are generalizations that are about configurations rather than immanence. If a tree is observed, it is a reasonable historical inference that it grew from a seed. That is an application of a generalization probable for the given case even though many exceptions are possible. (For example, tree ferns do not have seeds, and many trees can be grown from cuttings.) It is not a law of nature or an immanent feature of the universe that trees develop from seeds. Therefore this is not strictly an application of actualism, as previously defined, but it is an example of a necessary method in historical research. The method is similar when observations of erosional initiation of gullies and deepening of valleys are related to the classical problem of the causes of valleys.

In practice this aspect of uniformitarianism might be considered the most important of all because it is a major, while not the only, reliance for reconstruction of geohistory and biohistory. It might be called *procedural uniformitarianism*. It was in part this that Gould (1965) evidently had in mind when he applied the term methodological uniformitarianism to what, following Hooykaas, is here called actualism. However, actualism is a principle; its application, which is a method, is distinct from the principle itself. Further, the methods involved in historical inference are multiple, and what is here called procedural uniformitarianism is not altogether a simple application of actualism.

It has been claimed that growth of a tree or of a valley could in principle be wholly reduced to a microphenomenal basis and hence further to a strict interaction of immanence and configuration. In fact this has not been done at all in the example of a tree and only partially in the example of a valley. From the point of view of the historian (in any field of history) reduction to microphenomena and immanence would certainly be interesting and might be useful, but in the present state of the historical sciences that is never fully possible and it is rarely necessary. Reconstruction of the macrophenomena of geohistory and biohistory can be not only most practically but also most satisfactorily done by macromethods and macroprinciples. Explanation in this field is also fully practical only at the macrolevel. It is doubtful whether complete reduction to the microsciences of atomic physics and molecular chemisty is possible even in principle. (On the latter point, see for example Simpson, 1964, chapter 6; at a C.U.E.B.S. conference in June, 1968, Michael Scriven gave a paper, afterward distributed in manuscript form but not yet published, expounding the distinction between macro- and microsciences and the justification and need for macromethodology, macrologic, and macrophilosophy.)

Much of procedural uniformitarianism in geohistory and biohistory could be called detectival or of the "trout-in-the-milk school." A trout in that situation implies dilution; a tree implies reproduction; and a valley implies erosion. It is mostly this sort of interpretive procedure that may be called inductive, because it depends on generalization from a limited number of prior observations. As previously noted, the role of induction in science has been sharply questioned. That point need not be further discussed here, but it is noteworthy that induction in a strict and usual sense, or what is sometimes called enumerative induction, does not suffice in the examples given and seldom suffices for historical inference. Strict induction involves subjective formation of a class of observations and inference that some additional thing or process will have the characteristics of that class. That is involved in such examples as valley erosion, but there is also involved a gross extrapolation from the small to the large and the short to the long, from gullying in brief to valley formation in millennia or more. Inference from valley glaciation to an ice age or from annual changes in chromosome frequencies of *Drosophila* to the evolution of vertebrates involves so much extrapolation and so little strict induction that it can hardly be said to exemplify the latter.

Another aspect of historical inference is documentation. A document in itself is not necessarily historical. A fossil or a rock out of context is a document subject to description, analysis, and classification. It becomes historical only when fitted into a temporal sequence. The usual evidence consists of configurations formed in sequence and still, in the present that is here again the key to the past, retaining sequential relationships. In both geohistory and biohistory the classic and still the dominant method involves the principle of superposition, with which this essay began. It is supplemented in many ways, especially by the biohistorical principle that similar biotas are usually approximately contemporaneous. It does not follow that dissimilar biotas are of different ages. However, once different biotas have been established as sequential, their usefulness for geohistory and biohistory is supreme. Sequential evidence relevant to previous examples includes rings in trees and terraces in river valleys. There are many other kinds of sequential evidence (see, for example, Zeuner, 1950).

Other methods are not sequential in the same sense but depend on changes in configuration that are correlated with lapse of time. They thus afford, within widely variant limits of confidence, estimates as to when the changes began and thus dates to which historical configurations may be tied. Their accuracy depends, among other things, on how well their rates can be determined at the present time and on the relative constancy of that rate in the past. As regards configurational changes, we have seen that uniformity of the latter kind is often quite unreliable. Such early attempts as estimation of the age of the oceans by their saltiness gave grossly incorrect results, largely because determination of the present rate of increase was inaccurate and postulation of past constancy was false. At present, radiometric methods are incomparably the most important (reviewed in Faul, 1966). This is more directly and immediately an application of actualism than the historical methods previously exemplified, because it depends more directly and immediately on the immanent characteristics of matter, as seen in radioactive elements, and on postulation that those characteristics have been the same throughout measurable time.

Another extremely important historical method, also embodying an element of uniformity, is the comparative method. This depends on the fact that present configurations involving similar processes may represent different historical stages in the operations of those processes. That results when the pertinent processes began earlier in one case than in another, or when they acted more rapidly in one than in another, or both. A rill in a barnyard and the Grand Canyon represent, in the main, stages of valley erosion that began some millions of years apart. They can be connected because every intermediate configuration is now observable and can be interpreted as involving (among other factors for which allowance can be made) different periods of time since erosion began. This comparative method validates the previously mentioned extrapolation from the small amount of erosion now observable in action to the vast unobservable amount in the past of the Grand Canyon.

The comparative method is equally or even more essential in biohistory. For example, the reconstruction of phylogeny, an important although far from the only aspect of biohistory, necessarily brings in comparison of recent organisms. Different characteristics of organisms often evolve at different rates. If in one organism a given character has evolved more slowly than in a related organism, that particular character will be more primitive, that is, nearer the ancestral historical antecedent, in the former. It rarely happens that one organism is more primitive than a reasonably close relative in *all* respects. It also rarely happens that any organism has a character (above the molecular level, at least) *exactly* as in a remote ancestor. An example, all the plainer for being somewhat gross, is that human molar teeth have evolved more slowly and are therefore much more primitive than horse molar teeth, while human brains have evolved more rapidly and are less primitive than horse brains. But no molars just like human molars ever occurred in the horse's ancestry, and no brains just like horse brains in human ancestry. Moreover, in each case, teeth and brains are unlike any markedly earlier stage in their own ancestry. These problems by no means vitiate the method.

The comparative method requires the uniformitarian assumption that processes themselves have been the same in past as in present even if their rates have been different. Historical inference depends less on projection into the past of the immanent, construed in a static sense, than on projection of processes, which of course do depend on immanent characteristics. For the most part, these processes are recognized and characterized as they occur in the present. The record of the past and comparative results at present can then be interpreted as involving the same processes, on the uniformitarian principle that the processes have indeed been the same. That method also involves a confrontation of the record with knowledge of present processes. If known processes are evidently insufficient to explain the record, a minimal inference is that knowledge of present processes is incomplete. A maximal inference, which up to now, at least, has not proved to be necessary, would be that the uniformitarian principle is here incorrect, that there have been past processes not now operative. If the record should prove to be inconsistent with some supposed present processes, that, too, could be interpreted as indicating nonuniformity in the sense of being contrary to actualism. However, that conclusion is not necessary or acceptable in any instances known to me. In all such cases it turns out that one has a choice between alternative hypotheses or theories about processes. The uniformitarian viewpoint is then to reject the reality of supposed present processes that are inconsistent with the historical record.

In the geological field, this confrontation of record and present processes was initiated, in the main, by Hutton and fully developed by Lyell. That was their really essential accomplishment, and it made possible an effective, coherent science of geohistory. In the field of biohistory the confrontation was primarily performed by Darwin. Knowledge of present processes has considerably increased since Darwin, and later students have continued the confrontation that he began. One of my own earlier studies (Simpson,

1944) was an attempt to demonstrate that knowledge of the biohistorical record, vastly increased since Darwin, is consistent with one body of evolutionary theory (the synthetic theory), deriving in considerable part from Darwin but also with greatly increased knowledge of present processes. It was also maintained that the record is inconsistent with different, rival theories, such as those of Goldschmidt (1940) or of Schindewolf (1936).

In the total study of evolution, or indeed of any history, there are three phases: (1) obtaining and studying the historical data, for biohistory especially but not exclusively the fossil record; (2) determination of present processes, for biohistory especially but not exclusively those of population genetics; and (3) confrontation of (1) and (2) with a view to ordering, filling in, and explaining the history. Bock and von Wahlert (1963) have made almost the same distinction in somewhat different words. They emphasized the fact that (2) their (*b*) ("The study of the mechanisms of evolutionary modifications in organisms"), is definitely nonhistorical. It is a study of *present* ("actual") processes. That, more than incidentally, shows that the philosopher Marjorie Grene (1958) was quite wrong when she insisted that the study of evolutionay *theory* (processes) must be by historical methods. The main point in my present context is that (3) the explanation of evolution by confrontation of historical description and actual processes, is uniformitarian. It involves the principle that actual (present) processes are applicable and sufficient for interpretation and explanation of biohistory.

The element of explanation brings in another series of problems, both procedural and philosophical, lately much discussed by biologists and philosophers of science. Anything approaching adequate discussion of them cannot be included here and eventually would lead away from the topic of uniformitarianism. It is nevertheless necessary to indicate the existence of an issue and to refer to some treatments of it. Claims that the hypotheticodeductive method (e.g., Popper, 1959) is the only one allowable in science (e.g., Medawar, 1967) are almost absurdly extreme, but it obviously is *an* allowable method. Deduction from a hypothesis yields predictions, and the fulfillment of those predictions increases confidence in the hypothesis. If all evident predictions are fulfilled, the hypothesis becomes an accepted theory. That is a usual, perhaps for science as a whole the most usual, method of theory formation. In a paper now famous, Hempel and Oppenheim (1953) maintained that there is no difference in principle between prediction from (in layman's language) cause to effect, and explanation of the effect by retrodiction (or postdiction) from it to its cause.

Application of that supposed equality or temporal symmetry to nonhistorical procedures, for example in physical and chemical experimentation, may not involve serious difficulties. It breaks down badly, however, in historical application. By retrodiction, and especially by the method of uniformitarian confrontation specified above, it is possible to explain past events or present configurations that cannot be predicted. There is thus an evident asymmetry between prediction and *historical* retrodiction. That is certainly, indeed obviously, true in practice. Some philosophers (e.g., Scriven, 1959) and biologists (e.g., Mayr, 1961) maintain that it is also true in principle. On the other hand, some philosophers maintain that "Hempel-explanation" is applicable to biohistory, either in spite of temporal asymmetry (Grünbaum, 1963a, 1963b; see also rebuttal by Scriven, 1963) or by applying additional or "hedging" implications to the straight Hempel model (Beckner, 1967). "Hempel-explanation" seems to me quite in-acceptable in the practice of historical science, and it seems highly improbable that prediction and retrodiction are really equivalent even in principle. However, neither point of view necessarily involves contradiction of any aspect of uniformitarianism. It could be maintained that the view I consider *less* probable is *more* uniformitarian, but the point need not be pursued further here. (There is some further discussion of it in Simpson, 1964, Chapter 7.)

Throughout the present essay emphasis has been on the historical aspect of almost all the acceptable principles that have at various times gone under the loose designation of uniformitarianism. In closing, brief attention may therefore be given to the opinion that there is no such thing as a historical science or, at least, that there is no difference in principle between so-called historical and nonhistorical aspects of science.

An extreme example is provided by R. A. Watson (1966). He first argues that *only* particular events occur, with the implication that there is no difference in principle between formation of a particular mountain range (usually considered historical) and a particular performance of a chemical experiment (the *reaction,* not the performance of it, usually considered nonhistorical). However, particular events are (Watson says) trivial. The real stuff of science is said to be "types" of events, "abstractions" or "inventions," like the laws of geological tectonics or of chemical valence. But, having said that all real events are particular, that is, unique and nonrecurrent, Watson goes on to claim that there is no such thing as a "one-of-a-kind" event that cannot "recur." Using the Grand Canyon and *Homo sapiens* as a geological and a biological example, his reasons for claiming that they are not one-of-a-kind and can recur are that for all we know canyons just like the Grand Canyon and men just like us may be present elsewhere in the universe, and even if they do not exist Watson can *imagine* them in his mind! His general conclusion is that history is not science.

It is somewhat surprising that one thoroughly competent geologist, Siever (1968), has agreed with Watson to the extent of saying ". . . We do not really care . . . how the Grand Canyon of the Colorado River was formed. We only care how the generic class of Grand Canyons forms and has formed in the past, assuming that canyon-cutting was not a unique event." Here Siever, one might say "inadvertently," has put his finger right on a major distinction between historical and nonhistorical science. The historical scientist *does* care how the Grand Canyon of the Colorado River was formed. He takes that particular canyon's present configuration as a historical document; he also takes the nonhistorical knowledge of erosional processes, in which he and Siever are both interested; and then by confronting the two on uniformitarian principles he reaches a historical interpretation and explanation of the unique phenomenon. ("Unique," indeed, because no science is properly concerned with what is not known to exist, and only psychology is concerned with what may be imagined in Watson's mind.)

Along similar lines, extreme reductionists have denied the validity of historical explanation. Schaffner (1967), for instance, has maintained that only nonhistorical physicochemical laws are explanatory even in reference to unique phenomena such as constitute historical sequences. One of his examples is that "The Empire State Building is unique, nevertheless one would not expect that the laws of stresses and strains would not apply because of this uniqueness, and that the structure of the building would not be explicable on the basis of the principles of mechanics." Of course the "laws of stresses and strains" apply to this building, as they apply to every material object in relevant circumstances. Nevertheless it is surely obvious that those laws do not explain the actual structure of the building, the fact that it exists at all, and its uniqueness. Those are the interests of the historical scientist, and the needed explanaion is historical. Incidentally, in this example the explanation is not mechanical, as Schaffner assumes, but strictly biological. Historical causation of the Empire State Building is by actions of a species of animals.

In a footnote to the same study Schaffner (1967) tries to sweep the historical issue under the rug by saying "In neither the mechanical nor the biological case are we now concerned with the way in which the structure was formed; we only want to know why it is functioning as it does (the Empire State Building stands, the organism lives)." But historical principles and historical explanations are no less valid, necessary, and scientific just because Watson, Siever, and Schaffner are not interested in them.

Summary and Conclusion

Although Hutton is now usually considered the founder of uniformitarianism, he was not consistently uniformitarian and that was not the main issue in his theoretical discussions. Nevertheless he did banish preternaturalism from scientific inference, and he did show that the surface and crust of the earth can be treated as historical documents and interpreted on the principle that past geological processes were like those now observable. Those aspects of his work made the development of a valid natural science of geology possible. That development was carried out in great part by Lyell, who also made the first really full and explicit statement of uniformitarianism. In the most general terms, Lyell's proposition was that "the former changes of the earth's surface" can be explained "by reference to causes now in operation." That generalization has a number of facets and indeed involves several quite different principles still not always clearly evaluated and distinguished.

Lyell maintained not only that geological processes were the same in the past as at present but also that their average intensity has tended to remain constant. His was essentially a steady-state model of the earth, and his concepts of both geohistory and biohistory were at first anomalously nonhistorical. His opponents, among whom Conybeare is here taken as an example, proposed a historical model involving unidirectional change with continual decrease in the intensity of geological process. It was largely the latter point, which was in fact wrong, that Lyell was seeking to correct. However, in other respects the historical model of the catastrophists and some other nonuniformitarians was more nearly correct, and it was the background for the eventual acceptance of organic evolution. Although Darwin as a geologist was deeply indebted to Lyell, his evolutionary view of biohistory was a flat contradiction and not, as often stated, an outcome or application of Lyellian uniformitarianism.

As Hooykaas, Visotskii, Gould, and others have recently pointed out, much of the original and continuing confusion about uniformitarianism has been caused by failure to distinguish its two main aspects or divisions. One has to do with the immanent characteristics of the universe and the other with the configurations of matter and energy through time. Most important in the first category is actualism, the designation here accepted for the principle or postulate that the immanent characteristics of the universe are constant throughout the four dimensions of space and time. There is no apparent way to prove this proposition in a literal, absolute sense, but there is a great deal of observational evidence relevant to it, and all such evidence agrees with it. It is not correct to consider actualism as equally involved in all sciences, and it is not reducible to the method of induction and the principle of simplicity. Both the evidence for it and the applications of it are almost confined to the related and intergrading sciences of astrohistory, geohistory, and biohistory. The related principle or postulate that is relevant to all sciences is simply that at any one time (or specifically at the present time) the characteristics of the universe constitute a single, consistent system; that, in popular terms, the universe makes sense. Perhaps even more pervasive is the principle that scientific explanation must be naturalistic.

As regards configurations, the historical model that is essentially non-uniformitarian in origin is acceptable in one of its major points. The earth, moon, and sun constitute an essentially closed thermodynamic system, other sources or stores of energy being negligible as far as the earth is concerned. Configurational change follows a one-way course with increase of entropy from an energy-rich beginning toward an eventual end without available geological energy. Nevertheless Lyellian uniformitarianism was right in that unidirectional secular change in the intensity of geological processes is not evident in the course of recorded geohistory.

In biohistory a full steady-state model is completely untenable, as Lyell himself came to realize. Evolution is irreversible not only in the structural sense usually (but equivocally) ascribed to Dollo but also and especially in an overall historical sense. From the origin of life to the present day there has been great increase in the numbers and kinds of organisms and in various rates of evolution. Nevertheless these changes have not been even approximately constant or invariably in the same direction. Evolutionary acceleration continuing up to now is not shown by the record. Biomass and multiplicity of organisms in any one environment have probably tended to remain approximately constant once that environment was fully occupied. The directions of evolution have always been extremely multiple, without true central tendency toward unidirectionalism.

The old conflict between gradualism and catastrophism has taken a different form. Preternatural, worldwide catastrophes on the diluvial pattern have not occurred, but local catastrophes with actualistic and naturalistic causes have. Intensities and rates of processes have also varied greatly from time to time and place to place in both geohistory and biohistory. In geohistory a naturalistic doctrine of revolutionism involves belief that intensified, worldwide (but not ubiquitous in local detail), and relatively short episodes of tectonic activity have occurred. There now is, however, reason to think that supposed orogenic revolutions have been more long-continued, more multiple, and more local than previously supposed. In this respect, some reversion toward a Lyellian concept of an approximate statistical steady state may be justified.

In biohistory, each sufficiently recorded and studied major taxon has had times of accelerated evolution and times (not necessarily the same as the former) of greater diversity. Such times may be multiple within one taxon, and they are generally different for different taxa. Occasionally they do coincide for a number of large taxa and then there is a major episode of extinction, turnover, and proliferation, in that order. Notable well-documented examples are found in the Permian-Triassic and Cretaceous-Paleocene transitions. Such episodes may indeed be coincidental, but they probably represent intensification and broadening of some factors, especially those related to extinction, operative less strongly and less widely at other times. They probably do not represent factors operative only at these times. These episodes do not tend to be correlated with, and so are not caused by, the orogenic maxima of supposed geohistorical revolutions.

Darwinian gradualism in organic evolution is now well established as a general rule. There are minor exceptions among low toxonomic categories. Gradualism in evolution at the macrophenomenal level of populations is consistent with discontinuity in microphenomena within single organisms, that is, with the discrete nature of genes and mutations.

Actualism is an essential basis for historical inference, and that is the main reason for its interest and its acceptance. It is not, however, an adequate statement that uniformitarian method is simply the application of actualism to history. Procedural uniformitarianism requires that application and is not in any respect independent of it, but it does have other aspects. In its detective aspect it is partly inductive but more extrapolative, as it reasons from the small and brief to the large and long. It depends in considerable part on documentation, the observation of present configurations that can be put into a temporal framework. Sequential documents, like tree rings or river terraces, are present configurations retaining features that originated sequentially in the past. The most important sequential bases for both geohistory and biohistory are, first, the principle of superposition of sediments, and then the biotic sequence worked out on those grounds. Other important documents are those that, instead of retaining a past configuration, change configuration at more or less constant rates. The elements used in radiometric dating are most important in this respect.

The comparative method of historical inference depends on comparison of related present configurations that have been more or less affected by uniform processes. That may be because the processes began to operate on the different configurations at different times in the past, or acted at different rates, or both. That is an essential method in itself and also useful as a validation of previously mentioned extrapolation.

Interpretation of the past involves confrontation of its record and of comparisons of its present results with knowledge of relevant processes. The general procedure of historical research has three phases: (1) obtaining and ordering historical data; (2) determining present processes; (3) confronting (1) and (2). The result is largely retrodictive. It involves one kind of explanation of past and also of present configurations. This is not the same as hypotheticodeductive explanation or "Hempel-explanation," and it is not symmetrical with prediction. Retrodictive interpretation and explanation are almost unique to the historical sciences, astrohistory, geohistory, and biohistory, and are not characteristic of science in general.

303

They require application of uniformitarian principles in a broad, collective sense. Some of those principles, notably actualism, must be postulated as absolute. Others, such as gradualism, are relative and applicable only as warrented by the given data. The results often remain uncertain and subject to correction, sometimes but not always more so than in the so-called exact sciences, which also do remain subject to change even in their "exact" results. The difference in depth between a rill and a canyon can be objectively measured, and an n–dimensional coefficient of likeness between the teeth of men and horses has no uniformitarian presuppositions. But such nontheoretical quantifications of the obvious are trivial, indeed really meaningless, unless put into a historical framework on grounds largely uniformitarian.

References

ABEL, O. 1911. Grundzüge der Palaeobiologie der Wirbeltiere. Stuttgart, E. Schweizerbart'sche Verlagsbuchhandlung (Erwin Nägele).

ADAMS, F. D. 1938. The Birth and Development of the Geological Sciences. Baltimore, The Williams & Wilkins Co.

ALBRITTION, C. C., JR., ed. 1963. The Fabric of Geology. Reading, Addison-Wesley Publishing Co., Inc. [A useful annotated and indexed bibliography by the editor includes most of the basic references on uniformitarianism, and the chapters by Kitts, McIntyre, and Simpson, here cited under author's names, bear on the subject.]

———— 1967. Uniformity and Simplicity. Geology Society of America, Special Paper No. 89. [Chapters by Hubbert, Wilson, Newell, and Goodman are here separately cited.]

ARNOLD, C. A. 1947. An Introduction to Paleobotany. New York, McGraw-Hill Book Company.

BADASH, L. 1968. Rutherford, Boltwood, and the age of the earth: the origin of radioactive dating techniques. Proc. Amer. Philos. Soc., 112: 157-169.

BAILEY, E. E. 1967. James Hutton—the Founder of Modern Geology. New York, Elsevier. [A convenient summary of Hutton's "Theory of the Earth," with useful but not wholly satisfactory commentary.]

BEADLE, G., and M. BEADLE. 1966. The language of life. Garden City, New York, Doubleday & Company, Inc.

BECK, L. W. 1953. Constructions and inferred entities. In Feigl and Brodbeck, 1953, pp. 368-381. [Originally published in 1950; consulted by me in the 1953 .reprint.]

BECKNER, M. 1967. Aspects of explanation in biological theory. In Morgenbesser, 1967, pp. 148-159.

BLACK, M. 1967. The justification of induction. In Morgenbesser, 1967, pp. 190-200.

BOCK, W. J., and G. VON WAHLERT. 1963. Two evolutionary theories—a discussion. Brit. J. Philos. Sci., 14: 140-46. [A refutation of Grene, 1958.]

BRADLEY, W. H. 1963. Geologic laws. In Albritton, 1963, pp. 12-23.

BROUWER, A. 1959. Algemene Palaeontologie. Zeist, de Haan. [In Dutch; unrevised English translation by R. H. Kaye, London, Oliver and Boyd, 1966, and Chicago, University of Chicago Press, 1967.]

CANNON, W. F. 1960. The uniformitarian-catastrophist debate. Isis, 51: 38-55.

CLOUD, P. E., JR. 1968. Atmospheric and hydrospheric evolution on the primitive earth. Science, 160: 729-736. [Also discusses Precambrian biohistory.]

CONYBEARE, W. D. 1841. [See Rudwick, 1967.]

CUVIER, L. C. F. D. G. 1812. Recherches sur les Ossemens Fossiles de Quadrupèdes. Tome 1. Discours Préliminaire. Discours sur les Révolutions de la Surface du Globe. Paris, Deterville.

DARLINGTON, P. J., JR. 1957. Zoogeography: the Geographical Distribution of Animals. New York, John Wiley & Sons, Inc.

———— 1965. Biogeography of the Southern End of the World. Cambridge, Mass., Harvard University Press.

DARWIN, C. 1958. The Autobiography of Charles Darwin 1809-1882 with Original Omissions Restored. Edited with Appendix and Notes by his Grand-Daughter Nora Barlow. London, Collins. [The only complete edition of the autobiography written in 1876 and first published in extensively expurgated form in 1887.]

———— 1959. The Origin of Species by Charles Darwin. A Variorum Text Edited by Morse Peckham. Philadelphia, University of Pennsylvania Press. [Collates the original editions published from 1859 to 1878; Peckham supplies a system of reference to passages and editions that I have used here.]

DELEVORYAS, TH 1962. Morphology and Evolution of Fossil Plants. New York, Holt, Rinehart & Winston, Inc.

DOBZHANSKY, TH. 1951. Genetics and the Origin of Species. 3d ed. New York, Columbia University Press.

———— 1962. Mankind Evolving. New Haven, Yale University Press.

DOLLO, L. 1893. Les lois de l'évolution. Bull. Soc. Belge Géol., 7: 164-166.

DUNBAR, C. O., and J. RODGERS. 1957. Principles of Stratigraphy. New York, John Wiley & Sons, Inc.

EARDLEY, A. J. 1951. Structural Geology of North America, New York, Harper & Row, Publishers.

EMBERGER, L. 1944. Les Plantes Fossiles dans leurs rapports avec les Végétaux Vivants. Paris, Masson.

FARRAND, W. R. 1961. Frozen mammoths and modern geology. Science, 133: 729-735.

———— 1962. Frozen mammoths. Science, 137: 450-452. [A reply to Lippman, 1962.]

FAUL, H. 1966. Ages of Rocks, Planets, and Stars. New York, McGraw-Hill Book Company.

FEIGL, H., and M. BRODBECK, eds. 1953. Readings in the Philosophy of Science. New York, Appleton-Century-Crofts. [Authors cited from this collection are listed separately in this bibliography.]

GEIKIE, A. 1897. The Founders of Geology. London, Macmillan & Co. Ltd. [With a somewhat unsympathetic view of Lyell's uniformitarianism.]

GILLISPIE, C. C. 1951. Genesis and Geology. Cambridge, Mass., Harvard University Press. [A historical account of the period 1790-1850; see especially Chapter V, "The uniformity of nature."]

GILLULY, J. 1949. The distribution of mountain-building in geologic time. Bull. Geol. Soc. Amer., 60: 561-590. [This now classic presidential address is the anti-revolutionist manifesto.]

———— 1967. Chronology of tectonic movements in the western United States. Amer. J. Sci., 265: 306-331. [A richly documented detailing of data for a single area relevant to the general thesis of Gilluly, 1949]

GLAESSNER, M. F. 1962. Pre-Cambrian fossils. Biol. Rev. 37: 467-494.

GOLDSCHMIDT, R. 1940. The Material Basis of Evolution. New Haven, Yale University Press.

GOODMAN, N. 1967a. Uniformity and simplicity. *In* Albritton, 1967, pp. 93-99.

———— 1967b. Science and simplicity. *In* Morgenbesser, 1967, pp. 68-78.

GOULD, S. J. 1965. Is uniformitarianism necessary? Amer. J. Sci., 263: 223-228.

GRANT, V. 1963. The Origin of Adaptations. New York, Columbia University Press.

GRENE, M. 1958. Two evolutionary theories. Brit. J. Philos. Sci., 9: 110-127, 185-193. [The theories are Schindewolf's and the synthetic theory as expounded by Simpson.]

GRÜNBAUM, A. 1963a. Temporally asymmetric principles, parity between explanation and prediction, and mechanism versus teleology. *In* Induction: Some Current Issues, Middletown (New York), Wesleyan University Press, Chap. VI, pp. 114-149.

———— 1963b. Philosophical Problems of Space and Time. New York, Alfred A. Knopf, Inc.

HAPGOOD, C. H., and J. H. Campbell. 1958. Earth's Shifting Crust. New York, Pantheon Books, Inc. [Mr. Hapgood's unusual catastrophic hypothesis was also published in the Saturday Evening Post for 10 January 1959 and was endorsed by Ivan Sanderson in the same journal on 19 January 1960.]

HEMPEL, C. G., and P. OPPENHEIM. 1953. *In* Feigl and Brodbeck, 1953, pp. 319-352. [Originally published in 1948; consulted by me in the 1953 reprint.]

HENBEST, L. G., ed. 1952a. Distribution of evolutionary explosions in geologic time J. Paleont., 26: 298-394. [A symposium of seven papers with extended discussion; contributions by Henbest, Newell, and Simpson are here cited separately.]

———— 1952b. Significance of evolutionary explosions for diastrophic division of earth history. *In* Henbest, 1952a, pp. 299-318.

HERSCHEL, J. F. W. 1841. Whewell on inductive sciences. Quart. Rev. (London), 68: 177-238. [Ridicules the concept of configurational uniformitarianism.]

HOLMES, A. 1927. The Age of the Earth. Benn's Sixpenny Library, No. 102. London, Benn.

HOOYKAAS, R. 1956. The principle of uniformity in geology, biology, and theology. J. Trans. Victoria Inst., 88: 101-116.

———— 1959. Natural Law and Divine Miracle. Leiden, Brill. [Essentially an expansion of Hooykaas, 1956.]

———— 1963. The Principle of Uniformity. Leiden, Brill [Reissue and renaming of Hooykaas, 1959.]

HUBBERT, M. K. 1967. Critique of the principle of uniformity. *In* Albritton, 1967, pp. 3-33.

HUTTON, J. 1788. Theory of the earth. Trans. Roy. Soc. Edinburgh, 1: 209-304. [The first generally available form of Hutton's famous work, although it is now known that an abstract had been printed in 1785; the 1788 paper became the first part of the following edition.]

———— 1795. Theory of the Earth. 2 vols. Edinburgh, William Creech. [A facsimile was published in 1959 by Hafner Publishing Co., Inc., New York. A third volume was written but was discovered only in incomplete manuscript form and much later; it was published in 1899 by the Geological Society of London; it is summarized in Bailey, 1967.]

JASTROW, R. 1967. Red Giants and White Dwarfs. New York, Harper & Row, Publishers.

KELVIN, LORD [William Thomson]. 1899. The age of the earth as an abode fitted for life. Science, n.s., 9: 665-674, 704-711. [See also Thomson, 1862.]

KITTS, D. B. 1963a. The theory of geology. *In* Albritton, 1963, pp. 49-68. [Uniformitarianism is discussed on pp. 62-67.]

———— 1963b. Historical explanation in geology. J. Geol., 71: 297-313.

KULP, J. L. 1961. Geologic time scale. Science, 133: 1105-1114.

LADD, H. S., ed. 1957. Treatise on marine ecology and paleoecology. Vol 2. Paleoecology. Geology Society of America, Memoir 67, 2: i-x, 1-1077.

LAMARCK, J. B. M. de. 1809. Philosophie Zoologique. Paris, Dentu.

LIPPMAN, H. E. 1962. Frozen mammoths. Science, 137: 449-450. [An attack on Farrand, 1961.]

LOVEJOY, A. O. 1936. The Great Chain of Being. Cambridge, Mass., Harvard University Press.

LYELL, C. 1830-1833. Principles of Geology. 3 vol's. London, Murray.

MÄGDEFRAU, K. 1967. Die Geschichte der Pflanzen. *In* Heberer, G., ed. Die Evolution der Organismen. 3rd ed. Stuttgart, Gustav Fischer. Vol. 1, pp. 551-588.

MAYR, E. 1961. Cause and effect in biology. Science, 134: 1501-1506.

———— 1963. Animal species and evolution. Cambridge, Belknap Press.

McCrae, W. H. 1968. Cosmology after half a century. Science, 160: 1295-1299.

McIntyre, D. B. 1963. James Hutton and the philosophy of geology. *In* Albritton, 1963, pp. 1-11.

Medawar, P. B. 1967. The Art of the Soluble. London, Methuen. [See especially the (unnumbered) chapter on "Hypothesis and Imagination," pp. 131-155, for a violently negative view on induction as a scientific method.]

Merrill, G. P. 1906. Contributions to the history of American geology. Report U.S. Nat. Mus. for 1904 [published 1906]: 189-734 [From the point of view of this paper most interesting for its evidence of how little early American geologists were influenced by Lyell and uniformitarianism.]

Meyer, F. 1954. Problématique de l'Évolution. Paris, Presses Universitaires de France. [Includes an extensive but fallacious account of supposed evolutionary acceleration.]

Moore, R. C., ed., 1953. Treatise on invertebrate paleontology. Lawrence, Geology Society of America and University of Kansas Press. [Distribution of recognized genera of fossil invertebrates, publication continuing and approaching completion.]

Morgenbesser, S. 1967. Philosophy of Science Today. New York, Basic Books, Inc. Publishers. [An excellent collection of essays, marred by absence of references; essays most relevant to the present study are here cited separately by authors.]

Newell, N. D. 1952. Periodicity in invertebrate evolution. *In* Henbest, 1952a, pp. 371-385.

———— 1962. Paleontological gaps and geochronology. J. Paleont., 36: 592-610.

———— 1966. Problems of geochronology. Proc. Acad. Natural Sci., 188: 63-89.

———— 1967. Revolutions in the history of life. *In* Albritton, 1967, pp. 63-91.

Pap, A. 1953. Does science have metaphysical presuppositions? *In* Feigl and Brodbeck, 1953, pp. 21-33. [Originally published in 1949; consulted by me in the 1953 reprint.]

Piveteau, J. 1952. Traité de Paléontologie. Paris, Masson. [Distributional data, nearly complete for families of animals. Publication continuing since 1952.]

Playfair, J. 1802. Illustrations of the Huttonian Theory of the Earth. Edinburgh, Cadell, Davies and Creech. [Facsimile edition, 1956: Urbana, Univ. Illinois Press.]

Popper, K. 1935. Logik der Forschung. Vienna.

———— 1959. The Logic of Scientific Discovery. New York, Basic Books, Inc. Publishers.

Romer, A. S. 1966. Vertebrate Paleontology. Chicago, University of Chicago Press. [Includes distribution of all recognized genera of fossil vertebrates.]

Rudwick, M. J. S. 1967. A critique of uniformitarian geology: a letter from W. D. Conybeare to Charles Lyell, 1841. Proc. Amer. Philos. Soc., 111: 272-287. [Conybeare's letter is not only reproduced but also usefully discussed at length with its historical background.]

Russell, B. 1953. On the notion of cause, with applications to the free-will problem. *In* Feigl and Brodbeck, 1953, pp. 387-407. [Originally published in 1928; consulted by me in the 1953 reprint.]

Russell, L. S. 1951. Age of the Front-Range deformation in the North American cordillera. Trans. Roy. Soc. Canada, 45: 47-69.

Rutherford, E. 1904. The radiation and emanation of radium. Technics, for 1904: 11-16, 171-175. [Contains the epochal suggestion that the energy budget for the earth on a long time scale could be balanced by contributions from radioactivity.]

Schaffner, K. F. 1967. Antireductionism and molecular biology. Science, 157: 644-647.

Schindewolf, O. H. 1936. Paläontologie, Entwicklungslehre und Genetik. Berlin, Borntraeger.

———— 1950a. Grundfragen der Paläontologie. Stuttgart, Schweizerbart.

———— 1950b. Der Zeitfaktor in Geologie und Paläontologie. Rev. ed. Stuttgart, Schweizerbart.

———— 1954. Über die möglichen Ursachen der grossen erdgeschichtlichen Faunenschnitte. Neues Jahrb. Geol. Pal., 10: 457-465.

———— 1963. Neokatastrophismus? Z. deutsch. geol. Ges., 114: 430-445. [In "Jahrgang

1962," and usually cited as of that date, but issued in 1963; an example of confusion of different meanings of "catastrophism."]

SCHOFF, J. W., and E. S. BARGHOORN. 1967. Alga-like fossils from the early Precambrian of South Africa. Science, 156: 508-512. [Descriptions of some of the oldest known fossils and citations of the literature on others.]

SCHUCHERT, C. 1931. Geochronology, or the age of the earth on the basis of sediments and life. Bull. Nat. Res. Council, No. 80: 10-64. [Espouses a theory of acceleration of geologic processes.]

SCOTT, W. B. 1891. On the osteology of *Mesohippas* and *Leptomeryx,* with observations on the modes and factors of evolution in the Mammalia. J. Morph., 5: 301-406.

SCRIVEN, M. 1959. Explanation and prediction in evolutionary theory. Science, 130: 477-482.

———— 1963. Comments on Grünbaum paper. *In* Induction: Some Current Issues, Middletown (New York), Wesleyan University Press, pp. 147-149. [A rebuttal of Grünbaum, 1963a.]

SHKLOVSKII, I. S., and C. SAGAN. 1966. Intelligent Life in the Universe. New York, Dell Publishing Co., Inc. [Also paperback from the same publisher, 1968.]

SIEVER, R. 1968. Science: observational, experimental, historical. Amer. Sci., 56: 70-77.

SIMPSON, G. G. 1928. A Catalogue of the Mesozoic Mammalia in the Geological Department of the British Museum. London, British Museum (Natural History).

———— 1944. Tempo and Mode in Evolution. New York, Columbia University Press. [A facsimile edition was published in 1965 by Hafner Publishing Co., Inc.]

———— 1952. Periodicity in vertebrate evolution. *In* Henbest, 1952a, pp. 359-370.

———— 1953a. Life of the Past. New Haven, Yale University Press. [Also paperback, Yale, 1961, and Bantam Books, Inc. 1968.]

———— 1953b. The Major Features of Evolution. New York, Columbia University Press. [Also paperback: New York, Simon & Schuster, Inc. 1967.]

———— 1961. Lamarck, Darwin, and Butler. Amer. Scholar, 30: 238-249. [Reprinted as Chapter 3 in Simpson, 1964.]

———— 1963. Historical science. *In* Albritton, 1963, pp. 24-48. [Uniformitarianism is discussed on pp. 31-33.]

———— 1964. This View of Life. New York, Harcourt, Brace & World, Inc. [Chapter 3 is a reprint of Simpson, 1961, and Chapter 7 is a revised version of Simpson, 1963.]

———— 1965. The Geography of Evolution. Philadelphia, Chilton Book Company.

———— 1967. The Meaning of Evolution. Rev. ed. New Haven, Yale University Press. [On rates and directionalism see especially chapters 8 and 11.]

———— 1968. Evolutionary effects of cosmic radiation. Science [in press].

STEPANOV, D. L. 1959. [Neocatastrophsm in paleontology of these days.] Paleont. Zhurn. Akad. Nauk S.S.S.R., No. 4: 11-16. [In Russian; not read; cited from Schindewolf, 1963, and Newell, 1967.]

TEILHARD, DE CHARDIN, P. 1959. The Phenomenon of Man. New York, Harper & Brothers, Publishers. [First published in France as Le Phénomène Humain, Paris Editions du Seuil, 1955.]

THOMSON, W. [Lord Kelvin]. 1862. On the secular cooling of the earth. Trans. Roy. Soc. Edinburgh, 23: 157-169. [See also Kelvin, 1899.]

TOULMIN, G. H. 1783. The antiquity of the world. 2nd ed. London, Cadell. [A rarely noticed forerunner of Hutton; the first edition (not seen) was published in 1780.]

UMBGROVE, J. H. F. 1942. The Pulse of the Earth. The Hague, Nijhoff.

VELIKOVSKY, I. 1955. Earth in Upheaval. New York, Doubleday & Company, Inc. [One of the more radical modern versions of catastrophism.]

VYSOTSKII, B. P. 1961. Problema aktualizma i uniformizma i sistema metodov v geologii. Vop. Filos. Akad. Nauk S.S.S.R., No. 3: 134-145. [There is a large literature in Russian on uniformitarianism; this one paper is cited because it makes the same distinction as do Hooykaas and Gould.]

WATSON, J. D. 1965. Molecular Biology of the Gene. New York, W. A. Benjamin, Inc.

WATSON, R. A. 1966. Is geology different: a critical discussion of "The fabric of geology." Philosophy of Science, 33: 172-185. [A blast against Albritton, 1963, and especially Simpson's contribution to that work.]

[WHEWELL, W.] 1832. [Review of Lyell, 1830-1833, vol. 2.] Quart. Rev. 47: 103-132.

WHEWELL, W. 1837. History of the Inductive Sciences. 3 Vols. London, Parker.

WILMARTH, M. G. 1925. The geologic time classification of the United States Geological Survey compared with other classifications accompanied by the original definitions of era, period and epoch terms. Bull. U. S. Geol. Surv. 769.

WILSON, L. G. 1967. The origins of Charles Lyell's uniformitarianism. *In* Albritton, 1967, pp. 35-62.

ZEUNER, F. E. 1950. Dating the Past. 2nd ed. London, Methuen.

14

Copyright © 1970 by Royal Netherlands Academy of Sciences

Reprinted from *Koninklijke Nederlandse Akademie van Wetenschappen, afd. Letterkunde, Med. (n.r.),* **33** (7), 271–316 (1970)

Castastrophism in Geology, Its Scientific Character in Relation to Actualism and Uniformitarianism

REIJER HOOYKAAS

I. INTRODUCTION *

The history of geology has often been expounded, in the fashion of a fairy tale, as a battle between good and evil. Neptunism is black, Plutonism white; Catastrophism is black, Uniformitarianism white. In the 18th century darkness reigned until, through Hutton, suddenly all became light. In the beginning of the 19th century Cuvier, Buckland, c.s. fell back again upon deluges and catastrophes, until Lyell dispelled the clouds and definitively established uniformitarian orthodoxy.

Catastrophists are accused of giving free play to their phantasy, of rashly resorting to extraordinary events and supernatural causes, and of mixing up independent geological research with metaphysical beliefs.

In this paper [1] we will listen to the other side too. And the conclusion will be that, though there have been catastrophists who answer to the description just given, uniformitarians could be as metaphysical and perhaps even more dogmatical than their opponents, and that, quite apart from the resulting theoretical *system*, at least the *method* of the Catastrophists was a legitimate one.

II. CLASSIFICATION OF GEOLOGICAL METHODS AND SYSTEMS

In geological literature the 'anglosaxon' term 'uniformitarianism' and the continental term 'actualism' are generally used as perfectly synonymous, and both are put forward as the opposite of 'catastrophism'.

Uniformitarianism implies that ancient changes in the earth's crust were effectuated by causes of the same *kind* as those working

---

* The paragraphs I–III, IV *b* and *c*, VI *a*, *b*, *c* and *e*, contain the text of a lecture delivered before the Geological Society of Krakow on October 30th, 1967. Russian transl. in: Istorija Geologii, Acad. Sci. Armenian SSR, Erevan 1970, pp. 33–57.

[1] Reactions on my "The Principle of Uniformity in Geology, Biology and Theology" (Leiden 1959 [1]; 1963 [2]), urged me to further research on this topic.

The paragraphs II–V (pp. 5–25) replace the pp. 1–4 and 13–14 of the first and second editions of the book; par. VI (pp. 25–35) is an addition to pp. 33–42 of P.U.; par. VII, *a* and *c* (pp. 35–42) is an addition to pp. 11–12 of P.U.; par. VII, *b* an addition to pp. 90–92 of P.U.; par. VIII (pp. 42–45) an addition to p. 179 of the book.

As in discussions on actualism the relevant texts of Buffon sometimes are chosen onesidedly, and as Razumovsky and Dolomieu are never mentioned in this connection, we here give full quotations from their works. Also other geologists not dealt with in my earlier publications (Élie de Beaumont, Frapolli, Conybeare, Cotta, Bronn, Prestwich) have been brought to the fore. For authors already commented upon in "The Principle of Uniformity" we cannot but send back the reader to that book.

271

at present and that these causes had about the same *intensity* as their modern equivalents. Existing geological causes work rather slowly, so that strict uniformity of geological events requires an immensity of time: all past changes on the globe have been brought about by the *slow* agency of still existing causes (Lyell, 1830) [2].

That is to say, that uniformitarianism is antagonistic to catastrophism, which holds that causes now in operation (ice, water, winds, volcanism), if active with the now prevalent intensity, are not sufficient to explain the geological events of the past. Catastrophists, therefore, resorted also to the operation of extraordinary, violent causes: sudden elevations of whole continents, paroxysmal volcanic eruptions, and inundations of large areas of dry land by the ocean.

The usual contradistinction of uniformitarianism or actualism (by which a *method* as well as its resulting *system* was meant) and catastrophism (which is a geological *system* and not a method) has caused many misunderstandings.

In British and American literature the term "uniformitarianism" is always used. This term fits well to the methods and systems of Hutton and Lyell, who supposed a perfect similarity between the geological causes and effects of the past and the present,—a uniformity not only as to their kind but also as to their intensity.

In continental European languages, however, though the term "actualism" is considered as synonymous with the anglosaxon "uniformitarianism", it often has somewhat wider implications. For this term *in itself* implies only that the present (modern or actual) causes are sufficient to explain the events of the past; it does not necessarily include the idea that they operate with the same energy in the present as they did in the past. One could imagine that the geological causes of the past were of the same kind as the actual causes, but that they were much more powerful, so that they sometimes led to cataclysmic effects. In such a case they would be in the literal sense of the word "catastrophic" as well as "actualistic"; the *system* (the resulting historical description) would be catastrophist, whereas the *method* of constructing it would be actualistic.

Moreover, the quietness and slowness of change, which seem so characteristic of uniformitarianism over against catastrophism, are not sufficient to guarantee that a system is based on an "actualistic" method. It might be that totally different causes of change

---

[2] Ch. Lyell, Principles of Geology, being an Attempt to explain the former changes of the Earth's surface by reference to causes now in operation. sec. ed. London 1832, vol. I, pp. 72–73.

272

were active in the past and that they worked equally in a slow, non-catastrophic tempo: the result then would be neither "catastrophic" nor "actualistic".

The current division into catastrophistic and "actualistic" (or uniformitarian) *systems* or theories does not give an adequate representation of the present situation in geological science. It should be replaced by the more fundamental division in "conceptions based on an actualistic *method*" (strict uniformitarianism and actualistic catastrophism included) and "conceptions based on a non-actualistic *method*" (i.e. those recognizing ancient causes, whether catastrophal or not in their effects). This division, then, is determined in the first place by the extent to which the actualistic *method* is applied, and *not* by the uniformity or non-uniformity of the resulting descriptive *systems*.

Roughly speaking, then, at least four (or five) different conceptions of the history or the historiography of the earth may be distinguished.

a. *Non-actualistic conceptions*

1. The causes of some geological changes of the past *differ in kind and energy* from those now in operation.

This is catastrophism in the traditional sense (*non-actualistic catastrophism*), implying that forces which are not in operation at present, caused revolutions of an intensity much greater than that of the causes working now (Cuvier). Especially in paleontology, sometimes supernatural causes are introduced.

2. The causes of some geological changes of the past *differ in kind but not in energy* from those now in operation; their effects were not more violent, and the changes resulting from ancient causes occurred in the same slow tempo as is prevalent now.

b. *Actualistic conceptions*

3. The causes of geological changes in the past *differ not in kind*, though they may sometimes *differ in energy*, from those now in operation. This is actualism, though no uniformity of activity is assumed.

3*a*. In general, the background of this conception is the belief that the energy of geologic forces has gradually diminished, as the earth is cooling down (Hooke, Ray, von Buch, Breislak, and even the non-catastrophist actualists Scrope and Von Hoff). Therefore, when the earth was younger, the causes of change must have been more powerful and their effects more violent.

3*b*. A series of discontinuous outbursts of geological activity is assumed, and superposed upon the continuous changes. This

273

conception (Élie de Beaumont, Sainte-Claire Deville) is an *actualistic catastrophism* as to the resulting historic-descriptive *system*; it is a catastrophist *actualism* as to the *method* used, as it tries to interpret past phenomena as much as possible in terms of actually existing causes.

4.   The geological forces of the past *differ neither in kind, nor in energy* from those now in operation. This is *"actualism"* or rather *"uniformitarianism"* in the current sense. In this case the method largely determines the resulting theoretical system.

4*a*.   When using an actualistic method admitting strict uniformity in kind and energy throughout the ages, one may arrive at a geological system describing the situation of the earth in successive epochs in which the same circumstances and events are repeated with a large measure of uniformity.

The term *uniformitarianism* should be restricted to this subdivision. It may refer to a uniformitarian *system* or theory, propounding uniformity of material conditions and rates of change, as well as to a uniformitarian *method* (a subdivision of the actualistic method), asserting that the past should be reconstructed on the assumption that *all* geological causes (and not only the petrogenetical or even only the physical causes) of the past were of the same kind and intensity as those now in operation.

4*b*.   However, *uniformity* might refer not so much to uniformity of the *situation itself* as to uniformity of *change of the situation*. When a small rate of progressive or directed change prevailing now, is assumed to have prevailed always, a situation that is non-uniform throughout the ages is uniformly changed. In biology, the darwinistic protagonists of practically uniformly, or at least continuously, increasing complicatedness of animal structure, believed that they kept themselves to strict uniformitarianism. The resulting *system*, however, is *evolutionism* [3].

Perhaps one could say that both Lyell and Darwin used a uniformitarian *method*, but that Lyell (1830) arrived by its help at a uniformitarian geological *system*, whereas Darwin's theory of descent with modification is not a uniformitarian, but an *evolutionist system*.

5.   If the appearance of *new* causes in the course of the history of the earth is admitted, not *all* "actual" (present) causes could be used to explain past events. It would depend on the epoch concerned what part of them ought to be chosen. The *method* of explanation

---

[3]   Cf. R. Hooykaas, The parallel between the history of the earth and the history of the animal world. Arch. Internat. Hist. Sc. 10 (1957), pp. 1–18. Also: R. Hooykaas, Geological Uniformitarianism and Evolution. Arch. Intern. Hist. Sc. 19 (1966), pp. 3–19.

274

then would be actualistic (or perhaps even strictly uniformitarian), but the descriptive *system* would not be uniformitarian (cf. Johannes Walther's "Wüstenbildung")[4].

The above classification does not cover all differences of system and method and interpretation in geology. How far can we go back into the past in order to be able to speak of uniformity of the situation, or – less stringently – , of the applicability of "actual causes" in the explanation thereof? How long ought to be the period of change one takes into account for deciding whether a change is catastrophic or continuous?

Moreover, as to the identity of kind or the identity of energy of geological causes, a wide range of interpretation seems to be possible. It is difficult to establish what is meant by *geological* causes in contradistinction to *physical* causes. A good deal of confusion may arise through the ambiguity of the term "actual cause". It might be that a coincidence of fundamental primary physical causes (nuclear, atomic and molecular forces, gravitation etc.) in their combination gave rise to effects which were acting in their turn as causes of geological change (sedimentation, geochemical and petrogenetical phenomena), but that these ancient combinations or coincidences do not occur at present. In that case one might speak of *ancient* geological causes (dependent on an ancient geological situation and therefore only possible in the circumstances of the ancient world), and yet maintain that these are to be explained in an actualistic way, that is by physical forces similar to those active now. Thus actualism may be maintained on the level of physics, whereas, under the pressure of the evidence of geological observation, it is taken less strictly on the geological level.

When, however, the notion of "actual geological cause" has been widened then so far that it is practically considered as equivalent to *"physical* cause", systems based on a non-actualistic method become virtually non-existent. Only theories introducing supernatural, that is non-physical, causes would be non-actualistic then.

III. CATASTROPHISM

According to a widespread opinion, pre-scientific speculative systems, denoted as "catastrophism", prevailed in geology until, with Hutton's and Lyell's uniformitarianism and the overthrow of Cuvier's catastrophism, truly scientific geology triumphed.

It should, however, be borne in mind that uniformitarianism did

---

[4] R. Hooykaas, The Principle of Uniformity in Geology, Biology, etc. pp. 52–53.

275

not arise in a "catastrophic" way in the decades before and after 1800. Uniformitarianism and catastrophism already existed alongside each other in the 18th century. The cosmogonic systems of Burnet, Woodward and Whiston bore a strongly catastrophist character. Neither the kind, nor the energy of actual causes were considered sufficient to explain former changes. Moreover, these systems did not restrict themselves to changes in the *crust* of the earth, but embraced the genesis of the whole planet.

Over against them, less speculative, more scientific, systems, which were based on *observations* of the crust of the earth (the only part of the globe accessible to direct investigation), were put forward already in the late 17th and in the 18th century. Cuvier stated in 1821 that "since long it has been believed to be possible to explain past revolutions by actual causes" [5]. As a rule, those geologists who abstained from geogenic speculations and restricted themselves to explaining those changes the traces of which are still accessible to observation, tried to do so as much as possible with the help of causes they actually saw at work before their eyes.

a. *Buffon*

Buffon (1707–1788), having supposed the earth as detached from the sun by collision with a comet and then cooling off gradually, had no further use for this hypothesis. In his explanation of changes in the surface of the earth, he always referred to the actually existing causes.

In this "Théorie de la Terre" (written in 1744; published in 1749), he said that in the recent period (2000–3000 years) geological change was very small in comparison with "the revolutions which must have taken place in the first time after the creation", when the crust was much less solid than now [6]. So he had catastrophist ideas about the most ancient epochs, but at the same time he left no doubt about his actualistic conceptions, for he added that "consequently, the same causes which at present produce almost insensible changes in several centuries, must *then* have caused very great revolutions in very few years" [7].

On the other hand, when reconstructing the past situation of

---

[5] G. Cuvier, Discours sur les révolutions de la surface du globe, et sur les changemens qu'elles ont produits dans le règne animal. Paris 1826, p. 14.

[6] Buffon, Théorie de la Terre. Histoire Naturelle générale et particulière, Tome I, Paris 1749, p. 77.

[7] "... par conséquent les mêmes causes qui ne produisent aujourd'hui que des changemens presqu'insensibles dans l'espace de plusieurs siècles, devoient causer alors de très-grandes révolutions dans un petit nombre d'années; en effet il paroît certain que la terre actuellement sèche et habitée, a été autrefois sous les eaux de la mer, et que ces eaux étoient supérieures aux sommets des plus hautes montagnes ..." Buffon, o.c.p. 77.

276

what is at present dry land (but which for a long time has been covered by the sea), he starts from the assumption that it underwent "the same changes that the land *now* covered by the sea actually undergoes". "Therefore, in order to find what happened formerly on this earth, let us look at what is happening today at the bottom of the sea" [8]. The origin of the Atlantic Ocean may have been sudden (e.g. by the breaking down of a huge subterraneous cave and a subsequent universal deluge), or by slow action, but at any rate it was a *natural* event; "for deciding what has occurred and even what will occur, we have only to examine what is occurring" [9].

As "historians" we have, according to Buffon, to refuse to enter into vain and gratuitous speculations about the origin of the earth by the approach of a comet, etc. In order to have a firmer starting point, he intends himself "to take the earth as it is, to exactly observe all its parts and to conclude by inductions from the present to the past". He will not be affected by "causes whose effect is rare, violent and sudden", as "they do not belong to the ordinary course of nature", but he will use as "causes and reasons" only "effects which occur every day ... constant and always reiterated operations" [10]. Nevertheless, he recognizes that "sudden and rapid changes took place by inundations and earthquakes" [11], and he contrasts such "particular causes" (which produce upheavals, inundations and sinkings) with the continual and slow changes by the "general causes" (fire, air and water) [12].

In his later geological work, "Les Époques de la Nature" (1778),

---

[8] "... la partie sèche du globe que nous habitons a été longtemps sous les eaux de la mer; par conséquent cette même terre a éprouvé pendant tout ce temps les mêmes mouvemens, les mêmes changemens qu'éprouvent actuellement les terres couvertes par la mer. Il paroît que notre terre a été un fond de mer; pour trouver donc ce qui s'est passé autrefois sur cette terre, voyons ce qui se passe aujourd'hui sur le fond de la mer ..." Buffon, o.c., p. 81.

[9] "ce changement a donc pu se faire tout à coup par l'affaissement de quelque vaste caverne dans l'intérieur du globe, et produire par conséquent un déluge universel; ou bien ce changement ne s'est pas fait tout à coup, et il a fallu peut-être beaucoup de temps, mais enfin il s'est fait, et je crois même qu'il s'est fait naturellement; car pour juger de ce qui est arrivé et même de ce qui arrivera, nous n'avons qu'à examiner ce qui arrive". Buffon, o.c., p. 96.

[10] ",,, il faut le prendre tel qu'il est, et bien observer toutes les parties, et par des inductions conclure du présent au passé; d'ailleurs des causes dont l'effet est rare, violent et subit, ne doivent pas nous toucher, elles ne se trouvent pas dans la marche ordinaire de la Nature, mais des effets qui arrivent tous les jours, des mouvemens qui se succèdent et se renouvellent sans interruption, des opérations constantes et toujours réitérées, ce sont là nos causes et nos raisons". Buffon, o.c., pp. 98–99.

[11] Buffon, o.c., p. 605.

[12] Buffon, o.c., p. 609.

277

there is the same ambiguity. He emphasizes that the course of Nature is "not absolutely uniform", but that it undergoes "successive alterations, and is liable to new combinations", so that at present Nature is very different from what she was at the beginning and in the first periods [13].

Nevertheless, Buffon keeps to the actualistic method. In his opinion, when penetrating into the "night of time", one has to go back "only from existing facts to the historical truth of bygone facts"; one has to evaluate "not only the recent past, but also the most ancient past, by the present alone" [14].

The picture Buffon gives of the beginning of the Third Epoch is far from uniformitarian. He speaks of the "first moments of shock and agitation, of upheavals, irruptions and changes, which have given a second form to the greater part of the surface of the earth" [15]. "Nature was then in its first energy, and wrought the organic and living matter with a more active power and a higher temperature" [16]. Clay was produced in a shorter time than now, as the water was hotter, and, though this decomposition is still going on today, it is slower and less. [17]

----

[13] Buffon, Histoire Naturelle des Époques de la Nature. Histoire Naturelle, générale et particulière, Supplément V, Paris 1778, p. 3." . . . en l'observant de près, on s'apercevra que son cours n'est pas absolument uniforme; on reconnoîtra qu'elle admet des variations sensibles, qu'elle reçoit des altérations successives, qu'elle se prête même à des combinaisons nouvelles, à des mutations de matière et de forme: . . . et si nous l'embrassons dans toute son étendue, nous ne pourrons douter qu'elle soit aujourd'hui très-différente de ce qu'elle étoit au commencement et de ce qu'elle est devenue dans la succession des temps: ce sont ces changemens divers que nous appelons ses époques . . .".

[14] "ce n'est donc que de cet instant l'on peut commencer à comparer la Nature avec elle-même, et remonter de son état actuel et connu à quelques époques d'un état plus ancien. Mais comme il s'agit ici de percer la nuit des temps; de reconnoître par l'inspection des choses actuelles l'ancienne existence des choses anéanties, et de remonter par la seule force des faits subsistans à la vérité des faits ensevelis; comme il s'agit en un mot de juger, non seulement le passé moderne, mais le passé le plus ancien, par le seul présent . . .". Buffon, Époques, p. 5.

[15] "Quels mouvemens, quelles tempêtes ont dû précéder, accompagner et suivre l'établissement local de chacun de ces élémens! Et ne devons-nous pas rapporter à ces premiers momens de choc et d'agitation, les bouleversemens, les premières dégradations, les irruptions et les changemens qui ont donné une seconde forme à la plus grande partie de la surface de la Terre?". Buffon, Époques, p. 96.

[16] "La Nature étoit alors dans sa première force, et travailloit la matière organique et vivante avec une puissance plus active dans une température plus chaude . . .". Buffon, Époques, p. 99.

[17] "La décomposition des poudres et des sables vitrescibles, et la production des argiles, se sont faites en d'autant moins de temps que l'eau étoit plus chaude: cette décomposition a continué de se faire et se fait encore tous les jours, mais plus lentement et en bien moindre quantité". Buffon, Époques de la Nature, p. 103.

278

At any rate, in spite of his actualism, it goes too far to say that Buffon was "in advance of his time" by admitting only "actual and *slow* causes" [18]. There is always analogy, but not always identity of ancient and modern phenomena in his descriptions. His wavering attitude gives warning of the difficulties one meets when trying to classify geological theorists.

IV. NON-ACTUALISTIC CATASTROPHISTS (ad 1)

a. *G. Razumovsky*

In the ideas of the Russian geologist Count Gregor Razumovsky (1759–1837) we meet with another example of ambiguity. He was a neptunist and a catastrophist, that is, from the Huttonian standpoint he was as unorthodox as possible. But he was also an "actualist" in using physical and chemical *causes* which still are at work now for the explanation of "ancient" phenomena, whereas he was a non-actualist as well, in that he believed that these *phenomena* do no longer occur in nature today.

His catastrophism becomes evident e.g. when he says (1789) that the facts demonstrate that the environment of Lausanne has formerly been covered by waters, and that "the power of these waters is hardly conceivable by our imagination, as it was much superior to that of our most terrible modern waters in their effects" [19]. There are found there enormous boulders, which have a composition different from that of the surrounding rocks and similar to that of the most ancient Alpine rocks. This shows that they have been brought there by a strong torrent of water: "one cannot think of any power in nature today that would have been able to lift and to transport such large pieces so far from the place of their first formation" as has been done by those "ancient waters" [20]. The

---

[18] Cf. J. Roger, Mém. Mus. Hist. Nat. nouvelle série C, **X** (1962), p. 271. According to J. Staszewski (Kwartalnik Historii Nauki i Techniki IX (1964), p. 40) the actualistic principle was "a mere rudimentary conception in Buffon". The claim is then made that Hugo Kollontaj (1750–1812) was "the first actualist in geological history". Though it is recognized that he often depends on Buffon, the actualistic principle is said to be his "own, original creation". It is evident that Staszewski's under-estimation of Buffon's actualism matches the over-estimation by J. Roger and J. Piveteau.

[19] Cte G. de Razoumowsky, Histoire Naturelle du Jorat et de ses environs. Tome II, Lausanne 1789, p. 25. "... cette contrée ... a été couverte par les eaux, ... la puissance de celles-ci à peine concevable pour notre imagination, étoit bien supérieure à celle de nos eaux modernes les plus terribles dans leurs effets".

[20] "... on ne conçoit aujourd'hui aucune puissance dans la nature qui aye pu soulever et transporter des fragments de cette taille, si loin du lieu de leur première formation ..., on ne peut douter qu'elles ne se trouvent là encore à la place même où elles ont été déposées par les eaux anciennes". Razumovsky, o.c., p. 26.

environment of the Jorat unmistakably shows "incontestable monuments of the most astonishing catastrophes" [21].

In a later publication (1791) Razumovsky tackled the problem of the origin of the primitive rocks. In his opinion, granite was a product of crystallization from an "aqueous" fluid [22]. In order to find out from which solvent it has been crystallized, we have to take recourse either to the examination of the actions of still existing natural aqueous fluids on quartzeous matter, or to reasoning from analogy [23]. The only natural waters we know today (viz. fresh and salt water), do not dissolve quartz. Therefore, only the second way is open: chemistry teaches us that only "spar acid" [24] possesses this dissolving power. It seems, then, plausible that the globe originally was wholly covered by a sea containing this acid, whereas in more recent epochs salt water seas, resembling our modern seas, took its place. From the combination of the earthy, saline and acid principles, "according to the immutable laws of gravity, attraction and affinities", took rise this first crystalline kernel of the earth as well as the ancient fluid surrounding it [25].

Razumovsky, then, clearly states that the physical and chemical laws are immutable (and from this general viewpoint he could be called even a uniformitarian). In order to reconstruct the past, he starts from the *present* situation and from *actual* phenomena: he asks whether there are *now* causes active in nature that might give an explanation of an ancient event, and—when the answer turns out to be negative—he tries whether *now* experiments can

---

[21] "l'Histoire Naturelle du Jorat comme celle de toutes les Montagnes grandes ou petites, nous offre les traces non équivoques des révolutions successives des siècles les plus réculés. Chaque pas, nous y présente ces médaillons, ces monuments incontestables des catastrophes les plus étonnantes". Razumovsky, o.c., p. 228.
The interpretation of erratic blocks by Catastrophists and Uniformitarians will be dealt with in a forthcoming article.
[22] Comte de Razoumowski, Idées sur la Formation des Granites. J. de phys. **39**, (1791), p. 251.
[23] "Pour résoudre ce problème important d'une manière satisfaisante, nous ne concevons que deux voies: l'examen de l'action des fluides aqueux naturels que nous connoissons de nos jours sur la terre vitrifiable ou quartzeuse, qui forme la majeure partie des granits, et l'*analogie*." Razumovsky, l.c., p. 252.
[24] "Acide spathique", i.e. hydrofluoric acid. [24a] Razumovsky referred to the presence of fluor compounds in "primary" rocks as an argument in support of his hypothesis. As late as 1820 the Netherlandish scientist. H. C. van der Boon Mesch, in his "Disputatio geologica de Granite" (Lugd. Batav. 1820, p. 100) accepted Razumovsky's theory.
[25] "c'est . . . de la combinaison . . . de ces divers principes (terreuses, salines, acides) entre eux, selon les lois immuables de la pesanteur, de l'attraction et des affinités, qu'ont résulté cette première coagulation cristalline qui dès-lors a formé le noyau du globe, et ce fluide le plus ancien de tous qui ait jamais enveloppé notre globe". Razumovsky, l.c., p. 253.

be made in the laboratory which, by analogy, may reveal what possibly could have happened in the past. That is, he follows a truly *actualistic method*.

Nevertheless, he is also decidedly non-actualistic. Why do not we find remnants of this hypothetical fluid? The answer is: "one cannot compare the causes and the effects of such remote epochs, as one cannot compare their products; such rests do not exist and can no longer exist today, as neither granites nor fluor-spars are formed nor could be formed any longer, whatever may have been contended, ungroundedly, by a small number of naturalists" [26].

In this connection it is of no importance whether Razumovsky's hypothesis seems phantastic or not. What matters is, that he uses an actualistic method (comparison with phenomena occurring now; recognition of the immutability of physical and chemical laws), and that this leads him to conclusions that are decidedly non-actualistic. Moreover, the absence of any appeal to supernatural causes shows that catastrophism is not necessarily connected with "metaphysics".

b. *D. Dolomieu*

According to the system put forward (1791) by Déodat de Dolomieu (1750–1801) there has been a very slow sedimentation of the primitive rocks from the primeval ocean [27], during "thousands of centuries" [28]. After that period there occurred a worldwide catastrophe, which disturbed the horizontal layers by "a force of extraordinary violence" [29], and which gave rise to the primitive moun-

---

[26] "Que si l'on nous demande d'où vient qu'on ne trouve plus aujourd'hui des restes d'un fluide tel que nous le supposons, tandis qu'on trouve encore partout ceux qui ont formé les montagnes à couches? nous répondons qu'on ne peut pas plus comparer les causes et les effets d'époques si éloignées les unes des autres que leurs produits, que ces restes même n'existent ni ne peuvent plus exister de nos jours, puisque ni les granits, ni les fluors ne se forment ni ne peuvent plus se former, quoi qu'en aient prétendu, sans fondement, un petit nombre de naturalistes". Razumovsky, l.c., p. 253. Cf. p. 254.

[27] Commandeur Déodat de Dolomieu, Mémoire sur les pierres composées et sur les roches. In: Observations sur la Physique, etc. 39 (1791), p. 382: "Quel qu'ait pu être ce dissolvant, c'est avec M. de Saussure et M. de Luc que j'admets la précipitation comme première cause de la formation et de la consolidation des plus anciens matériaux de nos montagnes ... la précipitation s'est faite assez lentement ...".
Like Razumovsky, Dolomieu thought that the solvent which kept the siliceous matter that gave rise to the primitive rocks, in solution, finds no counterpart in nature now. In contradistinction to Razumovsky, however, he found no equivalent of it in the laboratory either. pp. 378–380.

[28] Dolomieu, o.c., p. 404.

[29] "La régularité du premier travail a été dérangée; une rupture a été produite par une cause quelconque, mais sûrement d'une force ou d'une violence extraordinaire ...". Dolomieu, o.c., p. 390.

281

tains [30]. After a long interval, an epoch of the formation of "couches de transport" started, in which enormous periodical inundations disturbed the regularity of the deposits of the first epoch. "No great antiquity" is supposed for "the actual order of things" [31].

Dolomieu energetically rejected the idea that during a very long period and with extreme slowness the sea could shape the surface of the earth: "When shaping the earth as we inhabit it, Nature has not spent time with such a prodigality as some famous authors did suppose" [32]. It seems probable that this is a thrust at Buffon, who—according to many contemporaries—was too lavish with thousands of years.

To Dolomieu, however, geological *facts* seem to point out the necessity of a catastrophist explanation: "Getting convinced that it is impossible that the sea, in its present circumstances, might operate anything similar to what exists on our continents ... the naturalist must imagine more powerful circumstances, capable of greater effects, in which, however, the sea must intervene, as there are certain proofs of its cooperation" [33]. "It is not by weak currents that I would open our valleys, but by all the power that the waters can receive from the uniting of the weight of a very large mass" increased by the acceleration through the impetus of their fall [34]: "it is not time that I will invoke, but it is force; one only relies on the first, when one does not know where to find the other" [35].

---

[30]   Dolomieu, o.c., p. 404.

[31]   ". . . je ne supposerois pas une bien grande antiquité à l'ordre actuel des choses". Dolomieu, o.c., p. 404.

[32]   "Que l'on ne me dise pas que la Nature ne compte pas avec le tems, que l'histoire des hommes est bien nouvelle; et que, dans le long période qui l'a précédée, la mer, quoiqu'avec une extrême lenteur, a pu faire tout ce qu'on lui attribue. Je conviendrai que le tems n'est rien pour la nature, mais cependant elle a placé au milieu de ses créations quelques bornes qui fixent différentes époques dans sa durée, et qui doivent modérer les élans de l'imagination. Tout me porte à croire qu'en façonnant la terre telle que nous l'habitons, la nature n'a pas dépensé le tems avec autant de prodigalité que quelques écrivains célèbres l'ont supposé". Dolomieu, o.c., p. 394.

[33]   "En acquérant la conviction de l'impossibilité où est la mer d'opérer, dans ses circonstances présentes, rien de semblable à ce qui existe sur nos continens, il (le naturaliste) ne peut plus supposer qu'elle y ait résidé long-temps; il doit imaginer des circonstances plus puissantes et capables de plus grands effets, où la mer doit cependant intervenir, puisqu'on a des preuves certaines de son concours". Dolomieu, o.c., p. 403.

[34]   "Ce n'est donc point la mer reposant tranquillement dans les bassins où elle est fixée par le centre de gravité de la terre, que j'appelle à la formation de nos couches, mais ce sont ses eaux dans le plus violent état d'agitation où elles puissent se trouver. Ce ne sera pas par de débiles courans que j'y ferai ouvrir nos vallées, mais par toute la puissance que l'eau peut recevoir de la réunion du poids d'une trèsgrande masse à une chûte précipitée". Dolomieu, o.c., p. 398.

[35]   "Ce n'est pas le tems que j'invoquerai, c'est la force; on ne place en général sa confiance dans l'un que lorsqu'on ne sait où trouver l'autre". Dolomieu, o.c., p. 399.

282

In a letter to H. B. de Saussure, for whom he had a great admiration, Dolomieu wrote that he would not have the slightest objection to abandoning his "hypothesis", if a more probable one could be put forward. He insisted, however, that this should present then "a cause sufficiently active for producing the required effects" [36].

The geological changes of the past, so he says, evidently are "outside the ordinary course of nature" [37], and therefore they cannot be explained by what is actually going on. It is precisely the comparison of the ancient phenomena with what the *actual* operations would effect if working in the past under the same circumstances as are prevailing now, which led him to the conclusion that actual causes are insufficient for explaining them: "At present the sea does not form similar strata; it does not excavate valleys; it does not bury lava currents under banks of calcareous rocks; it does not deposit salt mines, etc." He thinks that this will not be doubted by anybody who is free from "ancient prejudices" [38]. This dubbing the actualistic principle (according to which present causes must be sufficient for explaining past changes) an "ancient prejudice", demonstrates convincingly that at the end of the 18th century the actualistic method was not considered as something new.

To Dolomieu "the ideas of those who attribute an age of more than a hundred thousand years to our continents" [39] represents another prejudice, but this one he considers as of less importance, as it touches the system only and not the method.

c. *G. Cuvier*

Cuvier (1769–1832) was of the opinion that amongst those who have endeavoured to explain the present state of the globe, "hardly any one has attributed it entirely to the agency of slow causes, and still less to causes operating under our eyes" [40]. But, though approving of their non-actualism, he blamed these predecessors because this necessity of seeking causes different from those which we see acting at the present day, has made them "imagine so many extraordinary suppositions and lose themselves in so many erroneous and contradictory speculations, that the very name of their science

---

[36] Dolomieu à de Saussure, 26-4-1792. In: A. Lacroix, Déodat de Dolomieu, T. II, Paris 1921, p. 41.

[37] Dolomieu, l.c., p. 41.

[38] "Or, la mer ne forme point maintenant de couches semblables aux nôtres, ne creuse pas de vallées, n'ensevelit pas des courants de laves sous des bancs de pierre calcaires, ne dépose pas de mines de sel gemme, etc., etc. Je crois que pour ceux qui savent se défendre d'anciens préjugés, il ne doit rester aucun doute à cet égard". Dolomieu, l.c., p. 41.

[39] Dolomieu, l.c., pp. 42–43.

[40] G. Cuvier, Discours, etc., p. 21.

283

has long been a subject of ridicule" [41]. He deemed these early catastrophists too speculative and too ambitious, because they dealt with events (like the origin of our planet, or changes in the interior of the earth) of which, in his opinion, no trace has been left.

Accordingly, precisely like the uniformitarians before and after him, Cuvier rejected the cosmogonic systems of his predecessors, and he dated scientific geology from the moment that "it preferred the positive data furnished by observation, to fanciful systems, contradictory conjectures regarding the first origin of the globes" [42].

Evidently, the catastrophists of the school of Cuvier agreed with the uniformitarians of the Lyellian school in that they rejected the catastrophism of the cosmogonists, because it had not been built upon observations. But, for the same reason Cuvier and his disciples rejected uniformitarianism as well. They propounded their own theories not because of some prejudice in favour of catastrophes, but because they held that *observation* led to them.

Cuvier restricted his theory to those changes in the *crust* of the earth of which visible traces remain, and he was so successful, that, when Lyell in 1830 entered upon the scene, uniformitarianism had to fight its way with great difficulty.

In Cuvier's opinion the sudden transition of one kind of layer to another, and, in particular, the fossils they contain, testify to the revolutionary rapidity of certain changes which characterize the beginning of new geological and paleontological epochs. Consequently, the energy of these forces must also have been extremely great, "as no cause acting slowly could have produced sudden effects" [43].

Of course, the "ordinary" changes of the surface of the globe, caused by weathering, sedimentation and volcanic eruptions, were supposed to be common to all epochs. But geologists like Cuvier, Murchison and Élie de Beaumont deemed it impossible that any amount of these small agencies, though continued for millions of years, could have produced such results as the disruption and overturning of the mountain masses of the Alps, enormous dislocations which belong "distinctly to former epochs". The facts, according to Murchison, announce in emphatic language "how ordinary operations of accumulation were continued tranquilly during very lengthened epochs, and *how such tranquillity was broken in upon by great convulsions*" [44].

---

[41]   Cuvier, Discours, p. 21.
[42]   Cuvier, o.c., p. 145.
[43]   Cuvier, o.c., p. 21.
[44]   R. I. Murchison, Siluria, The history of the oldest known rocks containing organic remains. London 1854, p. 505.

That is to say, even in the past the more energetic causes were not always at work, but only during the relatively short periods of the catastrophes. Murchison and Sedgwick in England, and Élie de Beaumont in France, took their proofs of the greater *intensity* of former causation especially from the geological phenomena of the Alps, which, in their opinion, showed signs of former catastrophes, inexplicable by any reference to those puny oscillations of the earth which can be appealed to during the times of history [45].

But not only the *energy* of the causes of past geological changes was supposed to have been different from that of the causes now in operation,—these causes were sometimes also supposed to have been of a *different kind*. As Cuvier wrote: "It is in vain that we search among the powers which now act at the surface of the earth, for causes sufficient to produce the revolutions and the catastrophes, the traces of which are exhibited by its crust" [46]. . . "The thread of operations is broken; the march of Nature is changed, and none of the agents which she now employs, would have been sufficient for the production of her ancient works" [47].

But even Cuvier believed that, if not identity, at least some *analogy* with the physical phenomena of the present is indispensable for the reconstruction of the past. He deemed it the error of the cosmogonists that, inventing systems built upon "phenomena, which, having no resemblance to those of our actual physics, could find in it, for their explication, neither materials, nor touchstone"; "the geologists of whom I speak, neglected precisely the posterior facts which could alone have reflected some light upon the darkness of preceding times" [48].

And, finally, even catastrophists like Deluc and Cuvier distinguished between ancient causes which have ceased to act in the crust of the earth, and other causes, which have continued their activity up to the present day. Cuvier fully recognized the right of the actualistic method to be used for explaining phenomena which had occurred *between* the catastrophes or *after* the last revolution.

V. ACTUALISTIC CATASTROPHISTS (ad 3*b*)

a. *L. Élie de Beaumont*

Élie de Beaumont (1798–1874) was of opinion that those who would refuse to believe that the causes now in operation could ever have produced the great geological phenomena, would reason like

---

[45] Murchison, o.c., p. 476.
[46] Cuvier, Discours, p. 20.
[47] Cuvier, o.c., p. 14.
[48] Cuvier, o.c., p. 145.

people who, while having no experience of cold below the freezing point, would deny that water could ever become a solid body. According to his fundamental hypothesis, the irregularities of the crust of the earth, in its outward form as well as in its structure, result from the disappearance of part of the heat that the earth contained when its crust was still in a state of fusion. The "slow and continuous" phenomenon of cooling of the earth causes a slow and progressive diminution of its volume, from which ensues the rise of the mountains [49]. This cooling, which acts as a slow and gradual cause, has as its effects violent and sudden cataclysms, – "of a very short duration, and, as it were, instantaneous". Consequently, there are long periods of quietness, alternating with short periods of revolution [50].

Evidently, though being a catastrophist, Élie de Beaumont did not recognize ancient causes that are *essentially* different from the causes now in operation. The slow tectonic effects of the present day result from the same fundamental cause as the sudden and violent effects of the past. These latter, moreover, are not even essentially *ancient effects*, for it is possible that in the future a new catastrophe will strike the surface of the earth.

In Élie de Beaumont's system, however, this does not imply an eternal repetition of revolutionary and gradual effects in perfectly similar cycles. In his opinion, the effects of the causes presented by the phenomena of the past often differ from the phenomena of the present. He speaks even of the "gradual enfeeblement of the chemical agents which have been active on the surface of the globe" [51].

Petrogenesis

Élie de Beaumont established relations between the geochemical data and the sequence of events, thus indicating the way modern geochemistry would follow. He thought that in the formation of granite "extremely ancient phenomena, which must have been

---

[49] L. Élie de Beaumont, Notice sur le système des montagnes. Paris 1852. T. III, p. 1329.
Élie de Beaumont had already put forward his catastrophism in his "série de recherches sur quelques-unes des révolutions de la surface du globe", in Ann. sciences naturelles 19 (1829–'30).
[50] Élie de Beaumont, Notice, III, p. 1329.
[51] Élie de Beaumont, Note sur les émanations volcaniques et métallifères. Bull. Soc. géol. France [2], T. IV (1846–'47), p. 1331: "L'affaiblissement graduel des agents chimiques qui ont agi à la surface du globe, comparé à l'ordre suivant lequel y ont apparu les différentes classes d'êtres organisés, laisse apercevoir dans l'histoire de la nature un plan aussi harmonieux que celui qu'on admire dans la constitution de chaque être en particulier".

286

different from those occurring today on the surface of the globe", have been involved. There is an "enormous difference" between the phenomena characteristic of the epoch when granite was formed, and what happened later on in the formation of the other crystalline rocks. A large part of the chemical elements has been chemically bound in this first epoch, so that it could never reappear afterwards, and this fact alone indicates a gradual change in the course of geological phenomena [52]. In his petrogenetical conceptions, then, Élie de Beaumont was decidedly non-actualistic.

Summarizing, we may conclude that Élie de Beaumont's geology admits that there is a general decrease of the energy of geological causes, together with a gradual decrease of temperature and of the number of elements that participate in the formation of rocks. And, besides this qualitative and quantitative change of a continuous character, there are the cataclysms.

It goes without saying, that Uniformitarianism is energetically rejected by this catastrophist: if everything had always happened in the same way in perpetual geological cycles without essential change, in all mineral deposits the same elements would be found, so he says [53], whereas, in his opinion, this is not so in fact.

b. *L. Frapolli*

In the same year (1846–'47) L. Frapolli took over this distinction between *periods of tranquillity* (slow upheavals) and *epochs of*

---

[52] "... la concentration du *silicium*, du *potassium* et d'une classe nombreuse de métaux dans les granites ... remonte nécessairement à des phénomènes extrêmement anciens qui ont dû être différents des phénomènes qui se passent aujourd'hui sur la surface du globe; que lors de la coagulation de la première enveloppe du globe terrestre, il doit avoir existé une cause quelconque pour qu'un grand nombre de corps fussent retirés de la circulation; qu'il y a eu une énorme différence entre les phénomènes propres à l'époque où le granite s'est formé et ce qui s'est passé plus tard, lors de la formation des autres roches cristallines; d'où il résulte que les phénomènes qui se sont accomplis sur la surface du globe ont suivi une *certaine gradation*". Élie de Beaumont, Note, p. 1330.

[53] "Quelle qu'ait été la nature des premiers phénomènes géologiques, une grande partie des corps simples ont été alors séquestrés de manière à ne plus reparaître ailleurs, et ce fait seul indique un changement graduel dans la marche des phénomènes géologiques. On voit combien cela est contraire à certains systèmes dans lesquels on suppose que tout s'est constamment passé de la même manière sur la surface de la terre, et que l'origine du globe se perdrait dans la nuit d'une période indéfinie, pendant laquelle les phénomènes géologiques auraient tourné perpétuellement dans le même cercle. Si tout s'était toujours passé de la même manière, sans aucun changement essentiel, on trouverait dans tous les gisements de minéraux la même série de corps simples, et non pas une série plus nombreuse dans les gîtes formés les premiers que dans ceux formés les derniers". Élie de Beaumont, Note, p. 1330.

287

agitation (sudden upheavals, ruptures, inundations) [54]. In his opinion the problem of drawing an exact borderline between the products of the cataclysmic periods and those due to the "ordinary agents of the physical forces" and the activity of air and water during the periods of tranquillity, has been thrown into a regrettable confusion by "the substitution of phantastical agents for the real and actual causes" [55].

Evidently, with Frapolli a catastrophistic *system* does not exclude an actualistic *method*. He says that in the periods that are analogous to the present one, "similar causes produced effects resembling those we may observe in our time". There is, however, one restriction: the greater power of the chemical agents and the meteorological influences, which especially in the first epochs must have been modified by the higher temperature and by the different composition of the atmosphere, "must have made some difference" [56].

It goes without saying, that this difference from contemporary phenomena is more evident in the case of periods of agitation. But even for that case there is nothing to indicate that Frapolli would have resorted to "ancient" causes.

c. *Ch. Sainte-Claire Deville*

In the long run, through Lyell's triumph, Élie de Beaumont's system, which had been the orthodox one at least in France, became ridiculous in the eyes of the Uniformitarians. Nevertheless, he has had devoted partisans amongst later geologists, e.g. Charles Sainte-Claire Deville (1814–1881) [57] and, more recently, G. Simoens [58] (1907).

---

[54] L. Frapolli, Réflexions sur la nature et sur l'application du caractère géologique. Bull. Soc. géol. France [2], IV, pp. 623–625.

[55] "Malheureusement la substitution *d'agents fantastiques* aux *causes réelles et actuelles* a jeté dans ces derniers temps cette partie de la géologie dans une si déplorable confusion, qu'elle est à peu près encore à refaire". Frapolli, l.c., p. 626, note 1.

[56] (*"périodes de tranquillité"*:) "Dans ces périodes, espaces de temps analogues à celui où nous vivons, des causes semblables produisaient des effets pareils à ceux que nous pouvons observer de nos jours. Une plus grande puissance des agents chimiques, et les influences météorologiques modifiées, surtout dans les premiers temps, par la plus grande uniformité d'une température plus élevée, par la composition des eaux et de l'atmosphère de l'époque, par la disposition des mers et des continents, par l'existence probable d'une plus grande quantité de sources minérales et thermales, ont dû seules y apporter quelque différence, et réagir surtout puissamment sur la vie des végétaux et des animaux, en leur imprimant en général un cachet de contemporanéité respective; 2° des *époques d'agitation*, moments de soulèvement brusque et de *rupture*, marqués par l'arrivée des matières intérieures à la surface". Frapolli, l.c., pp. 624–625.
See below the paragraphs on Cotta and Bronn, showing similar conceptions.

[57] Ch. Sainte-Claire Deville, Coup d'oeil historique sur la Géologie. Paris 1878.

[58] G. Simoens, La Théorie de l'Évolution cataclysmique et de l'Évolution alternante. Paris-Bruxelles, 1936.

288

In his lectures delivered in the Collège de France in 1875, Sainte-Claire Deville is an actualist, though not a uniformitarian. He says that the opinion which Lyell, wrongly imputed to the geologists (viz. that ancient causes are wholly different from those that are behind the gradual changes we see today)[59], is a "geological heresy" [60], accepted only in *geogeny* and not in *geology*, with the exception "perhaps" of Cuvier [61]. Everybody agrees that the great geological causes, like the great astronomical causes, "cannot any longer be supposed to have ceased to exist at a certain moment" [62].

Even, in spite of Lyell's protests against Cuvier's wording, Sainte-Claire Deville thinks that one cannot reasonably suppose, that – when he spoke of "actual causes" – , Cuvier could have meant that there are two categories of forces of an essentially different nature, for such a proposition "would strike us by its absurdity" [63] Cuvier evidently meant that no agent "in its actual force and expression could have caused these ancient phenomena; he did not wish to say that the same agents, moved by incomparably superior forces, could not have produced the observed effects [64].

Sainte-Claire Deville himself, too, though recognizing with Lyell that "the ancient causes were the same as those we see active before our eyes", is not willing to admit that "the *intensity* of those forces has always, in all periods . . . been identical with that of the present time" [65]. Consequently, the final aim of geology precisely is to see how "essentially identical *causes*" could produce "exceedingly variable effects" [66].

Almost inevitably, such a conception must lead Deville into a non-actualistic direction (except for the elementary physical processes), as these effects in their turn become geological causes. And also, to Sainte-Claire Deville an actualistic *method* leads to a decidedly non-uniformitarian system.

The variability of effects goes so far, in his opinion, that there are phenomena that do not come back: mineral waters in the past deposited substances which are not (or almost not) formed in more recent deposits; the atmosphere has lost the substances that are harmful to the development of living beings [67/68]. The chemical

---

59   Deville, o.c., p. 251.
60   Deville, o.c., pp. 578, 208.
61   Deville, o.c., p. 252.
62   Deville, o.c., pp. 218–219.
63   Deville, o.c., p. 218.
64   Deville, o.c., p. 218.
65   Deville, o.c., p. 250.
66   Deville, o.c., pp. 379–380.
67/68   Deville, o.c.; p. 269. Élie de Beaumont had already made a similar remark on the more recent rocks, when saying that these are less harmful to the growth of plants and animals: "Cette marche graduée, suivant une

289

conditions of volcanic emanations have totally changed; modern lavas have no equivalent in the granite epoch [69].

A whole chapter [70] of Deville's book is devoted to an answer to the problem of "variation of the intensity of geological phenomena". In the Carboniferous period there has been in the atmosphere an enormous production and consumption of carbon-bearing gases, which finds no analogon in the recent period [71]; in some ancient periods the glaciers were larger than the present ones [72]; the causes remained, but their effects diminished.

Moreover, Sainte-Claire Deville, repeating an old catastrophist argument [73], is of opinion that we should emphasize over against Lyell, that a weak force would not always be able to perform in much time what a greater force can do in a short time. Consequently, "the explanations of gigantic phenomena by means of relatively microscopic forces that are still active before our eyes" crumble down [74]. And then Deville assumes exactly the same methodological position as Conybeare had taken 45 years earlier: "instead of torturing the facts in order to make them fit in with those precon- ceived ideas" (scil. of Lyellian uniformitarianism!), it would be better to "follow the really scientific way and to find out which are the phenomena of different order . . . and which are the variations their effects seem to have undergone, from ancient times of the earth up till the present time" [75].

Like Cuvier and Élie de Beaumont before him, Sainte-Claire Deville divides the *effects* of geological causes into two categories: *slow and continuous effects* (sedimentation, gradual elevation of continents), and *sudden and violent effects* (upheaval of mountains) [76]. He thinks that the slow and continuous causes have a tendency to lose intensity and that their effects, therefore, are becoming smaller: organic sedimentation is now practically restricted to tropical

---

progression décroissante, des phénomènes chimiques, est une des merveilles de la nature . . . Le globe terrestre était destiné aux êtres organisés qui ont peuplé sa surface, et l'ordonnance général des phénomènes inorganiques dont il a été succesivement le théâtre, était étroitement liée au plan général de la nature organique . . . les corps simples, qui, par leur nature, auraient pu exercer une action délétère sur les êtres organisés, ou qui devaient rester étrangers à leur composition, ont été retirés, en grande partie, de la cir- culation dès les premiers âges du monde". Élie de Beaumont, Bull. Soc. géol. 2, IV, p. 1331.

[69]   Deville, o.c., pp. 256–257.
[70]   Deville, o.c., pp. 241 ff.: Neuvième leçon: Y-a-t'-il eu variation dans l'intensité des phénomènes géologiques?
[71]   Deville, o.c., p. 253.
[72]   Deville, o.c., p. 254.
[73]   See below § VIb on Greenough, Cuvier and Conybeare.
[74]   Deville, o.c., p. 259.
[75]   Deville, o.c., p. 260.
[76]   Deville, o.c., p. 264.

290

regions; the gradual movements of the continents took place on a much larger scale when the solidified crust was thinner; volcanic eruptions and earthquakes are less energetic and less frequent today than in the past [77].

But, if, "attributing a greater importance to actuality in geology", one goes back to the great phenomena of elevation of mountains, one recognizes that the mountain chains that arose most recently, stand out in highest relief [78]. The more the thickness of the crust grew, the greater became the force necessary for breaking it. Consequently, the phenomena of dislocation have acquired a greater violence and the periods that separate them have grown longer [79]. That is to say, that there is a tendency of divergence between the ordinary geological phenomena and the cataclysms [80].

In our classification of geological methods, Sainte-Claire Deville's catastrophism, which admits nothing but *actual* causes, (causes which still are in operation), would be a kind of actualism. Sainte-Claire Deville himself, however, kept to the general use of terms. When speaking of the "actualistic method [81], or ,,the theory of actual causes" [82], he meant Lyell's strict uniformitarianism which he energetically combated.

VI. THE METHODOLOGICAL DIFFERENCE BETWEEN CATASTROPHISTS AND UNIFORMITARIANS.

a. *Physical causes. W. Conybeare*

We have distinguished non-actualistic from actualistic catastrophists, both standing in opposition to uniformitarianism. There is no hard and fast dividing-line between these two categories of catastrophists. The choice will depend on what kind of causes one takes into consideration: the more elementary causes, or the more complicated ones that are themselves the effects of the primary causes. It depends also on the willingness to regard difference of tempo as non-essential.

In general, both groups were actualists at least in so far as they supposed that the same *physical* causes as those prevalent today, were also behind the phenomena of the most ancient epochs and that the same physical laws describe the slow changes as well as the cataclysmic ones [83].

---

[77] Deville, o.c., pp. 267–270.
[78] Deville, o.c., p. 268.
[79] Deville, o.c., p. 269. Cf below Conybeare, l.c., p. 361.
[80] Deville, o.c., p. 268.
[81] Deville, o.c., pp. 253, 252.
[82] Deville, o.c., p. 571.
[83] This has been emphatically declared by catastrophists of the English school: Conybeare, Sedgwick, and Buckland.

291

Perhaps, one may even say that secondary, „geological", causes too were to a large extent considered to have been always essentially the same. From the beginning of the controversy between uniformitarians and catastrophists there was a misconception about the catastrophist position on this issue.

In 1830, immediately after the publication of Lyell's work, William Conybeare pointed out that Lyell's frequent use of the phrases "existing causes" and "uniformity of nature" seemed to imply that the catastrophists speculate on causes of a different order from any with which we are acquainted, and even on the supposition of different laws of nature. In his opinion, however, "both parties equally ascribe geological effects to known causes, viz to the action of water, and of volcanic power" [84].

After this actualistic statement, however, Conybeare immediately added that the catastrophists maintain that much which has resulted from aqueous action, e.g. the excavation of many valleys, "indicates the violent action of mighty diluvial currents" rather than effects "which do or can result from the present draining (of rain water) by the actual rivers ... to which Mr. Lyell looks exclusively" [85].

That is to say, Conybeare referred to geological causes which are *not* actually working now ("diluvial currents"), though he maintained that it is the power of water (an *actual* cause) which was then and now in operation.

Whether one would call Conybeare c.s. actualists or non-actualists, then depends on how far one is willing to go back in the series running from highly complicated combinations of causes up to simple, primary causes. With Conybeare the two most primitive steps—primary mechanical forces (of collision and gravitation of matter), manifesting themselves in the more special impact of water on rocks—, function in an actualistic way. From that point on, however, his and Lyell's ways part: one, in a non-actualistic way, refers to "combinations of forces" not occurring now (viz. "diluvial currents"), whereas the other, in an actualistic way, keeps to the still existing "slow excavation".

It depends on where one puts the accent, whether one would call Conybeare's geology actualistic or not. Dolomieu and Conybeare assumed about the same position on extraordinary catastrophes in the remote past, but Dolomieu,—thinking of the debacles—, spoke of a "different order", whereas Conybeare, referring to the

---

[84] W. D. Conybeare, An examination of those Phenomena of Geology, which seem to bear most directly on theoretical Speculations. Phil. Mag. 8 (1830), p. 360.

[85] Conybeare, l.c., p. 360.

activity of water and heat in general, could maintain that the
"same order" is still reigning.

b. *Multiplication of small effects*

The argument in favour of the non-actualistic aspect of the
catastrophist explanation, is that a cause not powerful enough to
have a certain effect (e.g. imparting of movement to a boulder)
in a short time, is neither able to do so in a long time. Dolomieu
did not observe even a small beginning in the present of some
effects of the past and, therefore, he supposed a different order of
events (greater intensity of operations) for the most ancient periods.
G. B. Greenough (1819) attacked the plutonists because they think
that slow action during a long time may perform the same thing
as violent action during a short time: "What profit can a man
expect from putting zeros out to interest?", he asked. "If seas
and rivers do not tend to produce within the period of human experience, any such effect as that which we are endeavouring to
account for (scil. mountain formation), they will evidently produce
no such effect in a million of years" [86].

In the same way Cuvier, when speaking of species transformation,
pointed out that what is produced on this issue in a *short* time, — to
wit *nothing*—, yields *nothing* in a long period [87]. Even Hutton
recognized the legitimacy of such a reasoning, when saying that
no change, when multiplied, remains no change. He added, however,
that a *small* change then becomes a large one! [88].

The question is: where is the „*nothing*" that cannot be multiplied? The most basic causes (gravitation, e.g.) are active always
and everywhere, but certain combinations of them are less general
the further one goes in the series running from general and simple
component causes to their complicated combinations. These latter
do not occur in all times and places. Consequently, being absent,
they cannot be "multiplied". But if we emphasize that finally
all phenomena have as their *primary* causes such fundamental and
immutable ones as gravitation, collision, etc., the controversy
between uniformitarians and catastrophists becomes meaningless,
as *all* geologists would admit the immutability of physical laws
as a working hypothesis.

Dolomieu stressed that certain causes and effects (violent outbursts of geological activity) do not occur now and, consequently,

---

[86] G. B. Greenough, A critical Examination of the First Principles of
Geology. London 1819, pp. 148–149. Greenough resorts to one universal
deluge, whereas earlier geologists (Pallas, James Hall) had invented a
plurality of partial debacles.
[87] Cuvier, o.c., p. 63. Cf The Principle of Uniformity, p. 71.
[88] R. Hooykaas, The Principle of Uniformity, pp. 94–95.

293

he could consider himself a non-actualist. Conybeare, on the other hand, looking to more general causes (water, heat), which are still active now, could maintain that he was true to the actualsitic method. Nevertheless, their methodological positions do not essentially differ.

Secondly, when the present causes are not deemed sufficient to explain ancient phenomena, one may try to find "ancient" causes of a different kind. But this difference is in general reduced to a simple difference in activity for causes still in operation today. *Identity* may be abandoned, but *analogy* remains. The catastrophists were willing enough to accept as much uniformity in nature as seemed warranted by observation, and to go as far as possible with the actualistic method, but they were of opinion that hard facts forced them to abandon, at a certain level, the uniformity of energy of causes or tempo of actions. It was because he judged it impossible to reconcile his *observations* with the uniformitarian hypothesis that Dolomieu resorted to ancient violent actions: "Getting persuaded that the cause of all he sees does not belong to the actual order of events, the naturalist will be authorized to seek it in a different order" [89]. He declared to be willing to give up his own theory as soon as it would be contradicted by reliable observations, but he was unwilling to stick to the methodological necessity of explaining past phenomena by what is actually going on [90].

Conybeare too emphasized that "the only question appears to be whether we prefer embracing an adequate or an inadequate cause" [91]. He tried to demonstrate that violent currents *must* have swept over our continents at several periods, and to him there can be "nothing unphilosophical in supposing that volcanic agency might have been capable of acting with greater energy" at the beginning of the formation of the crust of the earth, "than

---

[89] ". . . car, lorsqu'il (le naturaliste) sera persuadé que la cause de tout ce qu'il voit n'est point dans l'ordre actuel des événemens, il sera autorisé à la chercher dans un ordre différent". Dolomieu, Mémoire, etc., p. 403.

[90] "Mais comme les faits valent mieux que les systêmes les plus séduisans, je renoncerai au mien aussitôt que quelques observations bien faites y seront directement contradictoires". Dolomieu, Mémoire, p. 407.

In his letter to de Saussure, after having rejected the prejudice against ancient causes, he continues: "Je crois que pour ceux qui savent se défendre d'anciens préjugés, il ne doit rester aucun doute à cet égard. Autant donc que je crois devoir insister sur cette première partie de mes opinions, autant je tiens peu à l'hypothèse à laquelle j'ai dû recourir pour expliquer des faits en apparence contradictoires et qui sont certainement hors du cours ordinaire de la nature . . . Si un système plus vraisemblable m'est présenté, je l'adopterai volontiers . . .". Dolomieu à Saussure, Lacroix, o.c., p. 41.

[91] Conybeare, o.c., p. 361.

at present, when the whole weight and resistance of the actual crust opposes it" [92].

The present, then, to him turns out not to be a mere repetition of the past. Even Hutton's supporter, *James Hall*, though maintaining the cyclical, a-historical conception of the uniformitarian theory, felt himself urged by observation of the geological situation, to introduce periods of violence and revolution [93].

c.   *Sedgwick on method*

On the 18th of February 1831 the Rev. Professor Adam Sedgwick retired from the President's chair of the Geological Society of London with an address animated by the same spirit as Rev. W. Conybeare's article.

Sedgwick emphasized that "geology is a science of observations" [94], and his main objection against Hutton and Lyell is that they put forward arbitrary dogmas (the repetition of similar cycles) and *a priori* principles, already enounced on the title page of Lyell's book, which says that it is "an attempt to explain the former changes of the earth's surface, by reference to causes now in operation". Nevertheless, Sedgwick himself put forward also an a priori principle, viz. the constancy of the primary laws of physics (law of gravitation, laws of atomic affinity) [95]. This a priori belief, however he gave a more legitimate status when affirming at the same time that it is an empirically established fact [96].

The fundamental primary processes, then, are combined in "results of indefinite complexity ... which are removed far out of the reach of any rigid calculation". Volcanic forces, e.g., are the "irregular secondary results" of masses of matter obeying the primary laws of atomic action; in their turn they act as secondary, "geological" causes [97]. It seems to Sedgwick a "merely gratuitous hypothesis ... unsupported by the direct evidence of fact", that they have acted at all times and in each period with equal intensity.

---

[92]   Conybeare, l.c., p. 361.
[93]   R. Hooykaas, The Principle of Uniformity, pp. 20–23.
[94]   Rev. Prof. Sedgwick, Address to the Geological Society, delivered on the evening of the 18th of February 1831. Phil. Mag. 9 (1831), pp. 281–317; p. 298.
[95]   "I believe that the law of gravitation, the laws of atomic affinity, and, in a word, all the primary modes of material action, are as immutable as the attributes of that Being from whose will they derive their only energy". Sedgwick, o.c., p. 301.
[96]   "We show by the help of records, not to be misinterpreted, that during this vast lapse of time, in the very contemplation of which our minds become bewildered, the law of gravitation underwent no change, and the powers of atomic combination were still performing their office". Sedgwick, o.c., p. 300.
[97]   Sedgwick, o.c., p. 301.

Such a theory confounds the immutable primary laws of nature with the mutable results arising from their irregular combination, and it assumes that no elements have ever been brought together, which we ourselves have not seen combined [98]. In Sedgwick's opinion this is the prejudice of limiting the possibilities of nature by our own daily experience or by our own understanding [99].

Evidently referring to Lyell's parallel of the repetition of astronomical constellations and geological cycles, Sedgwick denied that the great phenomena of geology, "where the combinations are mutable and indefinite", where we have "no vestige of returning periods", and where the fixed elements of force are imperfectly known, could be compared with celestial movements which return in themselves and can be calculated. As in morals, so in physics, "the continued action of immutable (primary!) causes may and does coexist with mutable phenomena" [100]. Thus Sedgwick clearly pointed out the historical, non-repeatable aspect of geological events, which become the more historical the more complicated they are.

The limits of geological changes, so he went on, "may be studied in the records, but cannot be fixed by any a priori reasoning, based upon hypothetical analogies" [101]. This refers to the *kind* of geological causes, but Sedgwick protested also against their *energy* being submitted to apriori limitations [102]. If Lyell's principles be true, "there can be no great violation of continuity" [103], and this is again an unwarrantable prejudice.

Sedgwick wanted first of all to register facts and to build up his theory aposteriori: "We must banish all *a priori* reasoning from the threshold of our argument"; theory should only appear "as the simple enunciation of those general facts, with which, by observation alone, we have at length become acquainted" [104]. In his opinion, Lyell's "Principles of Geology" violates this proposition; the great objection against this book is that it *starts* from a hypothetical assumption and then interprets the phenomena in accordance with it: "from the very title page of his work, Mr. Lyell seems to stand forward as the defender of a theory" [105]. In this way he "vitiates all the great results of our observations", excluding beforehand a general cooling down of the earth, alternate periods of violence

[98]   Sedgwick, o.c., p. 301.
[99]   Sedgwick, o.c., p. 302.
[100] Sedgwick, o.c., p. 302.
[101] Sedgwick, o.c., p. 303.
[102] Sedgwick, o.c., p. 303.
[103] Sedgwick, o.c., p. 306.
[104] Sedgwick, o.c., p. 303.
[105] Sedgwick, o.c., p. 303.

and tranquillity, difference of mineral genesis in subsequent epochs, origin of new animal and vegetable types, and admitting only those interpretations in which "operations now going on, are not only the type, but the measure of intensity of the physical powers acting on the earth at all anterior periods" [106]. Thus Lyell in the general statement of his results has sometimes been warped by his hypothesis: instead of describing the history of nature, he has been defending his hypothesis; "in the language of an advocate, he sometimes forgets the character of an historian" [107].

Now Sedgwick, speaking for all catastrophists (Conybeare and Buckland included) [108], fully recognized that we have to apply an *actualistic method* of interpretation of geological phenomena in that we assume that the fundamental laws of physics did never change, and in that the present yields as it were the coordinates for the description and interpretation of the past:

"For we all allow, that the primary laws of nature are immutable — and that we can only judge of effects which are past, by the effects we behold in progress" [109].

But we cannot say *apriori* in how far the secondary combinations are the same in different periods:

"But to assume that the secondary combinations arising out of the primary laws of matter, have been the same in all periods of the earth, is ... an unwarrantable hypothesis with no *a priori* probability, and only to be maintained by an appeal to geological phenomena" [110].

Lyell's error, then, is in his opinion, that he did not decide *aposteriori*, from the phenomena themselves, which secondary combinations, and to what extent, were active in the past, but that he decided beforehand that *all* secondary combinations of the present were the same as those of the past and were working with the same intensity then. That is, Lyell's extremely actualistic *method* (putting forward the uniformity of kind and energy of secondary causes as a methodological principle) inevitably led to a *system* that was uniformitarian too, a system in which "all we now see around us is only the last link in the chain of phenomena arising out of a uniform causation, of which we can trace no beginn-

---

[106] Sedgwick, o.c., p. 304.
[107] Sedgwick, o.c., p. 303.
[108] Rev. W. Buckland, Geology and Mineralogy considered with reference to Natural Theology, vol. I. London 1836, p. 11: "Geology has already proved by physical evidence, ... that the ultimate atoms of the material elements, through whatever changes they may have passed, are, and ever have been, governed by laws, as regular and uniform, as those which hold the planets in their course".
[109] Sedgwick, o.c., p. 305.
[110] Sedgwick, o.c., p. 305.

ing and of which we see no prospect of an end" [111]. If Lyell's principles be true, the earth's surface ought to present "an indefinite succession of similar phenomena". Sedgwick, however, would enounce the inverse proposition and affirm that the earth's surface "presents a definite succession of dissimilar phenomena" [112]. As we know nothing of the secondary causes but by the *effects* they have produced, "the undeviating uniformity of secondary causes" [113] and other Lyellian phrases of like kind, only enunciate the proposition of a hypothesis, but do not "describe the true order of nature". We may, as Lyell does. *imagine* indefinite cycles and the indefinite succession of phenomena, but these things "do not belong to inductive geology", and "all I now contend for is, that in the well-established facts brought to light by our investigations, there is no such thing as an indefinite succession of phenomena" [114].

What, then, are those facts established by "inductive geology"?

d. *Sedgwick on the geological system*

Over against "one of the arbitrary dogmas of the Huttonian theory", viz. the doctrine of geological cycles, Sedgwick puts that there are indications of a primeval fluidity of the earth before the commencement of the typically geological phenomena, and of a great diminution of temperature before the earth was fitted for the habitation of organized beings [115]. Though the records show the constancy of the primary physical laws [116], the "evidence of fact" points out the non-uniformity of e.g. volcanic forces. Moreover, "inductive geology" demonstrates that there is no indefinite repetition of the same events: "in the well-established facts brought to light by our investigations, there is no such thing as an indefinite succession of phenomena". Between successive formations there is a mineralogical distinction as well as one of animal and vegetable forms, many types of which are now not living any more [117].

In particular the paleontological record shows, according to Sedgwick, "a series of proofs the most emphatic and convincing—, that the existing order of nature is not the last of an uninterrupted succession of mere physical events derived from laws now in daily operation: but, on the contrary, that the approach to the present system of things has been gradual, and that there has been a progressive development of organic structure subservient to the

---

[111] Sedgwick, o.c., p. 304.
[112] Sedgwick, o.c., p. 305.
[113] Cf. Lyell, Principles of Geology, vol. I, sec. ed., p. 86.
[114] Sedgwick, o.c., p. 305.
[115] Sedgwick, o.c., p. 299.
[116] Sedgwick, o.c., p. 300.
[117] Sedgwick, o.c., p. 305.

298

purposes of life". The recent appearance of man is "by itself absolutely subversive of the first principles of the Huttonian hypothesis" [118].

Not only equality of the average situation of the earth's surface, but also continuity of local change belonged to Lyell's system: "In the speculations I am combating, all great epochs of elevation are, and I think unfortunately, excluded", says Sedgwick [119]. Over against this exclusion apriori, he puts that structure and position of successive formations prove that there have been "enormous violations of geological continuity", produced by forces adequate to these effects [120]. Small wonder, then, that Sedgwick welcomes Élie de Beaumont's catastrophist theories as "little short of physical demonstration". He shares Beaumont's idea that "comparatively short periods of violence and revolution", during which the continuity was broken and elevation took place, were followed by changes in many of the forms of organic life, whereas they were separated by long periods of "comparative repose" [121].

It should be stressed that Sedgwick did not believe that catastrophes made the assumption of a very long geological time unnecessary: "in the phenomena of geology we are carried back . . . into times unlimited by any narrow measures of our own, and we exhibit and arrange the monuments of former revolutions, requiring for their accomplishment perhaps all the secular periods of astronomy" [122].

Sedgwick did not vote for Élie de Beaumont and against Lyell, in order to reduce geological time, but because the former's theory was in his opinion more conformable to geological data and sounder in its methodological basis: "because his conclusions are not based upon any *a priori* reasoning, but on the evidence of facts; and also, because, in part, they are in accordance with my own observations" [123]. With Élie de Beaumont, Sedgwick shared the conviction that facts demonstrate that not in all periods all geological events were of the same kind and intensity.

e. *Conybeare on method*

The question at issue between Uniformitarians and Catastrophists was only in the second place one of geological *systems* (uniformity

---

[118] Sedgwick, o.c., p. 306.
[119] Sedgwick, o.c., p. 307.
[120] The Scandinavian boulders found, even in Holland, can be explained, according to Sedgwick, as one of the effects of a period of intensive volcanic violence, (sudden elevation of the Scandinavian chain, enormous rush of retiring waters transporting these boulders), i.e. by a cause commensurate to the effects observed. Sedgwick, o.c., p. 306.
[121] Sedgwick, o.c., pp. 308–311.
[122] Sedgwick, o.c., p. 299.
[123] Sedgwick, o.c., p. 311.

and slow change, over against catastrophes); fundamentally it was one of difference of *method*: shall we start from the assumption that the geological causes at work in the past were precisely the same in kind and energy as those now in operation, or shall we try to make an unbiassed investigation of the relics of the past (under the supposition that the laws of physics have not changed), in order to find out in how far the secondary combinations of the physical causes (i.e. the so-called "geological causes,,) that are at work now, are sufficient to explain the phenomena of the past? As Conybeare put it:

"We may commence with the effects actually resulting from the causes still in operation and acting within their present power, and thus taking our departure from circumstances with which we are familiarly acquainted, we may proceed to the consideration of the geological changes produced at former periods".

This, according to Conybeare, is Lyell's method. But, alternatively we may choose the other method, that is:

"we survey the geological phenomena, in what may be called a chronological order . . . finally, comparing the whole together, with the view of observing whether they all indicate a uniform and constant operation of the same causes, *acting with the same intensity*, and *under the same circumstances*; or rather evince that there has been a gradual change in these respects, and that the successive periods have often given rise to such new circumstances as must have in a very great degree modified the original forces" [124].

The second method appears to him "more strictly philosophical", and he rejects the imputation that it implies that he and other catastrophists resort to unknown causes. One might add, that it is "more philosophical" from the methodological point of view, in that it does not exclude beforehand that the result might be a *uniformitarian* system.

Lyell, on the other hand, though recognizing that he had a *bias* towards uniformity, was of the opinion that the system *based* upon this assumption was "more philosophical" than a catastrophist one [125].

But his opponent William Whewell deemed it equally presumptuous to call in *time* to protect us from *force*, as to do the reverse; both are to him "superstitions": "the effects must themselves teach us the nature and intensity of the causes which have operated" [126].

---

[124] Conybeare, l.c., p. 360.

[125] Lyell to Whewell, 7–3–1837. In: Lyell's Life and Letters, vol. II, London 1881, p. 7.

[126] W. Whewell, History of the Inductive Sciences, 3d ed., London 1857, vol. III, p. 513. For Whewell's penetrating criticism of Lyell, cf. R. Hooykaas, The Principle of Uniformity, pp. 42–47.

300

This criticism of Lyell's method by his contemporaries underlines that what should be only a methodological principle of research, anticipated in fact part of a concrete theory (viz. the tenet of *strict uniformity*), which should at best have been a *result* of the method. The principle of actualism should be an empty form; Lyell's "principle of uniformity", however, possessed already concrete contents, i.e. it was also a working hypothesis. A methodological principle, may have a cogent (and at the same time, indefinite) character within a science, while a working hypothesis has only a tentative one. In its legitimate use, it should not force the results into conformity with itself. On precisely this point (that of giving the authority of a methodological principle to what legitimately should be but a working hypothesis), the arguments of Lyell's opponents were, more or less explicitly, concentrated.

It should be recognized, then, that the *method* of the catastrophists was a scientifically legitimate one. They emphasized that Uniformity, however "logical" and economical it may be, should not be maintained a priori, but that field research should be the basis of geological science. The method of explaining things in the simplest manner imaginable (uniformity!), should be subservient to, and not overrule the duty of "following Nature to whatever abysses it may lead you".

The method of the Catastrophists may be sound, quite apart from the question whether *their* method or that of the Uniformitarians led to the better geological theoretical *system*. Their merit remains that they refused to let their results be determined beforehand by the dogma of uniformity of the system of the earth, or that of uniformity of tempo and mode in geological change.

VII. Non-catastrophist actualism (ad 3 *a*)

It was, however, *not essential* for the catastrophists' conception of the actualistic method, that the resulting theory would be "catastrophism". It might as well be that geological investigation on this same methodological basis would lead other people to a non-catastrophist theory of slow changes of the surface of the earth as well as the organic world. Even a complete uniformity in the Lyellian sense should not be excluded a priori.

The theory of gradual decrease of temperature of the earth was not necessarily connected with catastrophism (Cf. Buffon, von Hoff, Poulett Scrope, Prévost) [127]. In its non-catastrophist version

---

[127] On von Hoff and Scrope, cf. R. Hooykaas, The Principle of Uniformity, pp. 4–12. About the non-catastrophist progressionism of Constant Prévost (1787–1856) in 1825, see R. Hooykaas, Geological Uniformitarianism and Evolution, Arch. intern. hist. sc. 19 (1966), pp. 12–17.

301

it was held that there is a rather slow and continuous change of character of mineral formation and surface building, and that gradually and slowly new types of plants and animals have developed.

a.   B. Cotta

The Freiberg professor Bernhard Cotta recognized valuable elements in Élie de Beaumont's elevation theory as well as in Lyell's uniformitarianism [128], but he deemed both standpoints onesided and wanted to unite them in a medium way [129]. Élie de Beaumont's theory was in his eyes an "artificial system", whereas against Lyell's tenet of the always uniform transformation of the earth's crust, he adduced the arguments that in ancient times volcanic action bore a different character and also that the most ancient eruptive rocks have not the same composition as the more recent ones [130]. Lyell assumes, "in contradiction to experience", that organic life did not develop by degrees, but was complete from the beginning. He was right in contending that the forces and laws of nature have always been the same, but it should be added that their effects are different in subsequent eras; they continually change with their objects; there is a "developmental history" (Entwicklungsgeschichte) of the earth and not only a sequence of changes of always the same energy [131].

As Cotta believed that the original state of the earth was that of a hot fluid mass which gradually cooled down [132], and that the original atmosphere contained much more carbonic acid than the present one, he had to hold also that the organic world underwent essential changes in the course of time [133]. In a mysterious way

---

[128]   B. Cotta, Der innere Bau der Gebirge. Freiburg 1851, p. 16; B. Cotta, Grundriss der Geognosie und Geologie (zweite Auflage der "Anleitung zum Studium der Geognosie und Geologie, 1842) Dresden, Leipzig 1846, p. 375.

[129]   Cotta, Der innere Bau, etc., p. 16.

[130]   Cotta, Grundriss der Geognosie und Geologie, pp. 378, 387, 376; Der innere Bau, etc., p. 60.

[131]   Cotta, Grundriss der Geognosie und Geologie, p. 376: "Gern wollen wir ihm zugeben, dass die Naturkräfte und Gesetze von Anfang an dieselben waren, und es dankbar anerkennen, dass er diese Idee lebhaft angeregt hat; aber die Wirkungen dieser Gesetze und Kräfte haben offenbar den gegenwärtigen Zustand, der kein ursprünglicher sein kann, erst aus einem früheren herausgebildet und sind sich folglich nicht durch alle Zeiten gleich geblieben, sondern haben sich mit ihren Objecten fortwährend verändert. Diese Hauptidee ist es, welche unserem Systeme zu Grunde liegt. Ich behaupte, dass man eine Entwicklungsgeschichte des Erdkörpers nachweisen kann, und nicht blos beständige Umänderungen von sich stets gleichbleibender Energie".

[132]   Cotta, Grundriss, p. 385.

[133]   ". . . es entstanden Organismen, angemessen jener hohen Temperatur, jener dichten Atmosphäre, welche beide auf der Erde damals noch alle klimatischen Unterschiede unbemerkbar machten". Grundriss, p. 391.

302

higher organisms arose, as is evident from the paleontological record [134].

In Cotta's system, then, there is, mainly attributed to the cooling down [135], an irreversible historical development (Entwicklung) of the earth (ancient eruptive rocks different from more recent ones; before the condensation of water, erosion was different; the rise of organic beings went together with decrease of carbonic dioxide in the atmosphere, etc.).

Cotta clearly saw that the results of geological change in their turn act as causes, so that a real change of the earth's crust leads to an accumulation of results, which inevitably introduces different and more complicated causes of further change, even when the degree or energy of geological activity would remain the same [136].

The basis of his geological theory he put forward (1850, 1858) as the "law of gradual development by summation of particular operations": "The multiplicity of the phenomenal forms is a necessary consequence of the summation of the results of all particular events" [137]. This law is, in his opinion, no hypothesis, but a logical necessity [138].

With the cooling down of the earth, then, goes together a growing diversity (at first only gaseous, afterwards also liquid, and finally solid bodies). In the earth's crust there has been an increase of

---

[134] "Es muss nothwendig auffallen, dass in diesen ersten Gebilden Reste von auf der Stufenleiter der Organisation ziemlich tief stehenden Geschöpfen gefunden werden, während in späteren Zeiten nach und nach immer höhere auftreten . . .". Grundriss, p. 392.

[135] Cotta, Grundriss, p. 388.

[136] "Hier erlaube ich mir nur eine ganz allgemeine Bemerkung gegen das Extreme dieser Ansicht. Vorausgesetzt, es sei wirklich nicht nur das Wesen, sondern auch der Grad (die Energie) aller geologischen Vorgänge von je her derselbe gewesen wie jetzt, so würde dennoch ihr Erfolg, ihr Resultat sich nothwendig beständig geändert haben, immer complicirter, mannichfaltiger geworden sein, da eine stete Summirung dieser Resultate stattfindet und nothwendig stattfinden muss, eine Summirung der Resultate, deren jedes auf das nachfolgende einwirkt. Alle Veränderungen der Erdoberfläche sind von dauernden Folgen begleitet, diese aber summiren sich, und jede frühere wirkt auf die spätere ein, macht dieselbe weniger einfach. Zu irgend einer Zeit müsste doch ganz gewiss ein erstes Gebirge erhoben worden sein. Dieses wurde durch keinerlei schon gegebene Unregelmässigkeiten seiner Art modificirt, sobald aber nachher ein zweites in der Nähe des ersten, wenn auch nur durch genau dieselben Kräfte, entstand, musste dessen Bau unbedingt durch das schon vorhandene erste beeinflusst werden . . . Das gilt aber keineswegs bloss für Gebirgsbildung, sondern für alle erdgeschichtlichen Vorgänge und ganz besonders auch für die Entwicklung des organischen Lebens, in welchem immer eine Form die andere bedingt". Cotta, Der innere Bau der Gebirge, pp. 4–5.

[137] B. von Cotta, Die Geologie der Gegenwart. 4 Aufl. Leipzig 1874, p. 185: "Die Mannigfaltigkeit der Erscheinungsformen ist eine nothwendige Folge der Summirung von Resultaten aller Einzelvorgänge, die nach einander eingetreten sind".

[138] Cotta, o.c., p. 186.

303

diversity of rocks and of texture [139], though it is not yet possible to find the chronological order of the first appearance of particular rocks and though it is not yet certain whether some of them do not take rise any longer (as is the case with extinct animals) [140]. The rise of organisms is a further step also in *geological* development: new materials are absorbed from the atmosphere and deposited afterwards. Moreover, there is a series of development of organic forms themselves [141]. No alteration has been completely reversible: every change left behind some permanent trace, and thus modified the next stage [142]. Especially the growing diversity of climate had a diversifying influence on the earth's surface and on the organic world [143]. Though, generally speaking, greater multiplicity goes together with higher forms (and greater complication), this ascent is no necessary consequence [144].

Cotta made efforts to prove the exaggeration (Ueberschwenglich-keit) of Lyell's proposition that always the same transformations took place as are now in operation, and that the degree of the transformations has always been the same as it is now. Nevertheless, he was of the opinion that Lyell liberated us from a hypothetical and miraculous primitive world (Vorwelt), by explaining all past change by natural laws still at work now [145].

In general, Cotta's sympathy for Lyell seems to have been greater than that for Élie de Beaumont, and he depicted the old catastrophists as phantastic miracle-mongers [146]. This demonstrates how soon Lyell's superficial and partial exposition of catastrophism was accepted, even by those who did not share his rigid views on uniformity.

b. *H. G. Bronn*

In one of his earlier works the German paleontologist H. G. Bronn made the idea of "development" the basis of his "history of nature". The "physiological" series of attraction (Attractions-Leben), affinity (Affinitäts-Leben), organic life (Organisches Leben) and mind (Vernunft-Leben) was to him also a chronological series. These subsequent degrees do not take rise suddenly, but imperceptibly and gradually: it is impossible to indicate the borderlines [147].

---

[139]   Cotta, o.c., p. 192.
[140]   Cotta, o.c., p. 194.
[141]   Cotta, o.c., p. 199.
[142]   Cotta, o.c., p. 203.
[143]   Cotta, o.c., p. 204.
[144]   Cotta, o.c., p. 208.
[145]   Cotta, Der innere Bau, p. 5.
[146]   Cotta, Der innere Bau der Gebirge, p. 4.
[147]   H. G. Bronn, Handbuch einer Geschichte der Natur, Bd I, Stuttgart 1841, pp. 5–6.

304

Bronn accepted the theory of the gradually cooling down of the earth [148]. Consequently, he rejected (with reference to Conybeare's critique) Lyell's tenet of the equality of intensity of geological forces and phenomena [149]. The plutonic forces diminished as the temperature decreased [150]. The chronological geological progression consists in the growing multiplicity of rocks, the increasing tectonic complication of the earth's crust [151] and the growing diversity of climate, soil and waters, and (dependent on these) animal and plant life [152].

In a later work, Bronn held the view that there has been a development of organisms from imperfect to more perfect forms. He assumed an "independent force of production" (eine selbstständige Produktions-Kraft), an "inner necessity" (eine innere Notwendigkeit), a law of inherent progressive development of the organic world, and at the same time a (more powerful) law of progression of external circumstances, in general running parallel to the former in its effects. Organisms that could not exist (bestehen) in certain circumstances, did not take rise (entstehen): "The conditions of creation and those of preservation. . ., then, must coincide to a certain extent" [153]. But the creative force of progression from lower to higher forms is continuous, whereas the progression of the external circumstances conditioning the existence of plants and animals, is sometimes rapid and at other times slow. The inherent law of development (progressive creation) would produce a continuous ladder of nature; as a result, however, of the external circumstances, in many cases only parts of such a series are to be seen [154]. Nevertheless, as the effects of the two causes run more or less parallel, there must be a rectilinear development also of external conditions. In fact, the main influence here is (in Bronn's opinion) the slowly cooling down of the earth (and the consequent change of the atmosphere), which causes a universal change of

---

148 H. G. Bronn, Handbuch einer Geschichte der Natur, I, pp. 75, 393.
149 "Wir bestreiten hiermit Lyell's Behauptung vom Gleichbleiben der Intensität geologischer Erscheinungen, sofern sie von astronomischen Kräften bedingt werden, eben so wohl als jene, die von geologischen Kraften selbst abhängen, welches letzte auch schon Conybeare (Jahrb. 1832, 324) u. A. gethan haben". H. G. Bronn, Handbuch, I, p. 52.
150 Bronn, Handbuch, I, p. 136.
151 Bronn, Handbuch, I, p. 246.
152 Bronn, Handbuch, I, p. 447.
153 H. G. Bronn, Untersuchungen über die Entwicklungs-Gesetze der organischen Welt während der Bildungs-Zeit unserer Erd-Oberfläche. Stuttgart 1858, p. 86; (cf. p. 352): "Die Schöpfungs-Bedingungen müssen daher mit den Erhaltungs-Bedingungen, die Schöpfungs-Kraft muss mit der Erhaltungs-Kraft in gewissem Grade zusammenfallen oder identisch seyn, obwohl die erhaltenden Bedingungen nicht immer nothwendig auch produzirende sind".
154 H. G. Bronn, Entwicklungsgesetze, p. 87.

305

population *"in one direction"* [155]. Thus the gradual change of geological conditions goes together with a gradual development towards higher forms of organic life [156].

Bronn thus, like Lamarck, introduced two main factors of development, but, in contradistinction with Lamarck, the ascending series is only a sequence of forms that are not necessarily connected by descent. Moreover, the change of external circumstances (gradual cooling of the earth; change of climate; alteration of composition of the atmosphere) [157], does not cause a change in animal forms *in response* to them. The external conditions sift out the viable forms and modify them [158].

In geology Bronn was actualistic, but not in the Lyellian uniformitarian sense: "the changes are still going on, but, as their result, the situation has become radically different "[159].

Though he did not put forward world catastrophes, but rather accentuated that there is a gradual cooling down of the earth, the effects of the latter are not wholly uniform: the rate of refrigeration and the intensity of plutonic eruptions is diminishing; some of the effects are continuous, others are periodical. In connection herewith, the changes of the animal world are, as a rule, not abrupt and not simultaneous over the whole world [160].

Bronn frankly confessed his ignorance about the way in which the rise of new species takes place. Like Lyell, he supposed that old species still continue becoming extinct, whereas cognate new ones arise [161]. But, in contradistinction to Lyell, he recognized that this goes together with a "development" towards more complicated forms.

Bronn's theory differed from catastrophism and progressionism

---

[155]   Bronn, o.c., p. 237.
[156]   Bronn, o.c., p. 115.
[157]   Bronn, o.c., p. 114.
[158]   Bronn, o.c., p. 354.
[159]   "Alle diese Bewegungen und Veränderungen dauerten mehr und weniger lange Zeit fort und dauern noch jetzt; ihre Wirkungen häuften sich daher immer mehr, und die Folgen jeder Art werden um so beträchtlicher und augenfälliger, in je späterer Zeit man sie zu summiren versucht; die Zustände der Erdoberfläche, der Wasser, der Atmosphäre sind von den anfänglichen um so verschiedenartiger, je weiter sie in der Zeit davon entfernt sind;—und so muss es auch die Bevölkerung der Erde seyn". Bronn, Untersuchungen über die Entwicklungs-Gesetze, etc., p. 114.
[160]   Bronn, o.c., p. 121, p. 114. "Die fortschreitende Vervollkommnung der organischen Welt ist daher in diesem Falle bloss eine Folge der fortschreitenden Vervollkommnung der äussern Lebens-Bedingungen und insbesondere der Wohnstätten der Organismen. Und wie die physische Ursache nur *allmählich*, Stück-weise und örtlich eintritt, so müssen auch die Folgen, muss auch das Fortschreiten der Bevölkerung im Einzelnen und im Ganzen *allmählich* geschehen . . .". Bronn, o.c., p. 129.
[161]   Bronn, o.c., p. 227.

as put forward by the British paleontologists (Buckland, Sedgwick, etc.) in that it rejected special periods of creation of new species in which a renovation of the whole animal world would have taken place. The extinction and rise of species took place at all times, dependent on the external circumstances, gradually or suddenly, locally or everywhere [162]. He called this doctrine "the theory of progressive development or systematic evolution" [163] (Theorie der progressiven Entwicklung oder der systematischen Evolution), and he considered it a result of induction. He wrote his book (in answer to a competition arranged by the Paris Academy of Sciences in 1855) under the motto: "Natura doceri" [164].

The theory of Étienne Geoffroy St. Hilaire, (which had been propounded many years before), was also "progressionist", but decidedly catastrophist at the same time. In periods of geological catastrophe, the rapidly changing external conditions (particularly of the atmosphere) cause saltatory change in animal types. In times of gradual and slow geological change, there has also been a gradual change of animal forms [165].

In Geoffroy's case, as in that of Bronn, there is "actualism" as to the method (Geoffroy even believed that the breeding of artificial monstrosities would give the clue to natural saltatory transformation). But in neither case there was "uniformitarianism" as to the resulting theoretical system.

On the other hand, there is more *uniformity of change* in Bronn's than in Geoffroy's theory. Bronn did not connect the components of the series of species by a hypothesis of descent. Yet, his ideas about the relation of the environment with the animal forms shows some affinity with Darwin's theory of natural selection, so that it does not seem strange that he introduced the "Origin of Species" in Germany.

c. *Evolutionism* (ad. 4b)

After Darwin, *Evolutionism* was put forward as a third way (at least in paleontology) beside *Catastrophism* and *Uniformitarianism* [166]. Now Darwin was strongly influenced by Lyell's *geological* uniformitarianism, but the *paleontological* basis of his

---

[162] Bronn, o.c., p. 237.
[163] Bronn, o.c., p. 355.
[164] Bronn, o.c., p. IV.
[165] For Étienne Geoffroy St. Hilaire, cf. The Principle of Uniformity, pp. 80–88, 117–118. Also: R. Hooykaas, The parallel between the history of the earth and the history of the animal world, Arch. intern. hist. sc. 10 (1957), pp. 9–13.
[166] Lamarck's theory was founded on the "ladder" of still existing animals and not on the paleontological record. Cf., The Principle of Uniformity, pp. 73–80, 88–89; R. Hooykaas, The parallel, etc. pp. 5–9.

307

theory he could only find with the progressionists of the catastrophist school (Buckland, Sedgwick, Conybeare, Murchison, etc.) or
with those of the more gradualistic type (Chambers, Bronn) [167].
That is to say, that, in contradistinction to progressionists of all
kinds (catastrophist as well as non-catastrophist), he did not
propound a parallel development of the earth and the animal world.
In the organic world of Darwin's system, there is development,
whereas in the inorganic world there is uniformity throughout
geological time. Darwin borrowed from Lyell the idea of slow and
imperceptibly small changes adding up to larger transformations
in the course of very long periods; saltatory transformation of the
kind advocated by Geoffroy St. Hilaire, was and is an arch-heresy
to all orthodox darwinists [168].

But the uniformity in his *evolutionism* is a uniformity of *becoming*
and not (as with Lyell) a uniformity of *being*. His organic world
is on the move in a certain direction.

VIII. Catastrophism, uniformitarianism and metaphysics.

Again and again the accusation of introducing non-physical,
supernatural, causes has been levelled against the Catastrophists.
It is said that they rejected the large geological time scale of
uniformitarianism because they held that the Biblical story of
Creation tells against it. This is one of the arguments supporting
the misconception that uniformitarianism is the only scientific
position over against the unscientific attitude of the catastrophists.

As far as geology proper is concerned, this charge is unjustified:
for the periods of quiet change between the catastrophes (i.e. for
the formation of sediments) many catastrophists too had recourse
to a long time. As a rule, when religion influenced their conceptions,
this happened rather by a general conception (e.g. that of "purpose"
in nature) than through the exegesis of a particular biblical text.
Dolomieu (and probably Razumovsky too) was an 18th-century
"philosophe", Cuvier a liberal protestant. Even so orthodox a
Low Church Anglican clergyman as Adam Sedgwick, when openly
disavowing in 1831 his former interpretation of "diluvial gravel",
said that he agreed with Francis Bacon [169] that one should not

---

[167] On Darwin, cf. The Principle of Uniformity, pp. 100–107; The parallel,
etc. pp. 15–16, and R. Hooykaas, Natuur en Geschiedenis, Mededelingen
Kon. Ned. Ak. Wetensch. afd. Letterkunde, nw. reeks 11, nr. 9, Noord-
Hollandse Uitgeversmij, Amsterdam 1966, pp. 46–50. On Robert Chambers,
cf. The Principle of Uniformity, pp. 90–92, and "The parallel, etc." pp. 13–15.
[168] This becomes evident in the controversy about saltatory evolution
between H. G. Schindewolf and E. Mayr. (Cf. The Principle of Uniformity),
pp. 121–133.
[169] F. Bacon,. "Of the Advancement of Learning", Bk II: "For to seek
heaven and earth in the Word of God . . . is to seek temporary things amongst
eternal: and as to seek divinity in philosophy is to seek the living amongst

seek for scientific data in the Bible [170]. His Oxford colleague, the Rev. Prof. William Buckland, in his Bridgewater Treatise (1836) abandoned the diluvial theory [171] which he had put forward in his famous "Reliquiae Diluvianae" (1823) [172].

It must be recognized, however, that in *paleontology* the embarrassing problem [173] of the creation of new faunas was either referred to some mysterious creative power God had laid into matter (Bronn, Cotta) [174], or to special divine interventions (Buckland, Miller). Especially in the latter instance there evidently was a mixing up of metaphysical and physical considerations [175].

With uniformitarians, however, no less metaphysical preconceptions and intrusions occurred. Hutton's "Theory of the Earth" (like his other works) is steeped in them, and even with Lyell they are not wholly absent. But, these two great geologists were soberminded enough not to propound an *eternal* repetition of cycles. They only declared that we *find* no vestige of a beginning and we *see* no prospect of an end in the cyclical course of events presented by the geological record.

Some uniformitarians, however, went much farther and made Uniformity into a kind of religious dogma. G. H. Toulmin (1780) dogmatically excluded the possibility of a beginning or an end of the earth. He tied uniformitarianism to the metaphysical belief in the eternity of Nature [176].

The geologist and mineralogist Otto Volger (1822–1897) wrote a book (1857) "Earth and Eternity" (Erde und Ewigkeit), the main

---

the dead, so to seek philosophy in divinity is to seek the dead amongst the living".

[170] He now recognized that not all the gravel that he had formerly attributed to the Noachian deluge, could stem from one and the same, so-called Diluvial period, and that the Noachian deluge, as far as we know, did not leave any geological traces. Having been a propagator of what he now called "a philosophical heresy", he showed the courage in this Presidential Address, as he said himself, "publicly to read my recantation". Sedgwick, o.c., p. 314.

[171] W. Buckland, Geology and Mineralogy considered with reference to Natural Theology, vol. I. London 1836, p. 95.

[172] W. Buckland, Reliquiae Diluvianae; or, Observations on the Organic Remains contained in caves, fissures, and diluvial gravel, and on other geological phenomena, attesting the action of an Universal Deluge. London 1823.

[173] Embarrassing not only to the catastrophists, but also to the uniformitarian, Charles Lyell.

[174] H. G. Bronn, Untersuchungen über die Entwicklungs-Gesetze der organischen Welt (1858), p. 81; B. Cotta, Die Geologie der Gegenwart, 4 Aufl. (1874), p. 270; B. Cotta, Grundriss der Geognosie und Geologie (1846), p. 408.

[175] R. Hooykaas, The Principle of Uniformity, pp. 199–206.

[176] On Toulmin, cf. R. Hooykaas, James Hutton und die Ewigkeit der Welt. In: Gesnerus 23 (1966), pp. 55–66 (on Toulmin and Hutton). Also Arch. intern. hist. sc. 19 (1966), pp. 10–12.

309

thesis of which is revealed by the subtitle, "The Natural History of the Earth as a circling development, in contrast with the unnatural Geology of Revolutions and Catastrophes" [177]. He unblushingly made gratuitous statements on "Eternity": the new formations once will be perfectly similar to those we now consider "ancient"; "who could doubt, that Nature always goes through the same course, from all eternity behind us to all eternity we are going to meet" [178]. As to the organic world, "here too we look backwards into infinity; here too the prospect of eternities opens itself to us" [179]. There is no indication of a cooling down of the earth from which we might conclude that the original and the future state of the earth would be different from the present one; there is always absorbed as much heat by the earth as is lost by her: "so it is now, so it has been always, so it will be in all future" [180]. The world will "in all eternity" be in destruction and reconstruction [181], the same minerals are formed and destroyed in all periods; all species of minerals we now find did exist in former periods wherever and whenever the circumstances were favourable; they come back like planetary constellations [182].

The "primitive" mountains require, according to Volger, the existence of animals and plants: without chalk no feldspar, no granite, – without plants and animals no chalk [183]. There has been no time in which inorganic nature was without organic nature; if no traces of living beings are found, this is because they have disappeared [184]. Though species died out and other species took rise, as a whole the same set of forms stayed on [185]. All types of animals existed always, though their fossil rests may not have been found as yet [186]. There is no progressive development of species, neither in the mineral, nor in the botanical or zoological world, but an eternal cycle [187]: the same species come back when the same conditions are fulfilled [188].

---

[177] G. H. Otto Volger, Erde und Ewigkeit. Die natürliche Geschichte der Erde als kreisender Entwicklungsgang im Gegensatz zur naturwidrigen Geologie der Revolutionen und Katastrophen. Frankfurt a.M. 1857.

[178] "Wer könnte zweifeln, dass die Natur stets den gleichen Gang gehe von aller Ewigkeit, die hinter uns liegt, bis in alle Ewigkeit der wir entgegenwalten!" Volger, o.c., p. 137.

[179] Volger, o.c., p. 148.
[180] Volger, o.c., p. 162.
[181] Volger, o.c., p. 474.
[182] Volger, o.c., pp. 497, 573.
[183] Volger, o.c., pp. 521, 526.
[184] Volger, o.c., p. 526.
[185] Volger, o.c., p. 537.
[186] Volger, o.c., p. 555.
[187] Volger, o.c., pp. 559, 573.
[188] Volger, o.c., p. 574.

310

In Toulmin's and Volger's theories, then, not only the dogmatic but also the a-historic character of Uniformitarianism has reached its extreme.

IX. THE HISTORICAL CHARACTER OF GEOLOGY.

Uniformity of geological *causes* on all levels, inevitably leads to uniformity of geological *effects*. If a change has occurred, this causes a different situation which will modify the secondary combinations that act as causes of further change. But when it is a priori held that the geological situation does not essentially change, this implies that the effect of change is compensated by reverse changes in other localities, so that a permanent equilibrium would be maintained, which establishes a global uniformity of causes as well as effects.

To the fathers of uniformitarianism, Hutton and Lyell, these two aspects were inextricably woven together. Quite apart from his geological theory, Hutton was strongly preoccupied with the idea of natural cycles. With him, the principle of uniformity and the uniformitarian system are hardly distinguished from each other; uniformitarianism is a method as well as a theory ensuing from it. He does not admit any causes in the past but those that are of the same kind and degree as those that are in operation now, and at the same time he confesses to find "no vestige of a beginning and no prospect of an end" of the geological cycles. As far as we can know, according to him, the geological situation (as well as the organic world) has been always the same and will remain always the same as it is now.

With Lyell, too, the uniformitarian method and the uniformitarian system are closely knit together. Even in paleontology (before his conversion to darwinism) he assumed that species may disappear, but similar ones will replace them; no "development" is admitted.

This cyclical and a-historic conception of the past, Sedgwick rightly called "one of the arbitrary dogmas of the Huttonian theory" [189]. It was dogmatically asserted for the mineral, vegetable and animal world by all true Uniformitarians (Toulmin 1780; Hutton 1785; Lyell 1830; Volger 1857).

From the methodological point of view, we made above the distinction between non-actualistic and actualistic methods, and,— within the latter —, between actualistic-empirical and uniformitarian-dogmatical methods. As to the resulting *systems*, then, one could make a distinction between *historical* and *a-historical systems*.

An a-historic system is cyclical; its actualism consists in the

---

[189] A. Sedgwick, o.c., p. 299.

311

events as well as in their elements. A historical system admits a sequence of unique events [190].

Catastrophism bears a *historical* character. If there would be a monotonous repetition of similar alternating periods of geological activity and tranquillity, there would be at least within such a cycle a kind of history. In general, however, catastrophists went even further: they did not put forward identical cycles. The idea of a continuously diminishing geological activity was one of their fundamental assumptions. This continuous course is interrupted now and again by outbursts of geological activity, and each of them bears its own individual character. Moreover, it is held that not all rocks have been formed at all times.

In paleontology catastrophism was almost always combined with organic progressionism, that is, with the idea that sudden geological outbursts run parallel with the rise of new (and also higher) animal types.

And even when the adherents of the gradual cooling down of the earth did not resort to catastrophes, they propounded a rectilinear development of the inorganic and the organic world, that is, a real, irreversible, *history* of the earth and the organic beings. Cotta (1842; 1846) contrasted the new geological theories, based on Lyell's supposition of continual equality of transformation and ever equal energy of forces—, the so-called "continuity theories" (Stetigkeitstheorieen)—, with the "development theories" (Entwicklungstheorieen), which start from a formerly different situation of the earth [191]. In his opinion "the essence of things is not known but when their coming-to-be is found therein" [192]. Geology, so he said, concludes from the structure of the earth's crust to the *history* of its formation [193]. As in the history of the development of organic beings a distinction is made between ancient and recent faunas and floras, there is also a series of ancient and recent mountains. These latter, too, have been "developed" (entwickelt); they show the "history of their becoming" by their structure [194].

---

[190] R. Hooykaas, Nature and History. In: Organon 2 (1965), pp. 5–16; R. Hooykaas, Natuur en Geschiedenis. Amsterdam 1966.

Perhaps one could say that among the historical systems, Darwinism shows strongly a-historic tendencies, whereas among the a-historic, uniformitarians, Lyell at least admitted two historical interventions in the ordinary course of nature: the rise of life and that of Man. Besides, though the "level" of the inorganic world did not change, the different epochs have in his system their individual characteristics, whereas on climate he held views that did not go far enough in the eyes of the uniformitarian diehard Rev. John Fleming (Principle of Uniformity, pp. 27–30 and 112–117).

[191] B. Cotta, Grundriss der Geognosie und Geologie (1846), p. 377.

[192] "Das Wesen der Dinge hat man erst dann erkannt, wenn man darin auch ihr Werden findet". B. Cotta, Der innere Bau der Gebirge, p. 1.

[193] B. Cotta, Der innere Bau der Gebirge, p. 1.

[194] B. Cotta, Der innere Bau der Gebirge, p. 10.

312

Finally, *Evolutionism*, as put forward by Charles Darwin, stressed, at least for the organic world, an upward movement from lower to higher forms. The uniformity in this system rather consists in the *rate of change* than in the resulting final effects, which form a sequence of unrepeatable unique phenomena. Thus, evolutionism may be methodologically close to uniformitarianism, but from the systematic point of view it is closer to catastrophism in that it is a *historical* system. As a system the evolution theory owes its historical aspects to the catastrophists and its uniformitarian aspects (slow changes, extremely long periods) to Lyell. The remarkable fact is, that, because of Lyell's (rather reluctant) conversion to Evolutionism, and Darwin's adherence to geological uniformity, evolutionism has wrongly been considered as necessarily connected with uniformitarianism [195].

X. EPILOGUE
a. *J. Prestwich's criticism*

Lyell's able advocacy, together with the triumph of Darwinism, gave to uniformitarianism, especially in Britain, "the charm of an infallible faith" [196]. On the European Continent, though catastrophism was generally abandoned, a moderate form of actualism (and, together with it, a more *historical* conception of geology) prevailed.

Yet, even Britain had its critics of Uniformitarianism. The London geological professor Joseph Prestwich (1812–1896) recognized the dogmatical character of the prevalent doctrine of uniformity, which, in his opinion, barred the advance of geology [197]. Though fully accepting the uniformity of *kind* of geological causes, he rejected the uniformity of *degree* [198]. In his opinion "Nature had

---

[195] In our article on "Geological Uniformitarianism and Evolution" (Arch. intern.hist. sc. **19** (1966), pp. 3–19) we opposed this thesis.

[196] "The argument in favour of uniformity of action has been put before us with so much skill and ability, and possessing as it does the charm of an infallible faith, that Uniformitarianism has become the accepted doctrine of the dominant school of geology". J. Prestwich, Collected Papers on some controverted questions of Geology. London 1895, p. 3.

[197] Prestwich, o.c., p. 1.

[198] Prestwich, o.c., p. 5. "In contradistinction to *kind* or *law*, where we are on common ground, no common scale on the question of *degree* is possible in judging of the past by comparison with the present" (p. 6). "The doctrine of uniformity in all time . . . still remains the creed of the majority, though I believe, in many cases, this arises from confounding *degree* with *kind*" (p. 6, note 1). "We would not for a moment contend that the forces of erosion, the modes of sedimentation, and the methods of motion, are not the same in *kind* as they have ever been, but we can never admit that they have always been the same in *degree*. The physical laws are permanent, but the effects are conditional and changing, in accordance with the conditions under which the law is exhibited" (p. 14).

313

greater forces at her command . . . than is admitted by Uniformitarians" [199]. As the shifting positions of uniformitarians show, their measures of time and change stand on an insecure basis and "they have probably done as much to impede the exercise of free inquiry and discussion as did the catastrophic theories which formerly prevailed" [200].

But not only did they shift the measure of time, they took also the freedom of lengthening the whole time scale at will. Sometimes, the protagonists of slow change (uniformitarian or evolutionist) made a virtue of this elasticity. Cotta (at that time an evolutionist), rather naively expressed his satisfaction about that "time was the only thing about which a geologist could dispose wholly freely, whereas in every other respect he is bound to natural laws, observations and experiences" [201].

It was precisely this freedom which Prestwich would not allow to the geologists, whereas, on the other hand, he criticized in them the lack of freedom they gave to change the degree of activity.

Whereas, in Prestwich's opinion, catastrophism found its own cure in the more accurate observation of geological phenomena, uniformitarian theories "hedge us in by dogmas which forbid any interpretation of the phenomena other than that of fixed rules which are more worthy of the sixteenth than of the nineteenth century. Instead of weighing the evidence and following up the consequences that should ensue from the assumption, too many attempts have been made—not unnaturally by those who hold this faith—to adjust the evidence to the assumption" [202].

Evidently, Prestwich, who certainly was no catastrophist, repeated the *methodological* objections to Uniformitarianism put forward by the Catastrophists half a century earlier, and he even recognized that, from the methodological point of view, the latter were more sound. He stressed how unfortunate it would be for any science to have free discussion and inquiry barred by assumed postulates, and not by the ordinary rules of evidence as established by the facts, however divergent the conclusions to which those facts lead may be from the prevailing belief [203]. The fact that he wanted to apply these remarks mainly to questions connected with the more recent geological periods, is the more remarkable because

---

[199]   Prestwich, o.c., p. 2.
[200]   Prestwich, o.c., p. 12.
[201]   "Die Zeit ist vielmehr das Einzige über welches der Geolog ganz frei zu verfügen hat, während er in jeder anderen Beziehung an Naturgesetze, Beobachtungen und Erfahrungen gebunden ist". B. v. Cotta, Die Geologie der Gegenwart, 4 Aufl., Leipzig 1874, p. 205.
[202]   Prestwich, o.c., p. 14.
[203]   Prestwich, o.c., p. 18.

314

even the catastrophists had always recognized the value of the uniformitarian approach for recent epochs. Prestwich hoped, that also the phenomena of these later periods would be judged "by the evidence of facts rather than by rules", so that the interpretation might "escape the dwarfing influence of Uniformitarianism" [204].

b. *Actualistic Method and Uniformitarian system*

As the matter stands now, geologists quite sensibly follow the catastrophists and other protagonists of the *historical* interpretation of nature in trying to be *as actualistic as possible*, but they do not push their actualism to the extreme of an almost absolute uniformity of causes. The principle of "uniformity", or actuality in general, serves as the methodological principle of trying to be as economic as possible with causes and notions in scientific explanation. Consequently, the conceptions of the scope and contents of the Principle of Actuality (Aktualitätsprinzip) are widely divergent: they run from strict uniformity of all geological causes (in the Lyellian sense) to such a trivial general verdict as that of the "immutability of the laws of physics".

Nevertheless, however much geologists are forced to adapt their contentions to the facts, generally speaking they all rally around the "Fetish of uniformity" [205], as adherence to it has become a token of scientific respectability. The holy names of Lyell and Darwin are connected with it, and, however widely one may deviate from its original meaning, one has to pay at least lip service to it. Catastrophism, on the other hand, remains the bugbear to the geologist.

Perhaps much confusion could be avoided if the anglosaxon term "uniformitarianism" were no longer translated by "actualism" in the continental European languages. The term "actualism" (Aktualitätsprinzip) has become automatically associated with the Lyellian system and method.

The term "uniformitarianism", however, should be restricted to theoretical *systems* like those of Hutton and Lyell, *and* to the rigid conception of the actualistic *method* as applied by those fathers of geology, that is, connected with the hypothesis of an almost perfect equality of causes at all times.

Actualism, on the other hand, covers a wide range of theories (from extreme catastrophism to uniformitarianism) that go together with the methodological principle of being as "actualistic" as the geological facts admit: a principle which finds a more rigid applica-

---

204 Prestwich, o.c., p. 18.
205 Prestwich, o.c., p. 8.

tion in a uniformitarian system than in that of catastrophism or in the systems of other protagonists of the "historical" conception of geology: a principle, however, that never should have its contents dogmatically fixed a priori [206].

Theses

1. The battle of Catastrophism versus Uniformitarianism, though revealing itself as that of two different geological systems, is essentially a controversy on method. Catastrophism held that the interpretation ought to be adapted to geological facts; Uniformitarianism tended to interpret data in conformity with the assumption of the immutability in kind and degree of all geological causes.

2. The Principle of Uniformity in the Lyellian sense implies the theory of the *identity* of the causes operating in past and present.

A moderately actualistic method implies the actualistic principle of being as uniformitarian as possible (i.e. as the *facts allow*); it admits the *analogy* of causes in past and present. That is, in itself it is "an empty form" [206], the contents of which depend on geological research.

3. The uniformitarian method leads to *a-historic* theories; the actualistic method (as defined above in the second thesis) may lead to *historical* theories of development of the inorganic and the organic world (either catastrophist or non-catastrophist). Evolutionism owes its historical character to the development theories.

4. Catastrophist theories may be based on actualistic or on non-actualistic principles; uniformitarian theories are based on an extremely actualistic "principle of *uniformity*".

5. The statement that the contrast between Catastrophism and Uniformitarianism is fundamentally one between explanations by supernatural and natural causes—though true in some cases— greatly oversimplifies the real situation. It overlooks the basic methodological controversy and the fact that many uniformitarians used metaphysical arguments and many catastrophists did not use them at all [207].

---

[206]   R. Hooykaas, The Principle of Uniformity, p. 161. One could compare this, perhaps, with the principle of economy of causes (or that of the simplicity of explanation): no more different causes should be assumed than is strictly necessary for explanation (or: explanatory systems should be as simple as possible). As soon as such a principle is transformed into the thesis that Nature *is* economic and simple (and this economy and simplicity get, moreover, a concrete formulation), it has acquired an ontological, instead of a purely methodological character.

[207]   The Principle of Uniformity, Part IV.

316

15

Copyright © 1966 by the University of Chicago Press

Reprinted from *Jour. Geol.*, **74**, 127–146 (1966)

GEOLOGIC TIME[1]

DAVID B. KITTS

University of Oklahoma, Norman, Oklahoma

ABSTRACT

In this paper an attempt is made to examine the foundations of the concept of geologic time. The temporal ordering of geologic events is based upon generalizations or laws which point to causal relations among classes of events. The measurement of geologic time entails serious theoretical and practical problems related to the fact that it is difficult to demonstrate that estimates of duration based upon successions of geologic events are in accord with estimates based upon standard physical clocks. Correlation is regarded as an attempt to extend the temporal ordering of events beyond the local section by means of causal chains or *signals*. It is not possible to confirm hypotheses regarding the simultaneity of geologic events because confirmation depends upon knowledge of the velocity of a particular signal, knowledge which, it is held, cannot be obtained. It is proposed that intervals of geologic time be defined by sets of events closed at one end.

I. INTRODUCTION

Historic time is not given in immediate experience. Because it lies beyond the range of our senses, we cannot regard it simply as an extension of common-sense time into the past. Historic time has its own special properties which are imposed upon it by the assumptions and procedures in terms of which we construct it. In this paper I have set out to examine the fundamental basis for our conception of geologic time.

In the course of this examination I have relied heavily upon two books, Hans Reichenbach's *The Philosophy of Space and Time* (1958), and Adolf Grünbaum's *Philosophical Problems of Space and Time* (1963). I hope that, if nothing else, the discussion that follows will induce geologists to consult these two great works.

II. THE ORDER OF EVENTS IN GEOLOGIC TIME

In an initial phase of historical investigation, geologists describe the present state of the earth. The term "present" does not refer to a point in time, but rather to an interval of time that covers tens, or even hundreds, of years. This extension of the geologic *specious present* is justified by the fact that the objects of geologic interest remain relatively unchanged as compared to other objects of our experience.

[1] Manuscript received October 4, 1965.

Even during the geologic present, however, significant change occurs and reference to this change is made in terms denoting temporal relationship. In the description of processes an essential reference to time occurs in terms such as "before" and "after." The "before" and "after" of these accounts often refers to temporal relationships between states in the awareness of an observer. Direct observation statements describing the processes and inductive generalizations that may be based upon them, even though they contain terms denoting temporal relationships, are not to be confused with descriptions of historical events and processes beyond the scope of immediate awareness.

Historic events and the temporal relationships among them must be inferred. The inference rests upon the conviction that the states which are observed now came to be what they are because of a causal relationship to states at some time in the historic past. The connection between a particular state in the present and a particular state in the historic past is justified by adducing some geologic and physical generalizations which point to a necessary relationship between states separated in time.

The term "state" implies a degree of precision which is not often attained in geology. In the discussion that follows, the term "event" will be used instead of the

127

term "state." An *event* is simply a set of particular conditions with limited spatial and temporal extension which is characterized by a descriptive sentence. Geologic events in this sense are to be distinguished from what Hempel has called *concrete events*.[2] A concrete event is not characterized by a descriptive sentence, but rather by a noun, or a noun phrase, which is usually understood to refer to the overwhelming complexity of *all* of the aspects of the designated event. When a geologist uses such a term as "The Appalachian Revolution" he usually means to refer to a concrete event. Hempel has pointed out (1962, p. 18) that to construe the term "event" in its concrete sense is self-defeating in scientific discourse, "for any particular event may be regarded as having infinitely many different aspects or characteristics, which cannot all be accounted for by a finite set however large of explanatory statements." The term "geological event" is meant to refer to what Hempel has called a *sentential event*. A sentential event refers to certain selected aspects of a concrete event. The aspects of a concrete event that are selected for inclusion in a sentential geologic event are just those aspects with which a geologist can deal, and with which he wishes to deal, in a particular inferential situation.

Geologic generalizations, reflecting the use to which they are ordinarily put, are framed in such a way as to point to causal relations among classes of temporally separated sentential events. This feature of geological generalizations permits the formulation of historical explanations for the present state of the earth's crust. Historical explanations proceed in terms of temporally ordered successions of particular events. The theory of geology may consequently be regarded as, among other things, a set of temporal ordering principles. Any two events, no matter what the spatial distance

separating them, may be temporally ordered provided a causal connection between them can be inferred in terms of some geologic or physical laws or generalizations. At this point I wish only to consider events which are relatively close to one another in space or, to put it into familiar geological terms, events which can be inferred on the basis of a single local section.

As an example of the temporal ordering of geologic events let us consider Moberly's account (1960, p. 1173) of the origin of the Morrison formation in the Bighorn Basin.

An extensive low-lying region, part of the present Western Interior of the United States, was exposed in late Jurassic time as a result of the regression of an epeiric sea in which the Sundance and related formations had been deposited. The present Bighorn Basin was a small part of this region. Aggradation commenced in rivers and on their floodplains. Fine-grained quartz sands and red muds, derived from lateritic weathering of Paleozoic marine shales and sandstones exposed to the west, were deposited together with minor additions of volcanic debris. Because the deposition was fairly continuous, the deposits were buried rapidly enough to inhibit modifying effects of the depositional environment except for destruction of most organic and the more labile volcanic components and reduction of part of the red coloring. Muds of the overbank deposits remained red. The climate was hot with fairly high rainfall interrupted by a brief, but pronounced, dry season during which small ephemeral lakes and swamps became highly alkaline, with deposition of calcium carbonate aided by algae.

In his account of the origin of the Morrison formation Moberly has inferred a succession of temporally ordered events. The inference upon which each of these events is based can be justified in terms of one or more familiar geologic generalizations, and indeed much of the discussion in the paper from which this passage was taken is concerned with such justifications.

In Moberly's detailed historical explanation there is a relatively close temporal association among the events involved. A very precise temporal resolution is unnecessary for many purposes and may indeed be det-

[2] Hempel draws the distinction between the two senses in which the term "event" may be construed in the paper quoted below and more fully in an as yet unpublished manuscript entitled "Aspects of Scientific Explanation."

rimental for some others. Geologists consequently often regard a particular sequence of events as a single complex event. We might meaningfully speak, for example, of the complex event "the origin of the Morrison formation in the Bighorn Basin." This is a perfectly permissible procedure for, as I have already suggested, geologists may conceive of events in such a way as to make them manageable within the total context of geologic history.

There is an important class of temporal ordering principles which permits the ordering of events of broad spatial and temporal extension, which I shall call *gross ordering principles*. The principle of superposition and the principle of crosscutting relations fall into the category of gross ordering principles. It must be emphasized that theoretically gross ordering principles do not differ from other ordering principles. The principle of superposition is based upon the notion that a particular sedimentary bed is a necessary temporal antecedent, although not properly the cause, of the bed lying immediately above it. Any gross ordering principle can be reduced by analysis to a number of geologic generalizations of more restricted scope.

A description of the hypothetical local section illustrated in figure 1 might consist of statements about the character and relationships of nine broadly defined entities; six sedimentary beds, two igneous dikes, and one unconformity. These singular descriptive statements together with general statements concerning superposition and crosscutting relations permit the inference of nine broadly defined events ordered in time in the relation "earlier than—later than."

If two events are causally related to one another, then they may be ordered *only* in the relation "earlier than—later than" because the causing event is invariably regarded as being temporally separated from the caused event. If two events are both regarded as having been caused by a single third event, then it *may* be possible to order them in the relation "simultaneous with." If, for example, a geologist chose to char-

acterize as two separate events the formation of the slickensides and the formation of the fault breccia resulting from a single increment of displacement along a particular fault plane, then it might be possible to infer that the two events had occurred simultaneously.

It should be noted here that the ordering of biological events is based ultimately upon geological ordering principles, which are themselves wholly independent of any biological considerations.

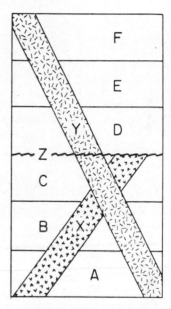

FIG. 1.—A hypothetical local geologic section

III. THE MEASUREMENT OF GEOLOGIC TIME

The ordering of events in time, or of points in space, is a non-metric or topological operation. The temporal ordering principles employed by geologists seldom contain reference, except in the most nebulous sense, to a metric. The great proportion of geologic ordering principles, including all of the gross ordering principles, are *purely* ordering principles. They cannot, consequently, be used in a measurement of the intervals of time between events. The measurement of geologic time involves great theoretical and practical difficulties.

In the world of more immediate experience intervals of time may be measured by comparing them with the uniform intervals of time marked off by a clock. How can we test the assumption that the successive intervals defined by any clock are uniform? In answer to this question Reichenbach stated (1958, p. 116): "There is only one answer: we cannot test it at all. There is basically no means to compare two successive periods of a clock, just as there is no means to compare two measuring rods when one lies behind the other. We cannot carry back the later time interval and place it next to the earlier one." And further (*ibid.*): "A solution is obtained only when we apply our previous results about spatial congruence and introduce the concept of a *coordinative definition* into the measure of time. The equality of successive time intervals is not a matter of *knowledge* but a matter of *definition*." Such a co-ordinative definition consists, in effect, of adapting some periodic process as a standard clock.

Let us now return to the hypothetical local section described above. Suppose that within each of the sedimentary beds of the section there is a very thin layer of volcanic ash, on the basis of which six ordered events —Aa, Ba, Ca, Da, Ea, and Fa—are inferred. Suppose further that someone were to claim that these events were separated by uniform intervals of time and justified his claim in terms of a co-ordinative definition. The claim would be rejected by almost any geologist. Attempts have been made, however, to introduce clocks based upon geologic systems for the measurement of geologic time. It has been said, for example, that the recurrence of events of the same type, such as orogenies or incursions of the sea, is regularly periodic. Such claims would be refuted if the defined intervals were found upon comparison with the intervals of a standard clock not to be uniform. To take, for the sake of simplicity, a non-geologic case, the attempt to falsify the claim that some individual human pulse defined uniform intervals of time would involve a comparison of the intervals marked off by the pulse with the intervals marked off by the second hand of a wrist watch. Even though the intervals defined by the movement of the second hand are of uniform length by definition only, we have not hesitated to reject the pulse as a reliable clock on the basis of a comparison with a wrist watch. Why is it then, in view of the definitional character of the uniformity of time intervals, that certain periodic processes are accepted as standards in terms of which the intervals defined by other periodic processes are measured? The answer is that clocks which permit the most comprehensive and simplest *total* system of scientific theory are accepted as standards. Thus a clock, though it permitted simple geologic descriptions, would be rejected if it required a change in the laws of physics.

There are a number of different physical systems whose periodic behavior permits measurements which yield more or less consistent estimates of duration. If, for example, a spring clock and a pendulum clock are used to measure the interval of time between two events, both clocks will yield nearly equal values for the interval if certain well-understood precautions have been taken in making the measurements. To determine whether or not two different periodic systems do in fact permit consistent estimates of duration is a matter for empirical test.

With one or two notable exceptions we have no grounds to support the contention that successions of geologic events yield estimates of duration which are consistent with estimates based upon standard physical clocks. Geologists are, however, confident that varves and tree rings are reliable standards by which intervals of time may be measured. This confidence stems from the fact that the processes which are thought to lead to the formation of varves and tree rings can be observed and their periodicity measured against standard clocks and that, furthermore, the processes which lead to the formation of varves and tree rings are held to bear a rather simple and direct relationship to regularly periodic events within the solar system.

It is probably fortunate for geology that the determination of the regular periodicity of events need not necessarily rest upon direct measurement of particular defined intervals, but may rest upon theoretical grounds. Justification for the contention that fluctuations in the obliquity of the ecliptic are regularly periodic is not to be sought for in direct measurement but rather in the theory of mechanics. If it could be shown that changes in the obliquity of the ecliptic resulted in geologically detectable events then we might have the basis for a geologic clock even though the periodicity of these events could not be shown by measurement to be regular.

There is another difficulty encountered in the attempt to devise a geologic clock. It has been found that clocks based upon *closed* periodic systems provide the most satisfactory standards for the measurement of time. Closed periodic systems are those whose periodic behavior, it may be concluded on the basis of physical theory, results from the effect of forces within the system and is free from the effect of forces external to the system. Jones (1950, p. 189) putting it somewhat differently, says, "any simple natural phenomenon which obeys one definite law without perturbation might be used to mark off equal intervals of time and therefore to serve as a clock." Reichenbach (1957, p. 119) points out that because no system is completely closed we can only speak of systems as being *closed to a certain degree of approximation.* In regard to its diurnal rotation the earth forms a very nearly closed system, and for this reason a clock based upon its period of rotation is considered to be highly reliable.

Physical and geologic theory require that we regard geologic systems as badly closed. This is, in itself, a sufficient basis for the rejection of almost any "clock" based upon a geologic system. If a geologist, unwilling to reject geologic clocks on theoretical grounds alone, should attempt to test the assertion that certain geologically defined

intervals are equal, he would encounter serious difficulties. Because practically *all* geologic systems are badly closed, there are almost no systems which might serve as standard clocks for the measurement of intervals. If geologists were to assume that *all* sequences involving the recurrence of geologic events of a certain type could be used to define uniform intervals of time, they could only be led to contradictory estimates of duration in geologic history.

It is possible to measure intervals of time by means of non-periodic processes. The assumption that a body moving in free space traverses equal distances in equal intervals of time leads to estimates of duration which are consistent with estimates obtained by standard physical clocks. The assumption that equal thicknesses of sediment are deposited in equal intervals of time, on the other hand, leads to estimates of duration which are inconsistent with estimates obtained by means of standard physical clocks.

The decay of radioactive minerals is a rectilinear, or non-periodic, process. If dike y of our hypothetical local section contained a mineral crystal containing Pb^{206} and U^{238} in a certain proportion, this fact could serve as the basis for an inference allowing the measurement, within the limits of probable error, of the interval of time between the crystallization of the mineral and the present. This inference is permitted because the rate of radioactive disintegration has been shown to be constant as judged by standard physical clocks.

The suggestion is sometimes made that the process of organic evolution, like the process of radioactive decay, can be used to measure intervals of time. Teichert, for example, says (1958, p. 116):

Measurements of past time periods depend on observational methods basically similar to those applied in the measurement of contemporaneous time. They are deductions from observations of rock properties or structures (including fossils) which may be interpreted as having resulted from the operation of processes which are either unidirectional and irreversible or occur with a fixed periodicity. The only terrestrial phenomena which fulfill these conditions and have proceeded universally for considerable periods are radioactive disintegration and the evolution of life.

And in the same paper, he writes (p. 102):

I do not wish to suggest that in the biological world genes mutate with the predictable precision of radioactive decay. Evolution is continuous only in the sense that life itself is a continuously operating chain of events. There are discontinuities and jumps, and it is these that are often particularly useful and valuable in the paleontologic record. However, the over-all effect of evolutionary processes which determine the composition and character of life communities was felt all over the world at the same time and at the same rate. On this uniformity of the over-all result depends paleontological correlation, a highly empirical science that can never take its own results for granted and yet has succeeded in accumulating a huge body of almost infallible guides to the interpretation of the geological past.

In the statement, "I do not wish to suggest that in the biological world genes mutate with the predictable precision of radioactive decay," Teichert clearly means to reject the claim that any temporal sequence of biological events could be said to define uniform intervals of time. There is indeed no empirical nor theoretical basis for supposing that any sequence of historically inferable biological events might mark off uniform intervals of time, and for this reason organic evolution cannot provide the basis for a clock which permits the assignment of numerical values to intervals of time.

In what sense, then, can the process of organic evolution be said to provide a means of *measuring* time? Teichert, and others who subscribe to this view of paleontological correlation, regard the evolutionary clock as one which measures time on the basis of events or *effects* which are "felt all over the world at the same time and at the same rate." This clock does not mark off uniform intervals of time, but this fact does not prevent measurement, because *measurement* in this sense apparently consists of nothing

more than pointing out that the interval of time between any two world-wide events is everywhere in space the same.

The grave defect of this view is that there is no reason to suppose that biological events are world-wide. The *evolutionary clock* not only does not define uniform intervals of time; it has not even been synchronized from place to place. I contended earlier that the notion of *geologic event* should be conceived of broadly enough to allow a greater or lesser restriction in time and space according to the function that an event was designed to perform in an inference. We cannot extend the concept of a geologic event to the degree that Teichert's view of correlation requires, for his view requires events which are extended all over the earth at a single instant in time and have identical consequences over every part of their spatial range. Every bit of evidence provided by evolutionary theory, and by the observations which may be adduced in support of this theory, compels us to conclude that the recognizable effects of evolution, far from being "felt all over the world at the same time" are restricted to limited areas on the surface of the earth.

Teichert is not naïve. He recognizes, implicitly if not explicitly, that biological events are local. In the paper from which the above quotations were taken, he presents so many qualifications to his contention that evolutionary events are world-wide that it loses all its force. The important point is that Teichert feels that it is necessary to *justify* correlation by means of fossils on the basis of a principle of world-wide organic events. To accept this principle is to obscure the only sound theoretical grounds upon which paleontological correlation can possibly rest.

Before presenting an alternative view of correlation I should like to discuss another matter relating to geologic time. I have not, up to this point, used either of the terms "relative" and "absolute" so frequently employed in discussion of geologic time. I should like now to examine these terms in the light of the foregoing discussion.

When a geologist uses the term "relative time" he is referring to the order of events. The sense in which order is relative is perfectly clear. If a set of events is temporally ordered then the elements of the set stand to one another in the *relation* "earlier than—later than," or the *relation* "simultaneous with."

The term "absolute time," as it is employed in geology, refers to the length of the time interval between events. The sense in which the magnitude of the time intervals between events is absolute is not at all clear. There is, however, an obvious and fundamental distinction, which has already been made, between the order of events in time and the length of the time interval between them. This distinction does not in any way hinge upon an understanding of the term "absolute." The measurement of time intervals involves quantification. A certain numerical value is assigned to each interval in question. The ordering of events in time need involve no quantification whatever. The significant distinction to be made is not between absolute and relative but rather between quantitative and non-quantitative, or even more simply between metric and non-metric.

My construal of the geologic meaning of "absolute time" may be too narrow. Eighteenth-century geology borrowed the term from the classical physics. Perhaps some contemporary geologists mean to use it in its full Newtonian sense. Newton defined absolute time in the *Principia* in the following way:

1. Absolute, true, and mathematical time, of itself, and from its own nature, flows equally without relation to anything external, and by another name is called duration: relative, apparent, and common time, is some sensible and external (whether accurate or unequable) measure of duration by means of motion, which is commonly used instead of true time; such as an hour, a day, a month, a year [Cajori edition, 1947].

In regard to Newton's concept of absolute time Withrow (1961, p. 34) remarks:

In practice we can only observe events and use processes based on them for the measurement of time. The Newtonian theory of time assumes, however, that there exists a unique series of moments and that events are distinct from them but can occupy some of them. Thus temporal relations between events are complex relations formed by the relation of events to the moments of time which they occupy and the before-and-after relation subsisting between distinct moments of time.

There is no method for discovering the "unique moments" of geologic time, nor is there any way of showing that geologic time "flows." According to Grünbaum (1964, p. 229-230): "*Physically*, certain states are later than others by certain amounts of time. But there is no "flow" of *physical* time, because physically there is no egocentric (psychological) transient *now*." Flow is not a feature of physical time, but of common sense, or psychological, time. The notion of the flow of time is suggested by, as Grünbaum has put it, the steady shift of the "now contents" of our consciousness. How then are we to impose *flow* on geologic time? The answer is that there is no way to impose it on time beyond the reach of immediate experience. *Geologically* "certain states are later than others but certain amounts of time" but there is no flow of geologic time.

Does the idea of Newtonian absolute time enter significantly into the geologist's conception of time? I think it does. The idea is certainly a significant factor of Jeletsky's conception of geologic time even though he calls it by another name. He says (1956, p. 701-702): "There is a certain advantage in using geologic time units as ideal time units based solely on the unknowable *abstract* time, and independent from all material criteria of geochronologic division. So interpreted, the ideal geologic time units would be strictly theoretical; they would serve solely the purpose of keeping us aware of the existence of the uninterrupted *abstract* geologic time behind its partial biological and physical standards."

Jeletsky decides, for various practical reasons, not to define geologic time units in terms of abstract time. It is absurd to even suggest, however, that there might be an advantage in using terms in such a way as to remind us of the *existence* of something which is *unknowable*.

Allusions to Newtonian absolute time are not usually so explicit as they are in the passage from Jeletsky. I suspect, however, that the notion of a *flowing, unknowable* time lies hidden in many discussions of geologic time and rises occasionally to make its contribution to the prevailing confusion.

In any case, I propose that we abandon once and for all the term "absolute time." There are perfectly familiar terms that can do all the work of "absolute time" and do it better by avoiding the ambiguity that has become attached to this term.

IV. The Extension of Local Temporal Ordering

A. INTRODUCTION

Up to this point the discussion has been largely confined to the temporal ordering of events at a single point in space or within a restricted geographic area. A problem of overriding importance for geology is, of course, correlation or the extension temporal ordering among events which are separated from one another in space.

B. CORRELATION BY RADIOMETRY

Inferences of geologic time based upon the laws of radioactive decay depend, in principle, upon the *measurement* of intervals of time between events. Not only can the interval of time between two events be measured in terms of rates of radioactive decay, but events, either in the same place or in different places, can be ordered in terms of these rates. Given three points on a line, the distances from each point to each other point and the direction of increasing magnitude, the order of the points on the line is uniquely determined. The relationship of distance and order holds equally well for *points*, or instants, in time and the *distance*, or amount of time between them. The problem of inferring order from distance

is considerably simplified in geology because one of the points in time, whose distance from other points may be known, is the present. The *other side* of this point, that is, the future, can be ignored. It is consequently possible to order three points in historic time if one of the points is the present and the distance to the other two points is known, which is simply a roundabout way of saying that if it is known how long before the present two events occurred then the order in which they occurred can be inferred.

Radiometric determinations may permit the ordering of events separated in space in the relation "earlier than—later than" and it is obvious how this same method can be used to infer simultaneity of events separated in space. If two events separated in space occurred at a time equidistant from some other event and on the same "time side" of that event then they may be said to have occurred simultaneously. In principle this method provides the most promising means of establishing the simultaneity of two events separated in space. Unfortunately the rather considerable probable error attending the measurement of time by means of radioactive decay prevents us from assuming that any two geologic radioactivity clocks are ever precisely synchronized. This method does, however, allow us, upon occasion, to talk meaningfully about the *probability* that two events occurred simultaneously.

C. CORRELATION BY SIGNAL

1. INTRODUCTION

The radiometric method of extending local temporal ordering is in most respects almost ideal. The applicability of this method is, in practice, very restricted. Two events separated from one another in space may, however, be temporally ordered with respect to one another in the complete absence of knowledge of the length of time between either of the events and any other events if one of the events in question has an influence on the other event, that is to say, if the two events are causally connected.

Because the causal connection between two events separated in space is usually a complex one involving what it is convenient to regard as several events, it is customary to speak of connection by means of *causal chains*. In geology, no less than in astronomy, the extension of local temporal ordering usually depends upon a causal chain, or *signal*, of finite velocity traversing the distance between two points separated in space.[3]

The transportation of a volcanic ash provides an example of a simple geologic signal. Certain generalizations may be adduced to support the conclusion that the deposition of a volcanic ash represents one end of a causal chain or, to put it somewhat differently, the terminus at a particular point of a signal. The arrival of the signal cannot be observed. It must be inferred in the same way that an astronomer might infer the arrival of a signal from the condition of a photographic plate. The signal whose arrival has been inferred from a volcanic ash bed, it might be further inferred, was initiated by a volcanic eruption at a point *A* separated in space from the site of deposition of the ash bed at point *B*.

Any geologic causal chain which has spatial extension may, in principle, be employed as a signal. Radioactive decay has, for geological purposes at least, no spatial extension and cannot, consequently, be employed as a signal. Most physical geologic signals are of limited use because their spatial range is more or less restricted. This restriction of range is *de facto* or contingent. It is imposed not by the generalizations or laws covering the events comprising the signals but rather by, to borrow a phrase from Grünbaum (1963, p. 211), "spatial-

ly ubiquitous and temporally permanent boundary and initial conditions." Thus the *running-water signal* might, in principle, have a much greater range than it in fact ever has. The drainage basin, however, provides *contingent* barriers to the extension of a running-water signal as surely as a room with no windows provides *contingent* barriers to the extension of a light signal.

There is general agreement that organic remains preserved in sedimentary rocks provide a particularly effective means of extending local time scales. Because we must reject the notion of measurement in terms of universal biological events, the correlation of two sedimentary beds separated in space and containing similar fossil assemblages rests upon precisely the same grounds as the correlation of a flow basalt and a bed containing volcanic ash of similar chemical composition. A fossil assemblage provides the basis for the inference of a complex local biological event which may be regarded as the terminus of a causal chain consisting of many local biological events. Under certain circumstances it may be possible to perform another inference that permits the recognition of another local biological event which may be regarded as the other end of the same causal chain, and thus to recognize a signal.

2. THE RECOGNITION OF HISTORIC SIGNALS

The recognition of historic signals as such depends upon our ability to recognize unique associations of events in causal sequences. We have at our disposal a vast body of well-confirmed geological and physical generalizations which permit us to associate *certain kinds of events* with *certain other kinds of events* in causal sequences. It is quite another matter, however, to causally associate a *particular instance of some kind of event* with *a particular instance of some other kind of event*. It is, for example, a familiar causal generalization which allows a geologist to conclude that a particular ash fall is to be associated with a volcanic eruption, but how does he then proceed to associate the particular ash fall with a particular volcanic

---

[3] After having completed a draft of this paper, I examined a manuscript on stratigraphy by John A. Wilson in which the concept of a signal was introduced into a discussion of correlation. It is interesting to note that this treatment of correlation was suggested to Wilson by a reading of Reichenbach's *The Philosophy of Space and Time*. Alfred Fischer has recently called my attention to a paper by Wegmann (1950) in which the geological *Zeitsignal* is discussed (see esp. p. 128, 129).

eruption chosen from all the volcanic eruptions of which he has knowledge? The answer to this seemingly trivial, but nonetheless important, question is that particular associations are made on the basis of the well-supported conviction that the unique characteristics of a concrete event are reflected in the unique characteristics of a concrete event to which it is supposed to be causally related. Thus a particular ash fall may be associated with a particular volcanic eruption on the basis of the unique mineralogical composition of a flow basalt on the one hand and of ash shards on the other.

It is assumed that historic geologic events, and consequently the causal chains composed of them, are non-repeatable. We conclude, in effect, that a signal of particular character is transmitted only once in history. Although in some contexts it is important to emphasize the fact that in a general way history may repeat itself, it is in the present context necessary to point out that it never does so precisely. If the assumption of the non-repeatability of geologic events were not made, correlation by means of a signal would be impossible.

There is no theoretical reason why some event should not be precisely repeated nor can it be claimed that the geologic record by itself supports the view that events are not repeated. We can, however, justify our contention of non-repeatability by pointing out that in any geologically significant situation the number of possible combinations of pertinent boundary and initial conditions is so high that the probability of precise repetition is to all intents and purposes zero.

Nor is the non-repeatability of sequences of organic events easily demonstrated. Suppose that someone were to claim that two identical paleontological sequences separated in space originated during different, non-overlapping intervals of time and presented this claim as an argument for the repeatability of organic sequences. The only effective argument against such a claim would be to point out that the two sequences were correlative and had consequently originated during the same interval of time. But paleontological correlation presupposes the non-repeatability of organic sequences. To avoid the circularity of this argument, independent justification for the assumption of non-repeatability must be presented. Contemporary evolutionary theory, in fact, supports the contention that sequences of organic events are contingently non-repeatable. Thus evolutionary theory, while it cannot provide a basis for the measurement of time, provides a firm foundation for the assumption that a biological signal of particular character was transmitted only once in history.

Finally, it may be possible under certain circumstances to distinguish structurally similar causal chains from one another in much the same way a physicist might distinguish different signals from one another, that is, on the basis of their distinct arrival and departure times. If, for example, a geologist were to encounter two bentonites of identical composition, he would certainly regard them as the termini of two separate causal chains if undisputed stratigraphic evidence indicated that one had been formed from an ash deposited during Cambrian time and that the other had been formed from an ash deposited during Cretaceous time. It must be borne in mind, however, that "undisputed stratigraphic evidence" must be ultimately based upon signals which were distinguished upon grounds *other than* their unique departure and arrival times.

Although the contingent non-repeatability of geologic events provides a theoretical basis for the recognition of particular historic-geologic signals, some difficulty usually attends an attempt to delineate a signal. Causal chains which are in principle distinguishable from one another may in practice be confused with one another because they are so similar in their characteristics. Two causal chains, each one of which began with an orogeny and ended with the deposition of a suite of sediments, for example, might be confused with one another if the sedimentary suites marking their ends were very similar.

Another practical difficulty attending the attempt to recognize particular causal chains stems from the fact that no geologic causal chain may be regarded as a well-closed system. Physical and biological causal chains are subject to the influence of forces and circumstances which may alter their character during transit. It might be relatively easy, for example, to associate the erosion of a particular igneous body with the deposition of a particular gravel, while it will usually be difficult to associate the erosion of a particular igneous body with the deposition of a particular clay. A biological signal may be so altered during transit that the association of two events widely separated in the chain could be inferred only in cases where evidence was unusually abundant.

3. DIRECTION OF SIGNAL TRANSIT AND "EARLIER THAN—LATER THAN" ORDERING

If the direction of transit of a signal of finite velocity is known, then the ordering of two events connected by it in the relation "earlier than—later than" can be immediately inferred. Once a particular signal has been defined in terms of two or more events, the determination of its direction of transit will rest upon a knowledge of the temporal direction of the processes comprising the signal and upon the assumption that the processes are temporally irreversible.

In almost every case the temporal relationships referred to in geologic generalizations are asymmetric. The geologic processes covered by these generalizations are thus held to be irreversible. It is interesting to note that the laws of dynamics, upon which many geologic generalizations are based, relate to processes which are fully reversible. At this point it will be useful to introduce a distinction, made by Grünbaum, between two senses in which a process might be held to be irreversible. He states (1963, p. 210):

There is both a weak sense and a strong sense in which a process might be claimed to be "irreversible." The weak sense is that the temporal inverse of the process in fact never (or hardly ever) occurs with increasing time for the following reason: certain particular de facto conditions ("initial" or "boundary" conditions) obtaining in the universe independently of any law (or laws) combine with a relevant law (or laws) to render the temporal inverse de facto non-existent, although no law or combination of laws itself disallows the inverse process. The strong sense of "irreversible" is that the temporal inverse is impossible in virtue of being ruled out by a law alone or by a combination of laws.

Clearly geologic processes are held to be irreversible in the weak sense rather than the strong sense. When geologists apply the laws of dynamics to geological situations, they, in effect, add restrictive initial and boundary conditions which make these laws applicable to situations of geologic interest. It is restrictive initial and boundary conditions that render most geologic processes irreversible. Processes of sedimentation, for example, are not irreversible because the laws and generalizations covering the processes forbid reversibility, but because the special initial and boundary conditions under which reversal might conceivably occur are in fact never realized.

Grünbaum points out that the distinction between the two kinds of irreversibility may be difficult to make in sciences in which the distinction between boundary conditions and laws is not sharp. He states (1963, p. 211):

The distinction between the weak and strong kinds of irreversibility has a clear relevance to those physical theories which allow a sharp distinction between laws and boundary conditions in virtue of the repeatability of specified kinds of events at different places and times. But it is highly doubtful that this distinction can be maintained throughout *cosmology*. For what criterion is there for presuming a spatially ubiquitous and temporally permanent feature of the universe to have the character of a boundary condition rather than that of a law?

Geologic theory, in addition to containing references to boundary and initial conditions that are regarded as limited to particular times and places, contains reference to boundary and initial conditions that are assumed to be "spatially ubiquitous" and "temporally permanent" features of the geologic universe. The value for the acceleration of gravity at the earth's surface is specifically assumed in a number of geologic generalizations and its assumption imposes *de facto* irreversibility upon a number of geologic processes. The distinction between law and boundary condition is, however, sharp enough in geologic theory to allow the distinction between the two kinds of irreversibility to be meaningfully made.

In cases where two events can be shown to be directly connected by a physical signal, the determination of the direction of transit seldom involves any difficulty because it usually depends only upon a knowledge of some causal generalization. In the case of a signal beginning with a volcanic eruption and ending with the deposition of a volcanic ash, for example, the direction of transit is immediately evident.

The determination of the direction of transit of a biological signal does not depend simply upon knowledge of some causal law. Occasionally conclusions concerning the direction of dispersal of a fauna or a species are reached. Such conclusions, it turns out, are based upon a correlation which in turn rests upon another signal. The direction of dispersal then, is knowledge which can seldom be invoked in a correlation but almost always depends upon a previously determined correlation. An examination of the stratigraphic literature indicates that the direction of organic dispersal is, in fact, seldom invoked in paleontological correlation. Knowledge of direction of dispersal may, on the other hand, assume great significance in discussions of paleobiology.

4. SIGNAL VELOCITY AND "SIMULTANEOUS WITH" ORDERING

Cases in which correlated events are not directly connected by a signal, or cannot be shown to be so connected, are of particular interest because so many correlations depend upon indirect signal connection. Suppose that a geologist is presented with two

sedimentary beds, A and B, widely separated in space and containing fossil assemblages of nearly identical characteristics. In the absence of very convincing evidence to the contrary the conclusion would be drawn that the two points had been indirectly linked by a signal which had originated at a third point, C. Unless the location of the point C could be inferred, there would be no basis for a conclusion as to whether the bed at A had been deposited before the bed at B, or vice versa, or whether they had been deposited simultaneously. The fact that so may situations like the one described above are known has seemed to lend support to the concept of universal biological events. Universal biological events must be precluded, however, for the reason that I have already given. We cannot consistently hold to contemporary evolutionary theory and the doctrine of universal biological events at the same time.

Suppose now that we were presented with two volcanic ash beds of identical physical and chemical characteristics, one located at a point A and the other located at a point B separated in space from A. Let us further suppose that we were able to infer that the ash which had been deposited at A and B had originated at a third point C, the site of a volcanic eruption. Knowing the distances C–A and C–B we could determine whether or not the deposition of the ash at A and at B had occurred simultaneously if we knew the velocity of ash transport between C–A and C–B. Could we measure these velocities?

It turns out that the measurement of signal velocities involves some intriguing problems as contemporary physicists are able to testify. Grünbaum (1963, p. 344) states:

It might be thought that distant simultaneity can be readily grounded on local simultaneity as follows. Suppose that two spatially separated events E_1 and E_2 produce effects which intersect at a sentient observer so as to produce the experience of sensed (intuitive) simultaneity at himself. Then this local simultaneity of the effects E_1 and E_2 permits us to infer that the distant events occurred simultaneously, if the influence chains that emanated from them had appropriate *one-way velocities* as they traversed their respective distances to the location of the sentient observer. But this procedure is unavailing for the purpose of first characterizing the conditions under which the separated events E_1 and E_2 can be held to be simultaneous. For the *one-way* velocities invoked by this procedure presuppose *one-way transit times* which are furnished by *synchronized* clocks at the locations of E_1 and E_2 respectively.

It cannot be assumed that two clocks remain synchronized after being separated in space. An attempt to demonstrate that they had remained synchronized would involve the use of a signal which would again presuppose a one-way transit time.

But it is not necessary to have prior knowledge of simultaneity to measure certain kinds of signal velocity. A light signal may be transmitted from a point A to a point B and reflected back to A. If the distance between A and B and the elapsed one clock time for the round trip are known, then the average round trip velocity can easily be calculated. In order to base a determination of the one-way velocity of light upon the measured round-trip velocity, it is assumed that the time required to travel from A to B is the same as the time required to travel from B to A. Einstein pointed out that the conception of the simultaneity of distant events depended critically upon this assumption. He stated (1920, p. 4): "We have hitherto an A-time, and a B-time, but no common time to A and B. This last time (i.e., common time) can be defined, if we establish by definition that the time which light requires in traveling from A to B is equivalent to the time which light requires in traveling from B to A."

There is no standard signal in geology whose round-trip velocity has been measured and whose one-way velocity has been inferred on the basis of the assumption of constant velocity. Geologic theory does not permit the assumption that the average velocity of any two historic-geologic signals

was the same nor that the velocity of any particular signal was constant. A judgment of simultaneity in historical geology must rest upon a knowledge of the average velocity of the *particular* signal employed. In no case, however, can the average velocity of a particular historic signal be determined because the determination would presuppose the very thing that would make the determination of velocity superfluous, that is, a knowledge of simultaneity at the two points connected by the signal.

If a geologist were to attempt to demonstrate that an historical signal had been transmitted instantaneously, he would encounter the difficulties alluded to above. A world-wide event which occurred at an instant in time or during a very restricted interval of time and which had detectable geological consequences at widely separated points on the earth's surface would constitute, in effect, an instantaneously transmitted signal linking those points. The possibility that physical events of this kind have occurred cannot be precluded in principle. To avoid the presupposition of simultaneity, however, the grounds for supposing that such an event had occurred would necessarily rest upon the use of conventional signals of finite, but unknown, velocity. The assumption that orogenies have occurred simultaneously at widely separated places on the surface of the earth has been used from time to time as the basis for correlation. The underlying assumption seems in this case to be that whatever the sufficient conditions for the occurrence of an orogeny may be, these conditions tend to be, in any given instance, widely extended in space and narrowly extended in time. Attempts to demonstrate the simultaneity of widely separated orogenies on the basis of paleontological signals has been largely unsuccessful. At the same time the theoretical grounds for supposing that orogenic episodes were world wide have become less and less tenable. The assumption that other world-wide events have occurred during relatively short intervals of time, for example, climatic changes, is plausible but nonetheless difficult to substantiate.

Because a geologist cannot determine the velocity of individual historical signals, he cannot say of any two historical events that they occurred simultaneously. This conclusion will startle no one, for very few, if any, geologists have ever claimed that they could determine that two historical events occurred at precisely the same time. Geologists do claim, however, that they can determine approximate simultaneity, and that given the length of geologic history, these approximations are significant and useful. To deny the legitimacy of such claims on technical grounds would constitute unjustifiable quibbling. The question of whether we can meaningfully speak of even approximate simultaneity, however, is bound to arise. It certainly arises in the case of intercontinental correlations. Intercontinental correlations almost invariably invoke biological signals. Biological signals are, of course, employed because of their greater spatial range as compared to physical signals. We have reason to suppose, however, that biological signals have on the whole, relatively low velocities, that the velocity varies greatly from signal to signal, that the velocity is far from constant, and finally that independent judgments of direction of transit are extremely difficult to come by. The characteristics of biological signals, together with the great distances involved in intercontinental correlation, conspire to undermine our confidence in the extension of *time planes* on a world-wide scale.

This is not to say that hypotheses of the simultaneity of events occurring on different continents might not be more or less well supported. Those who are concerned about the problem of historic-geologic signal velocities attempt to place certain broad limits upon velocities on the basis of the plausible assumption that the measured velocity of a signal in the present is more or less similar to the velocity of a historical signal of similar character. On the basis of such considerations judgments about the reliability

of particular signals are reached. Eames, Banner, Blow, and Clark (1962, p. 130), for example, state, in justifying the stratigraphic usefulness of the Globigerinaceae: "The planktonic habit of the Globigerinaceae ensures rapid faunal (and genetic) migration; the observed density of living populations, which are free to migrate, indicates rapid reproduction in this superfamily (as also observed in other foraminiferal groups) and easy genetic exchange. All these factors reduce impediments to faunal migration and genetic interchange to a minimum."

We must always bear in mind, however, that an hypothesis that two events are ordered in the relation "simultaneous with" can never be confirmed because the confirmation of such an hypothesis would rest upon the determination of the average velocity of a particular historic-geologic signal. Hypotheses involving "earlier than—later than" relations can on the other hand be confirmed, because they do not rest upon knowledge of the velocity of a signal, but only upon the knowledge of the direction of transit of a signal, knowledge which is in some cases available.

To regard geologic correlation as primarily a method for establishing simultaneity, while not necessarily detrimental to the practice of stratigraphy, tends to obscure the fundamental nature of the methods available to geology for the spatial extension of local time. Correlation is, in its most precise form, a method which permits the ordering of spatially separated historical-geological events in the relation "earlier than—later than." This conclusion in no way diminishes the importance of correlation. A very high degree of ordering in the relation "earlier than—later than" can be achieved. Because this relation is, in addition to being irreflexive and asymmetric, transitive (if *A* earlier than *B*, and *B* earlier than *C*, then *A* earlier than *C*), it is not necessary that events be directly connected by a signal to be ordered according to it. Furthermore, it is possible to base a very elaborate ordering of *intervals of time defined by events* in the relations "included in," "partially included in," and "excluded from" upon the "earlier than—later than" ordering of events.

V. PERIODS OF GEOLOGIC TIME

A. ROCKS AND TIME

A geologist would be very surprised, upon attending a meeting of historians, to find the participants engaged in a discussion of the problem of how to develop a set of terms which would allow a distinction to be made between divisions of historic time and the documents on the basis of which the divisions of historic time had been formulated. Yet one of the most striking features of geology during the past twenty-five years is the extent to which stratigraphers have become involved in a parallel discussion which raises the question of the distinction between "time terms" and "rock terms." The discussion has, for the most part, centered about terms, yet it is clear that the points at issue cannot be dismissed, as some have tended to do, as "merely questions of semantics." The pressing need for a separate set of terms for rock intervals and time intervals grew out of a fundamental conceptual confusion which, despite the heroic efforts of a large number of stratigraphers, persists in some quarters today. It is remarkable that a distinction which, once it has been made, seems so natural and necessary should have resulted in so much difficulty. The vigorous discussion, begun about twenty-five years ago, was precipitated by the recognition that so intimately entwined had become the concept of divisions of geologic time with the objects which constituted the basis for their inference that it could indeed be said that no distinction whatever was being made.

The confusion between rock and time was permitted by the assumption that the physically defined boundaries of rock intervals represented "isochronous surfaces," and that for this reason rock intervals could be said to somehow "correspond with" or be "equivalent to" certain intervals of geologic time. This assumption is clearly unwarranted on the basis of geologic theory, and, more than this, it is incompatible with the general historical point of view which leads us to conclude that the observable, describable objects of our immediate experience came to be what they are as a result of a series of events in historic time. It makes exactly as much sense to say that the boundaries of the Woodford Formation are isochronous as it does to say that the boundaries of the Great Wall of China are isochronous.

Opponents of the view that the boundaries of rock units are invariably isochronous have sometimes employed a vocabulary in their arguments which has tended to perpetuate the very confusion between object and time which they were striving to remove. Is not the confusion evident, for example, in the statement: "A rock-stratigraphic unit may possess approximately isochronous boundaries, or its boundaries may transgress time horizons" (from the Code of Stratigraphic Nomenclature, 1961, p. 649)? What does it mean to say that the boundaries of rock-stratigraphic unit transgress time? One who was unaware of the particular conceptual and historical context in which this statement was made might understandably conclude that the statement meant no more than that rock boundaries persist in time. It is clear to a geologist that the statement is meant to convey more than this. It is not meant to say that the boundary which we observe in the present transgressed a time horizon, but rather that some events which account for the presence of the boundary transgressed a time horizon. To say that a boundary transgressed a time horizon when what is meant is that events transgressed a time horizon is profoundly confusing. When a geologist says, "the boundaries of a rock-stratigraphic unit may transgress a time horizon" he only means, "the event which marked the origin of a particular point (or limited area) on the surface of a rock-stratigraphic boundary may not have been simultaneous with the events marking the origin of every other point (or limited area) on the same boundary." Compare this statement to "the event which marked the ori in of a particular point (or limited area) on the surface of the Great Wall of China may not have been simultaneous with the events marking the origin of every other point (or limited area) on the surface of the wall." Neither of these statements is particularly illuminating except to a person who might, for some reason, have believed otherwise. This explication, incidentally, makes no recourse to the notion of a *time horizon*.

B. OUR KNOWLEDGE OF THE PAST AND OUR KNOWLEDGE OF THE PRESENT

To make a distinction between rocks and defined intervals of geologic time, however useful and illuminating it may be, does not by itself remove all of the conceptual difficulties that surround the notion of geologic time. It is true that rocks form the ultimate basis for the derivation of intervals of geologic time, but it is important to realize that the inferential gap between rocks and time intervals is a very wide one indeed. As a matter of fact, rocks on the one hand and defined units of geologic time on the other are at opposite ends of an inferential spectrum. Between these ends lies nearly all of our knowledge of the earth.

We may conceive of our knowledge of the geologic past as a knowledge of states or events which are ordered with respect to one another in space and in time. We may conceive of our knowledge of the geologic present in the same way. The events of the past and the events of the present are characterized in terms of the same descriptive vocabulary. By the use of this vocabulary to characterize both the geologic past and the geologic present, geologists clearly mean to convey the conviction that the past was composed of the same *kinds of events* as the present. If we were to conclude that one million years ago a violent volcanic eruption had occurred on a particular island, we should have concluded that an event had occurred which had some characteristics in common with a volcanic eruption that might be going on at this moment. If we had been standing off-shore in a ship one million years ago at the time of this historic event, we might have turned to a companion and remarked, "Ah, a violent volcanic eruption!"

The point made above may seem to be entirely trivial, yet I think that in the present context it is necessary to be reminded that the events of the past are supposed to differ from the events of the present primarily in being located in a different region of time. Historic events are so located temporally as to be beyond the reach of direct observation by a sentient observer, and our knowledge of them must consequently be acquired by inference.

It can be said that, in general, our knowledge of the past is inferentially derived from our knowledge of the present. If we do not keep this important fact in mind, we may come to believe that our knowledge of historical events is much more immediate than it actually is. Our knowledge of the present is itself, however, in part inferentially derived. We can legitimately and usefully distinguish between knowledge that is acquired by direct observation and knowledge that is acquired by inference. This distinction is not, however, a sufficient basis for distinguishing between our knowledge of the past and our knowledge of the present.

When we speak of "our knowledge of the present" we refer to, among other things, descriptions of the state of the earth which are based upon more or less direct observation. Descriptive statements referring to direct observation do not, however, encompass all that we claim to know about the present. We claim, for example, to know something about the state of the earth's interior. This knowledge, though ultimately grounded in observation, is acquired by means of an inference involving the mediation of physical and geological laws. Our knowledge of the present is not only extended in space beyond the range of immediate observation, but it is also extended in time. Much of the knowledge we claim to have of the "present" is in fact knowledge of the proximate past and like our knowledge of the distant past and of distant places, is acquired inferentially.

It is not even true that we can claim that our knowledge of the *extended present* or *proximate past* is always more complete and certain than our knowledge of the distant past. We might, for example, be much more confident that some particular event had occurred at some point on the surface of the earth in the distant past than we are that some event is occurring in the interior of the earth at this moment. And we may have much more confidence in our knowledge of an event which occurred ten million years ago than in our knowledge of an event which may have occurred in our own lifetime.

The completeness and certainty of our knowledge of an event do not depend directly upon the amount of time which has elapsed since the event is supposed to have occurred. They depend, among other things, upon an observable trace which the event has left in the present, upon the availability of laws in terms of which the trace may be seen to have relevance to the past, and upon independent confirmation that the hypothetical event did occur. These factors are only in part dependent upon the amount of elapsed time involved.

C. THE CLASSIFICATION OF GEOLOGIC TIME

Primary historical inferences in geology lead to statements about events ordered in space and in time. Events are the stuff of which our systematic knowledge of the earth is woven. Consequently the basis for any notions we may have about the divisions of geologic time and about the classification of these divisions must lie in a knowledge of events.

We may conceive of geologic time as being defined by the set of all the events of the geologic past. The events of the set are spatially ordered and temporally ordered according to the relations "earlier than—later than" and "simultaneous with." Whatever our notion of geologic time may be, our knowledge of it does in fact consist of knowledge of a set of events which *may be* regarded as a proper subset of the set of all of the events of the geologic past. This is a way of saying that our knowledge of the past is incomplete. Events have transpired of which we have no knowledge. The relations of the known events of the past to one another in terms of "earlier than—later than" is incompletely known. Only rarely is it possible to assign some numerical value to the interval of time between events. And very rarely indeed is it possible to say of any two known events that they occurred simultaneously.

If geologic time is regarded as being defined by a set of ordered historic events, then it follows that the defined intervals of geologic time may be regarded as being defined by subsets of this set. An interval of geologic time might be defined by a set of events that occurred after, and includes a designated *limiting event* and occurred before, a designated limiting event included in the next later interval of time. To define an interval of geologic time by a set open at one end would avoid the well-known difficulty described in the following quotation from the stratigraphic code (p. 659–660).

The period comprised an interval of time defined by the beginning and ending of the deposition of the system. To define periods rigorously in this manner is to create unnamed time units between periods, in other words, gaps in formal geologic time. By later work supplementary sections largely or wholly filling the hiatuses have been found elsewhere in the world and their rocks, by common consent, have been assigned to one or another of the contiguous systems. Many of the gaps have thereby been essentially filled. Today it is probable that formal geologic time as referred to actual rocks is continuous or even (as now classified) duplicated.

The problem of unnamed time intervals between designated periods and the overlapping of designated periods arises because a period defined by the "beginning and ending of deposition of a system" amounts, in effect, to defining the period by two events both of which are included in the period. The period so defined constitutes a set closed at both ends. If the event marking the end of period A and the event marking the beginning of the period B which is supposed to be the temporal successor of A did not occur simultaneously then a gap between A and B or an overlap of A and B must result.

named gap or to both periods at the same time.

A *defining time plane* or *time boundary* could be introduced by defining it as a set of simultaneous events which includes a designated limiting event. It follows from this definition that an overriding consideration in the selection of limiting events would be the possibility of including such an event in a set of simultaneous events. To put it somewhat differently, the system of classification of geologic time ought to be based upon time

TABLE 1

Rock-stratigraphic (Lithostratigraphic) and Biostratigraphic Units	Time-stratigraphic (Chronostratigraphic) Units	Geologic-Time (Geochronologic) Units
Article 4. A rock-stratigraphic unit is a subdivision of the rocks in the earth's crust distinguished and delimited on the basis of lithologic characteristics. Article 19. A biostratigraphic unit is a body of rock strata characterized by its content of fossils contemporaneous with the deposition of the strata.	Article 26. A time-stratigraphic unit is a subdivision of rocks considered solely as the record of a specific interval of geologic time.	Article 36. Geologic-time units are divisions of time distinguished on the basis of the rock record, particularly as expressed by time-stratigraphic units. They are material units.
No change.	A time-stratigraphic unit is a set of observable things (rocks) whose descriptions serve as the basis for the inference of a particular set of historic events included in a unit of geologic time.	A unit of geologic time is defined by a set of historic events which occurred after and includes a designated limiting event and occurred before a designated limiting event included in the next later unit of geologic time.

If, however, the set of events defining a period is left open at one end then the problem of gaps and duplications simply does not arise. If the Cambrian period were to be defined by the set of events occurring after a particular event included in Cambrian time and before a particular event included in Ordovician time then no gap between the two periods could occur. Such a device does not, of course, serve to remove the uncertainties connected with correlation. It might not be possible to assign a particular event to one period or the other, but it would not be possible to assign an event to an un-

planes that can be defined with the highest possible degree of precision. Often other considerations, for example, historical tradition, are given precedence over precision of definition in the selection of time boundaries.

Because time boundaries or horizons depend for their definition upon a set of at least two simultaneous events, they can, even under the most fortunate circumstances, be defined only indefinitely. We are reminded, therefore, that time boundaries are only "approximate" or "ideal." An unfortunate consequence of basing a stratigraphic system upon the notion of simultaneity is that it imparts to the system a high degree of instability simply because simultaneity is a relationship among events that is very difficult to confirm and rather easy to falsify. Time boundaries are shifted about, and a principal basis for this shifting lies in the falsification of hypotheses of simultaneity.

Is it necessary that something so indefinite and unstable as a time plane should bear the burden of the system of geologic time classification? Many obviously consider it desirable, but it is certainly not necessary. It is a simple matter to eliminate the time boundary, and along with it simultaneity, as a basis for the division of geologic time. That geologists, or for that matter anyone else, could conceive of time without recourse to the notion of simultaneity, is almost impossible to imagine. It does not follow from this, however, that the scheme in terms of which we characterize our knowledge of geologic time must critically depend upon planes which presuppose for their specification the demonstrable, or, worse still, the undemonstrable simultaneity of events.

It is logically permissible to define the divisions of geologic time by individual limiting events with no requirement that they be simultaneous with any other events. The events of geologic history would be ordered with respect to these limiting events. Most of the order would be in the relations "earlier than" and "later than." Simultaneity would not be eliminated from consideration, but only from the definition of units of geologic time.

Such a system seems to have no fatal defects and to have one or two features which positively recommend it. In table 1, definitions of time-stratigraphic units and time units in terms of events are compared with definitions from the stratigraphic code.

ACKNOWLEDGMENT.—This paper was written during the academic year 1964–1965 while I was a visiting fellow in philosophy at Princeton University. I wish to express my gratitude to the faculty of the Department of Philosophy at Princeton for extending to me every courtesy and assistance during this period, and to the University of Oklahoma for granting me sabbatical leave. During my leave I was aided by a grant from the University of Oklahoma Research Foundation. Professor Carl G. Hempel read several different versions of this paper and offered me encouragement as well as valuable criticism and advice. Professor Adolf Grünbaum, of the University of Pittsburgh, read an earlier draft of the paper, and I have attempted to incorporate his valuable suggestions into the final version. I have benefited from conversations with many geologists, particularly from those with Alfred G. Fischer, Hollis D. Hedberg, and Heinrich D. Holland, all of the Department of Geology at Princeton. I need hardly remind the reader that the opinions expressed above are my own. The philosophy faculty at Princeton awarded me the 1965 J. Walker Tomb prize for this paper. I deeply appreciate this honor.

REFERENCES CITED

AMERICAN COMMISSION ON STRATIGRAPHIC NOMENCLATURE, 1961, Code of stratigraphic nomenclature: Am. Assoc. Petroleum Geologists Bull., v. 45, p. 645–665.

EAMES, F. E., BANNER, F. T., BLOW, W. H., and CLARKE, W. J., 1962, Fundamentals of mid-Tertiary stratigraphical correlation: Cambridge, Cambridge University Press, vii+163 p.

EINSTEIN, A., 1920, On the electrodynamics of moving bodies, in The principle of relativity, Original papers by A. Einstein and D. Minkowski, translated into English by M. N. Saha and S. N. Bose: Calcutta, University of Calcutta, xxiii+186 p.

GRÜNBAUM, A., 1963, Philosophical problems of space and time: New York, Alfred A. Knopf, xi+448 p.

—— 1964, The anisotrophy of time: Monist, v. 48, p. 217–247.

HEMPEL, C. G., 1962, Explanation in science and in history, in Frontiers of Science and Philosophy, ed. R. G. Colodny: Pittsburgh, University of Pittsburgh Press.

JELETSKY, J. A., 1956, Paleontology, basis of practical geochronology: Am. Assoc. Petroleum Geologists Bull., v. 40, p. 679–706.

JONES, H. S., 1950, The determination of precise

time: Ann. Report Smithsonian Inst. (1949),
    p. 189–202.
MOBERLY, R., 1960, Morrison, Cloverly and Sykes
    Mountain formations, northern Bighorn Basin,
    Wyoming and Montana: Geol. Soc. America
    Bull., v. 71, p. 1137–1176.
NEWTON, I., 1947, Principa, ed. F. Cajori: Berkeley,
    University of California Press.
REICHENBACH, H., 1957, The philosophy of space
    and time: New York, Dover Publications, xvi +

295 p. (English trans. of Philosophie der Raum-
    Zeit-Lehre published in 1928).
TEICHERT, C., 1958, Some biostratigraphic con-
    cepts: Geol. Soc. America Bull., v. 69, p. 99–120.
WEGMANN, E., 1950, Diskontinuität und Kontinuität
    in der Erdgeschichte: Geol. Rundschau, v. 38,
    p. 125–132.
WHITROW, G. J., 1961, The natural philosophy
    of time: London, Thomas Nelson & Sons, xi +
    324 p.

Addendum by David B. Kitts

I am happy to have the opportunity to comment upon my paper eight years after its publication. The response to my views on geologic time has been gratifying, and I am pleased that my paper has been selected for inclusion in this volume. I admit to some disappointment that the suggestions I made have gone unheeded by stratigraphers. But I did not, after all, write a work devoted to the practice of stratigraphy. I attempted to clarify the conceptual foundations of geologic time in such a way as to remove what appeared to me to be some significant difficulties. The recommendations that I made did not concern the day-to-day practice of stratigraphy but the formal system in terms of which the knowledge acquired by stratigraphers is finally organized.

My most radical proposal was that simultaneity and the "time plane" be abandoned as the basis for defining intervals of geologic time because reasonable conclusions concerning simultaneity cannot be achieved. Intervals of time may be defined by events with no requirement that the events be simultaneous with any other events. Events that mark the end of one period may mark the beginning of the next period in succession. By this simple *convention* the gaps and overlaps of geologic time that appear in the *Code of Stratigraphic Nomenclature* (1961) may be avoided. But the latest pronouncement of the stratigraphic community perpetuates the traditional dilemmas. In the *International Guide to Stratigraphic Classification, Terminology, and Usage* (1972), a "chronostratigraphic unit" is defined in the following way (p. 28):

> A chronostratigraphic unit is a body of rock strata which is unified by representing the rocks formed during a specific interval of geologic time. It represents all rocks formed during a certain time-span of earth history and only those rocks formed during this time-span. It is best defined as corresponding to the stratigraphic interval between two designated boundary-stratotypes. The relative magnitude of a chronostratigraphic unit is a function of the length of the time interval which its rocks subtend rather than its physical thickness.

For decades stratigraphers have insisted that a distinction be made between rock terms and time terms, yet a confusion still exists. Rocks, according to the *Guide*, subtend periods of time. Yet time intervals must be defined by events, not by physical objects. When statements about chronostratigraphic units are translated into statements about events, the source of confusion is revealed. It is a peculiar consequence of selecting bodies of rock strata as the basis for defining intervals of geologic time that the event that begins an interval of time must occur in the *same place* as the event that ends that interval. If it cannot be demonstrated that an event that ends one period is simultaneous with the event that begins the next period in succession, then a gap between, or an overlap of, these periods inevitably results.

How remarkable that historians have been able to avoid the difficulties alluded to above. Books do not "subtend" historical periods. An event that marks the end of one period may mark the beginning of the next period. An event that begins a period need not be located in the same place as the event that ends that period. And the concept of simultaneity does not enter into the definition of intervals of historical time at all.

My colleague Sabetai Unguru begins the Middle Ages with the closing of the Platonic Academy in Athens by Justinian. He may end Antiquity with the same event and thereby assure the absence of a gap between the two periods. An event that occurred just before the closing of the Academy is an event of Antiquity, and an event that occurred just after the closing of the Academy is an event of the Middle Ages. These conclusions in no way rest upon an "isochronous surface" that extends from Athens to the corners of the earth. And finally, a historian who begins the Middle Ages with the closing of the Academy in Athens may end it with the death of Nicolas of Cusa in Umbria. Historians may disagree about which events are to be used in defining intervals of historical time. They may even, on occasion, end up with gaps and overlaps between periods. But unlike geologists they are not directed by a code to define intervals in such a way as to require gaps and overlaps.

Why are geologists so willing to accept a set of rules that leads to such inelegant consequences? Why is any set of rules necessary when the means of adequately defining periods of time is provided by ordinary language? The answer to these questions lies as much in sociology and history as it does in geology. Part of the answer is, I think, provided by a recognition that the relationship between the geometry of rock bodies and the geologist's concept of geologic time is intimate. It is the *plane* boundaries of sedimentary bodies that suggest the *plane* boundaries of geologic time. If our world had contained only igneous rocks, "time planes" would probably not have become a part of our conceptual baggage.

There is probably a lesson in all of this. Rational reconstruction has little effect upon the practice of science. Stratigraphers continue to acquire knowledge of the past and to organize it unaware of, or at least undisturbed by, the difficulties that have loomed so large in my consideration of geologic time.

References

American Commission on Stratigraphic Nomenclature, 1961, Code of stratigraphic nomenclature: Amer. Assoc. Petroleum Geologists Bull., v. 45, p. 645–665.
International Subcommission on Stratigraphic Classification, 1972, An international guide to stratigraphic classification, terminology, and usage: Lethaia, v. 5, p. 283–295.

Author Citation Index

Subject Index

383